A First Course in Applied Statistics

SECOND EDITION

with applications in biology, business, and the social sciences

MEGAN J. CLARK AND JOHN A. RANDAL

www.pearsoned.co.nz

Your comments on this title are welcome at
feedback@pearsoned.co.nz

Pearson
a division of Pearson New Zealand Ltd
67 Apollo Drive, Rosedale, Auckland 0632, New Zealand

Associated companies throughout the world

© Pearson 2004, 2011
First published by Pearson Education New Zealand 2004
Reprinted 2006, 2007, 2008 (twice), 2009 (twice), 2010
This edition published by Pearson 2011
Reprinted 2011, 2012 (three times)

ISBN: 978-1-4425-4151-1

All rights reserved. No part of this publication may be
reproduced, stored in a retrieval system, or transmitted, in any
form or by any means, electronic, mechanical, photocopying,
recording, or otherwise, without the prior permission of the
publisher.

Produced by Pearson

Cover design: Marie Low
Cover image: © Marilyn Volan/Dreamstime.com

Printed in Malaysia via Pearson Malaysia (CTP -VVP)

This Pearson Originals edition has been published directly from the authors' manuscript in order to get the book to
you as soon as possible. It has not gone through the rigorous editorial and production processes normally afforded to
Pearson titles. While every effort has been made by the authors to ensure the accuracy of the text, Pearson does not take
responsibility for the editorial quality of this edition.

Contents

Preface ix

Acknowledgements x

1 Introduction 1
   1.1 What is statistics, the subject? 1
   1.2 Using this book 2
   1.3 Course structure 3

I Summarising data 4

2 Data collection, organisation and graphical representation 5
   2.1 Classifying variables 5
   2.2 Sampling 9
   2.3 Stemplots, dotplots and histograms 13
       2.3.1 Stemplots 13
       2.3.2 Dotplots 16
       2.3.3 Histograms 16
   2.4 Bar charts 20
   2.5 Summary 22
   Exercises 22

3 Summary statistics — 31

- 3.1 The centre of a data set 31
  - 3.1.1 Mean 32
  - 3.1.2 Median 33
  - 3.1.3 Percentiles 34
- 3.2 Spread or variability 36
  - 3.2.1 Range 36
  - 3.2.2 Interquartile range, IQR 37
  - 3.2.3 Variance and standard deviation 38
- 3.3 Sample statistics for repeated or grouped data 40
- 3.4 Summarising the features of sample data 42
  - 3.4.1 Boxplots 42
  - 3.4.2 Outliers 44
  - 3.4.3 Modified boxplots 45
- Exercises 46

4 Describing bivariate relationships — 54

- 4.1 Scatterplots 55
- 4.2 Correlation 57
  - 4.2.1 Pearson's linear correlation coefficient 57
  - 4.2.2 Spearman's rank-order correlation coefficient 62
- 4.3 Regression 66
  - 4.3.1 Choosing the regression line 66
  - 4.3.2 Estimating the regression line 67
  - 4.3.3 Prediction 71
  - 4.3.4 Assumptions 73
- 4.4 Optional technical material 75
  - 4.4.1 Pearson's linear correlation coefficient 75
  - 4.4.2 The method of least squares 76
- Exercises 78

5 Time series data — 90

5.1 Displaying time series data 92
  5.2 Time series components 94
  5.3 Summarising time series data 97
       5.3.1 Linear time trend 98
       5.3.2 Moving averages 100
       5.3.3 Sample autocorrelation 104
  Exercises 106

II Introduction to probability   114

6 Working with probabilities   115
  6.1 What is a probability? 115
  6.2 Probability rules 117
  6.3 Conditional probability 119
       6.3.1 Probability trees 120
       6.3.2 Bayes' Rule 123
  6.4 Probability distributions and random variables 124
  Exercises 127

7 Proportions and the binomial distribution   134
  7.1 Properties of the binomial distribution 135
  7.2 Finding binomial probabilities 137
  7.3 Mean and variance of a binomial random variable 141
  7.4 Proportions 142
  Exercises 143

8 The normal distribution   146
  8.1 The normal distribution and its uses 146
       8.1.1 Using normal tables to compute probabilities 150
       8.1.2 Using the inverse normal table 153
       8.1.3 Relative standing 155
  8.2 The central limit theorem 156
       8.2.1 A demonstration of the CLT 156

		8.2.2 How large is large?	158
	8.3	Sampling distributions	159
	8.4	Normal approximation to the binomial	161
	8.5	Summary	164
	Exercises		165

III Estimation and testing — 169

9 Single population — 170

	9.1	Large sample inference for a mean	171
		9.1.1 Confidence intervals for μ	171
		9.1.2 Hypothesis tests for μ	173
		9.1.3 p-value approach to hypothesis testing	180
		9.1.4 Type I and type II errors	183
	9.2	Small and normal sample inference for a mean	184
		9.2.1 Confidence intervals for μ	186
		9.2.2 Hypothesis tests for μ	187
	9.3	Small and non-normal sample inference for a median	188
	9.4	Inference for a proportion	191
		9.4.1 Confidence intervals for p	191
		9.4.2 Large sample hypothesis testing for p	193
		9.4.3 Small sample hypothesis testing for p	195
	9.5	Finite populations	197
	9.6	Margin of error and sample size	199
	Exercises		201

10 Two populations — 212

	10.1	Large sample inference for means	212
		10.1.1 Confidence intervals for $\mu_1 - \mu_2$	214
		10.1.2 Hypothesis tests for $\mu_1 - \mu_2$	215
	10.2	Small and normal sample inference for means	217
		10.2.1 Confidence intervals for $\mu_1 - \mu_2$	219

10.2.2 Hypothesis tests for $\mu_1 - \mu_2$	220
10.3 Tests for variances	221
10.4 Small and non-normal sample inference for medians	223
10.4.1 The Mann-Whitney U-test	223
10.4.2 The normal approximation to the U-test	227
10.5 Paired comparisons	228
10.6 Tests for proportions	229
10.6.1 Confidence intervals	230
10.6.2 Hypothesis tests	231
10.7 Finite populations	233
Exercises	233

11 Many populations   247

11.1 Small and normal sample test for means	247
11.1.1 The ANOVA test	248
11.1.2 The model-based approach	254
11.1.3 Interpreting the ANOVA result	256
11.1.4 Residuals and assumptions	258
11.2 Small and non-normal sample test for medians	260
Exercises	262

12 Tests for categorical data   269

12.1 One-way chi-square	269
12.1.1 The general test	269
12.1.2 Assumptions	272
12.1.3 The goodness of fit test	272
12.1.4 Optional advanced material	274
12.2 Contingency tables	275
12.2.1 The general $r \times c$ case	276
12.2.2 Testing in the 2×2 case	281
Exercises	283

13 Inference and the regression line   292

13.1 Testing the slope 292

13.2 Testing slope and intercept simultaneously 296

13.3 Prediction intervals 297

Exercises 298

14 Review of estimation and testing   301

14.1 Large samples 301

14.2 Small samples 302

    14.2.1 Normal data 302

    14.2.2 Non-normal data 303

14.3 Bivariate data 304

IV Appendices   305

A Self-assessment guide and mathematical basics   306

B Calculator use   310

C Spreadsheet use   315

C.1 Summary statistics 315

    C.1.1 Sorting data 317

    C.1.2 Case study: Spearman's correlation 317

    C.1.3 Case study: time series moving averages 318

C.2 Probability distributions 318

C.3 Inference 319

    C.3.1 Case study: contingency table test 320

D Tables and table use   323

E Solutions to selected exercises   337

Index   345

Preface

This book was written for New Zealand first-year undergraduate students of statistics. We realise that many of you are required to do statistics to support your major subjects, primarily biology, business, and psychology. Because of this, we have strived to create a text that is relevant, accessible, and streamlined. Examples and exercises are drawn from the local (New Zealand) context, or from work by local researchers. Techniques are explained carefully, and for the non-expert, without the distraction of information over and above that which is required to sensibly apply the techniques. The material has been limited to fit within the scope of a 12-week single-semester course.

Both authors have worked in an environment where people who need to use statistics in their research and workplace have come for consultation. Inevitably, the same questions arise time and again, and it is these recurring questions that this book attempts to address. Our ongoing relationships with people working in the fields of biology, criminology, ecology, econometrics, economics, finance, geography, marketing, psychology and others, have provided a wide range of datasets, and this range, in turn, demonstrates the widespread applicability of the techniques covered.

It is our belief that learning statistics is like learning a musical instrument. To become proficient, the student must not simply watch their instructor perform, they must try it themselves. Ideally, this practice will be frequent, but not too long. Thus, we recommend doing a small number of problems from this book a few times a week. And, as with a musical instrument, it doesn't hurt to play a piece you mastered long ago.

Acknowledgements

This second edition would not have happened without the help and support of many.
- Special thanks to Pearson Education, and particularly Zoë Haws, for providing the impetus and day-to-day support for this update. Much of the variety you see is directly due to Zoë.
- Richard Arnold, Walt Davis, John Haywood, Leigh Roberts and Cushla Thomson gave feedback which helped us greatly improve parts of the original book. Suze Randal identified hundreds of grammatical errors in the draft manuscript.
- Khan (Joey) Au played a crucial role in the production of the first edition of this book, and because of this, producing the second has been almost fun!
- The following people shared fantastic data with us:
  - Colleagues from Victoria University of Wellington: Laurie Bauer, Kevin Burns, Simon Carey, Catherine Duthie, Laura Green, Stephen Hartley, Dean Hyslop, Paul Jose, Martin Lally, James Liu, David McKee, Rachel McKee, Phillip Morrison, Nicole Phillips, Ken Ryan, Susan Schenk, Jessie Wilson and Marc Wilson.
  - Peter Ferguson, Malaghan Institute of Medical Research; Kelly Hare, Celia Lie, University of Otago; George Major, Macquarie University; Andrew Martin, University of Tasmania; Les Oxley, Canterbury University; Michelle Peterson, Lloyd Stringer, The New Zealand Institute for Plant and Food Research Limited; Jeff Radford, Vancouver Island University; Stephen Stannard, Massey University.
  - Paula Acethorp, Meteorological Service of New Zealand Limited, Rebecca Craigie and Carly Harker, Reserve Bank of New Zealand, Jonathan Kennett, Kennett Brothers, Simon Kennett, Greater Wellington, Clio Reid, Department of Conservation, Andrew Robertson, Colmar Brunton and Ross Findlay, Rongotai College.
- A special thank you to those who have contributed to two marvellous projects, without which this book would never have been written: the typesetting system LaTeX (ctan.org) and the R language and environment for statistical computing and graphics (r-project.org).
- Finally, to the thousands of past students in STAT 193 and QUAN 102 have helped us fine-tune the explanations that feature in this book. We will continue to learn.

For our children: Jasper, Max and Isabel; and Kaitlyn. Thank you all for your never-wavering patience and support.

Chapter 1

Introduction

1.1 What is statistics, the subject?

The subject matter of statistics is numerical data, its method of collection and analysis. There is always more than one way to analyse any set of data, and we need to understand which techniques are legitimate in any particular situation. As in most disciplines, if you start from different assumptions, you get different answers.

The word *statistics* comes from the description of a state (political entity). We still use it in this context in those official statistics where complete records are kept, e.g.

- census data
- registers of births, deaths and marriages
- registers of endangered species
- prison populations
- companies register, etc.

The more modern usage is concerned with data that have some element of *uncertainty* associated with them, and the methods we use for drawing conclusions from these data.

Uncertainty in data can be due to:

- Intrinsic randomness in phenomena, e.g. variation in the maximum daily temperature in Wellington according to the time of the year, and also due to local weather patterns.
- Sampling, where we access only a small part of the entire population, e.g. 30 randomly selected New Zealanders will have an average age different from that of a second group of 30 New Zealanders.

- Experimental errors, e.g. our inability to control experimental conditions fully.

Not until the 17th century did reasoning from data occur in a formal way, i.e. generalising patterns observed in sample data to the population from which they were drawn. In the life and social sciences, this reasoning from data began in the 19th century with Florence Nightingale, and Adolphe Quetelet.

Recently, the use of statistics has spread to disciplines such as law, especially in balance of probability arguments. Here, and in other areas, we take a conservative approach. When uncertainty in data is high, we are cautious when we draw conclusions.

1.2 Using this book

Dealing with numerical data requires us to use basic mathematical tools frequently. In this book, we do not make any attempt to teach basic mathematical concepts; however, through practice, the user's confidence and familiarity with these tools will increase. Appendix A outlines the *minimum* mathematical requirements students should have before embarking on this course of study.

We recommend purchasing a calculator that is capable of *bivariate* statistical analysis. A cheap and easy-to-use calculator that satisfies this requirement is the Casio *fx-82MS*. We give worked examples using this calculator for all the relevant statistical analyses in Appendix B. If you have a different calculator, use your manual to follow the examples in Appendix B to ensure you can correctly use your calculator to do the calculations. Used properly, your calculator will be a great time saver, and will also protect you against mis-calculations due to sloppy arithmetic mistakes.

While a calculator is very convenient, it is hardly a "modern" computational tool. Appendix C describes how many of the techniques in this book can be implemented in a spreadsheet – a common computer application. A spreadsheet is a much more powerful computation tool than the calculator, but brings with it some drawbacks, namely portability and transparency.

Each chapter has a set of exercises which are drawn from the major disciplines of biology, business and psychology, as well as general examples drawn from current New Zealand society. In the early chapters the exercises primarily use societal data, much of which have been obtained from official New Zealand government sources. Towards the end of the book the exercises have a much higher academic content, and many are from specialist academic journal literature. Selected exercises have sketch solutions in Appendix E.

The book itself is divided into four major parts.

Part I: Summarising data In this first part, we move from a data set containing numbers to a graphical representation of the data, or numerical summaries. Graphs include histograms, stemplots, bar charts and boxplots, as well as scatterplots for representing paired data. Numerical summaries include means and medians, standard deviations, quartiles and ranges, which we also calculate from grouped data. For paired data, we present Pearson's and Spearman's correlation coefficients, and look at estimating a linear relationship using regression.

Part II: Introduction to probability The second part contains a very limited introduction to probability. Some basic rules of probability are stated. We encounter two probability distributions: the binomial, and the normal. These, and the basic notion of a probability, are crucial prior knowledge for the third part of this book.

Part III: Estimation and testing The third part contains the most important material in this book. It covers the linked ideas of confidence intervals and hypothesis testing. We first present these ideas for a single population, progressing from the most standard case, where we test a mean based on a large sample, to more advanced situations. Inference for means, medians, and proportions is covered. Following this, we extend the tests to two populations, and then to many populations. Inference for categorical variables is also covered, and we return to the regression line with related tests. The final chapter summarises the tests and intervals covered.

Part IV: Appendices The final part of the book contains useful supplementary material. Much of the content has already been mentioned: a self-assessment guide for basic mathematical skills, calculator and spreadsheet instructions, and solutions to selected exercises. Also given are the probability tables required to complete examples and exercises throughout the book.

1.3 Course structure

It is unlikely that all the material in this book will be covered in a twelve-week university statistics course. There is also no guarantee that topics will be presented in the same order as here – there is scope for reordering chapters, or material within chapters. We hope the layout of the book will make such changes a relatively painless exercise for instructor and student alike. We recommend spending roughly three weeks on Part I, two weeks on Part II, and seven weeks on Part III.

Part I

Summarising data

Chapter 2

Data collection, organisation and graphical representation

The purpose of collecting data or experimenting is to make a discovery about a situation. The overall aim is to extract information from data in order to understand what is happening in the situation you are investigating. This understanding is normally difficult to get from a data set which consists of a large group of numbers. We first try to "clean" the data, i.e. get rid of noise in the data. Noise is variation in the data that is not intrinsic to the data but caused by outside factors such as errors in reading instruments. We look for sources of variability by looking for patterns in the data. Then we ask questions like:

- Is there a trend in this mass of data?
- What is a typical value?
- Are there clusters in the data and what do they mean?

It is difficult to answers questions like these just by studying pages of numbers or a computer screen. The techniques we use to describe the overall characteristics of the data are called descriptive statistics and the process of coming up with these statistics is sometimes called data mining.

2.1 Classifying variables

To get a feel for our data we usually start by collapsing it into a table or displaying it in a graph in order to see any overall pattern or any oddities in the data. Then we summarise its characteristics. In order to do this we need a few technical terms, especially the notions of *population*, *sample* and *variable*.

> The *population* is all the values, people or objects we are interested in.

We need to be very clear what the population of interest is. If we are too vague we may end up wasting our time taking a sample from a slightly different population. For example, suppose we specify the population we are interested in as: "the population of all 20-year-old psychology students," but then collect data only from Victoria University. The definition of the population is not specific enough: do we mean psychology *majors*, students at this point in time, students at our university only, registered students, first years, second years, third years, any level? When we give a clear definition of the population we are interested in, and then collect data from that population, any information we gain from the sample can then be used to make statements about the population we have defined.

Once we have clearly identified the population we are interested in then we can investigate it. For convenience, populations are often labelled with capital letters, e.g. X, U, W,... and the number of individuals in the population is usually labelled N. Any single individual in a population is denoted by a lower-case letter. This is shown in Figure 2.1.

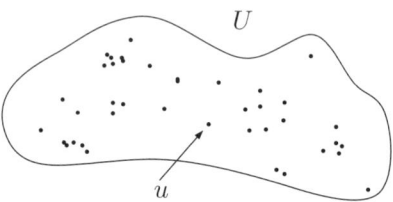

Figure 2.1   The population U, and an individual u

Note that population members do not have to be alive (unlike common usage of the word). They could be people, possums, therapies, fungi, policy options, the corpus of colloquial English, stocks, companies, etc. Each member of the population will have at least one characteristic that we are interested in measuring, for example: cost, number of syllables, anxiety level, size, etc.

Example 2.1

We might be interested in the population of Samoans born in and living in New Zealand. Let U = the population of Samoans born in and living in New Zealand. Then $N \approx 90{,}000$. X might be age, anxiety level, occupation, religion or some other characteristic of this population. The first two of these have numerical values while the second pair is not numerical.

The measurement we are interested in is usually labelled X or Y, and any data we collect on this measurement is labelled with the corresponding lower-case letter, e.g. x or y.

2.1 Classifying variables

> *Variables* are the characteristics of the population.

Example 2.2

If we are interested in the population of New Zealanders, there are many variables we could measure, including: age, sex, marital status, religion, occupation. These variables, and more, are collected from the whole New Zealand population in national censuses every five years by Statistics New Zealand.

Specific ages and sex are *values* of the variables age and sex. Depending on the actual variables we are dealing with, these values may or may not be numbers.

Example 2.3

- Typical values of the variable *age* are: 25, 32, 47, etc. We usually round our age down to a whole year figure.
- Typical heights of people are: 1.65 m, 1.78 m, etc. We usually round our height to the nearest centimetre.
- Typical numbers of people living in the same house are: 1, 2, etc. These are counts and do not need to be rounded.
- Values of the variable sex are "male" and "female".
- Values of the variable occupation are: professionals, clerks, service and sales workers, trades workers, etc. Many individual occupations will fall into these broad categories.

In order to make more sense of the different types of variable, we classify them according to the tree shown in Figure 2.2.

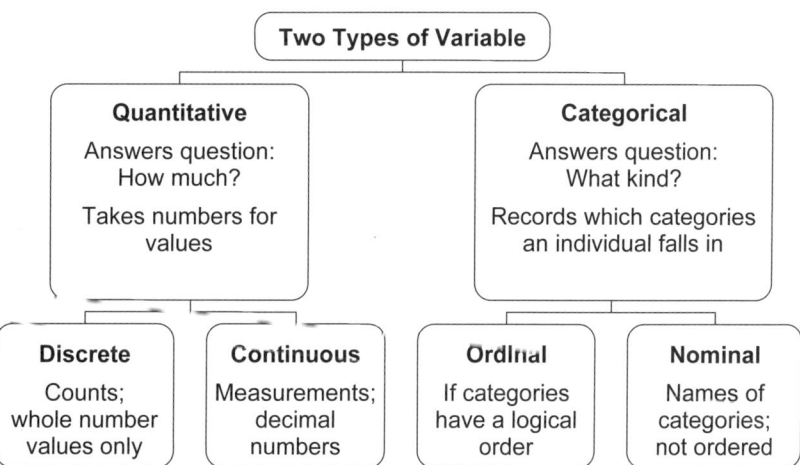

Figure 2.2   Classification of variables

8 Chapter 2 Data collection, organisation and graphical representation

The first sort of variable indicated is a *quantitative* variable.

> *Quantitative* variables are typically *measurements* or *counts*. They are numerical variables.

Quantitative variables answer the question "how much?" or "how many?" Almost all numerical variables are quantitative, and they can be further divided into *discrete* variables or *continuous* variables. Discrete variables are counts, or numbers that can be measured exactly, with no rounding. Typical examples of discrete variables are: numbers of individuals, prices of goods at the supermarket ("measured" in cents), etc. Continuous variables cannot be measured exactly, and if we could improve our measurement device, we could also improve our measurement, i.e. the value of the continuous variable. Typical examples of continuous variables are: heights, weights, etc.

The second sort of variable is a *categorical*, or *qualitative*, variable.

> *Categorical* variables describe the category which population members fall into.

Categorical variables answer the question "what kind?" Their values are typically words, e.g. "male" and "female". The values of some categorical variables are numerical, but these numbers do not have numerical meaning; they are just labels associated with the particular categories. It is important that we distinguish between categorical variables with numerical labels, and quantitative discrete variables. Some categorical variables have a built-in order, e.g. opinions "strongly agree", "agree", "disagree" and "strongly disagree", and these are called *ordinal* variables. If the categorical variable has no built-in order, e.g. sex, or occupation, then it is a *nominal* variable.

Example 2.4

- The number of refugees accepted by Australia in a given year is quantitative (answers the question "how many") and discrete as we expect integers.
- The weights of babies born in a given city in a given week is a quantitative variable ("how much?") but this time we have a continuous variable as we expect to get decimal numbers such as 3.27 kg for our data. The weights must be measured, rather than counted.
- The number of new emigrants in each of the categories Somali, Vietnamese, Chinese, is a situation with two variables: the number, and the ethnicities. The number of people is a count, and is a quantitative discrete variable. The ethnicities are categories without order, so this variable is nominal.
- The data gathered from a questionnaire in which people respond to the question *how do you like your coffee?*, by checking the appropriate category below

    Strong ☑   Average ☐   Weak ☐

    is categorical and ordinal (the category "strong" is stronger than "average", which in turn is stronger than "weak"). Although we should not put a number value on these categories, any numbers allocated to "strong", "average" and "weak" would have to be in order, e.g. 1, 2, and 3, or 3, 2, and 1. Because this variable is not numerical, which sequence we choose is not important.

A categorical variable is a special sort of *qualitative* variable. Qualitative variables focus on characteristics which are *qualities* rather than numerical *quantities*. Other qualitative variables include information gathered in interviews and diaries, and other non-quantitative data. We will not be covering such data in this book.

2.2 Sampling

In order to gain a better understanding of a population, we collect data from the members of that population. In a *census*, information is collected from every member of the population. For example, we could think about taking a census of pupils in a school, or of birds in a colony.

A census is not always the sensible thing to do, for a number of reasons. It may be too expensive, or may be so time consuming that the results can't be reported before the time they are needed. Also, a complete census may be unnecessary, e.g. there is no need to drink a whole cup of coffee before you can make a decision on whether it is sugared or not. Collecting the whole population may be impractical for other reasons: we do not require your entire blood supply in order to test for certain antibodies.

In most situations instead of taking a census we collect data from a subgroup of the population. This process of data collection is called *sampling*, and the data we collect is called a *sample*.

> A *sample* consists of n measurements from a population on a variable X. The sample values are denoted
>
> $$x_1, x_2, \ldots, x_n$$
>
> where n is called the *sample size*.

In a census, the sample size is equal to the population size: $n = N$.

Having obtained a sample, we then use the properties of the sample members to describe the properties of the whole population, e.g., the proportion of people who are unemployed in a sample of working-age people may be used to estimate the unemployment rate in the whole population. However, an important question immediately arises: how should we select a subgroup from the population to form our sample?

We want the sample to be in some way *representative* of the population, and so it is important that there be no bias in our selection; we don't unconsciously slant our sample by selecting subjects on the basis of ethnicity, appearance, or any other variable. Even worse would be the sample members themselves deciding whether or not to be in the sample. This leads to *self-selection bias* which occurs in phone-in or text-in surveys. In such surveys only people with the most extreme viewpoints will phone or text in to express their opinions, and their views will not be representative of the population as a whole.

The solution is to take the selection process entirely out of the hands of both the sampler and the subjects being sampled, and use *random sampling*. This means that every member of the population is given a chance of being selected, and the actual sample members are selected by generating a sequence of random numbers.

Example 2.5

Q: Select a random sample of $n = 10$ individuals from a population of $N = 852$.
A: We begin by numbering the 852 individuals. (A convenient way to do so would be using a spreadsheet; see Appendix C.) The individuals then have an ID number from the list $001, 002, 003, \ldots, 852$. We now need to randomly choose ten of these IDs and we can do so using the random number table on page 324.

First, we must randomly select a starting position from this table. For this example we will start from the 3rd digit on the 10th row and move along the row and down, giving:

```
621   94478   18444   55082   35446   20242   12008   13452   82374
88817 93562   09253   97167   95526   94725   99285   49859   ...
```

Grouping these into three-digit numbers, we have

```
621  944  781  844  455  082  354  462  024  212  008  134  528
237  488  817  935  620  925  397  167  955  269  472  599  285
498  ...
```

and these can now be compared with the three-digit IDs allocated to the population individuals (001-852). The first ten valid numbers are: 621, 781, 844, 455, 082, 354, 462, 024, 212 and 008. Note that the second number (944 in the table) is discarded since there is no population member with this ID. There are no repeats in the list so the sample would be individuals with ID: 008, 024, 082, 212, 354, 455, 462, 621, 781 and 844.

The example above outlines selection of a *simple random sample*, where each member of the population has an equal probability of being selected. Notice that this method assumes that we know in advance that there are exactly 852 individuals in the population. With a large simple random sample we can usually be confident that the properties of the sample will be similar to those of the population. However, some care is needed even so. Consider the following issues.

- Can we actually get valid measurements or responses from each of the individuals we select? What should we do if someone refuses?

- In a trial of a new antibiotic, we take a random sample of patients from one city. Do the results we find generalise to people from elsewhere?

- In the example above, are the ten in the sample representative of the population? What if it turns out that all individuals in the sample are males, when only 50% of the population are male? Should we replace some of the sample?

The first of these issues is not simple to resolve. A practical example of its occurrence includes collecting data using a telephone – not everyone owns a telephone, and not everyone who owns a telephone is prepared to answer surveys on it. If the measurement being sampled is completely *independent* of the reason the individual is excluded from the sample, then this isn't a big deal. However, it is very difficult to be confident that this is true, e.g. people who will not respond to phone surveys are unlikely to be a random subgroup of the population.

The second issue relates to the *assumptions* that researchers make when setting up surveys and experiments. The researchers' interests may be in a large population (all patients everywhere), but since they can't possibly sample that whole population, they *assume* that the local population of patients are similar to those elsewhere. They may or may not be. There may be genetic differences between the local population and people elsewhere which may influence the results. When sampling a cup of coffee to see if there is sugar in it, we may need to *assume* that it is well mixed (and that there isn't a pile of undissolved sugar at the bottom of the coffee cup).

Throughout any statistics course or use of statistics we need to always have at the back of our mind the questions: *what am I assuming here?* and *what limitations do my data have?*

The third issue, i.e. the sample characteristics being non-representative of the population, exposes a risk of random sampling; namely, that we might get, purely by chance, a sample that really is rather different from the population. For example, if we were to take a simple random sample of pupils from a school, there is a chance, though a very small one, that we might select no students from the youngest year group in the school.

One way to avoid such an eventuality is *stratified sampling* which we illustrate in the following example.

Example 2.6

Q: A high school has 1191 pupils enrolled in five year groups (Years 9 to 13).

Year Group	9	10	11	12	13	Total
Pupils	242	231	257	259	202	1191
Percentage	20.3%	19.4%	21.6%	21.7%	17.0%	100%

If we were to take a simple random sample of 100 pupils from this school we would expect to end up with roughly, but not exactly, 20% of the sample (i.e. approximately 20 pupils) from each year group. Every pupil in the whole school has the same chance of selection: $\frac{100}{1191} = 8.4\%$.

In order to ensure exactly 20 pupils are selected from each year group we can take a *stratified* sample. We select *separate* simple random samples of exactly 20 pupils from each year group using the simple random sampling procedure in Example 2.5. In this procedure the Year 9 pupils would be given IDs 001 through 242, and two would be selected randomly; the Year 10 pupils would be given IDs 001 through 231, and selected using a *different* sequence of random numbers, and so on for all five year groups.

Notice that in the stratified sample the probability of being selected depends on which year group a pupil is in: a Year 9 pupil has a $\frac{20}{242} = 8.3\%$ chance of selection whereas a Year 13 pupil has a $\frac{20}{202} = 9.9\%$ chance.

Stratified sampling is not without its own challenges. In particular the fact that different individuals can have different chances of selection means that we have to use special statistical methods when working with stratified samples. Also, the sample in the example above may be stratified by year group, but it has not been stratified by gender, ethnicity, weight, etc.

In summary, collecting a "good" random sample can be difficult to do, and sampling should always be done with care and with methods that are well thought through. Beyond trying our best, we must be sure to accompany any analysis with a clear warning about the limitations of our results. They may not be generalisable to the whole population, but rather a sub-population determined by aspects of our sampling process.

Example 2.7

Q: Comment on the limitations of the following samples.
(i) 20% of adults in a single colony of penguins.
(ii) A random sample of first year psychology students is used to learn about anxiety levels in young adults.
(iii) A survey conducted on the street at lunchtime is used to measure attitudes towards a recent government announcement.
(iv) Health of trees adjacent to a dirt road is used to indicate health of trees within a large block.
(v) A "researcher" asks his friends and neighbours to fill in a survey for a paper on consumer habits.
(vi) Media focus attention on a couple who won Lotto in two consecutive weekends.

A: (i) The single colony of penguins that was sampled will be likely to have much less genetic variation than the species as a whole. Findings from this sample might not generalise to the species.
(ii) The students have at least two things in common – they are all university students and are all studying psychology. They might be systematically different from law students, from polytechnic students, from young adults in the work force, etc.
(iii) The people who answered this survey all stopped, and they might prove to have different attitudes to those who did not. The interviewer may also have subconsciously stopped people who looked like her, thereby eliminating diversity from the sample. The sample excludes people who aren't able to wander the streets at lunchtime.
(iv) While convenient to observe, these trees may face different light, soil and weather conditions due to their location.
(v) This reseacher's friends may have similar habits to his own, and those in his neighbourhood are likely to share the same socio-economic status, and therefore have similar shopping habits.

(vi) This couple is a sample of the Lotto-playing population. It is very unlikely that anyone would win Lotto in two consecutive weekends. But, very many people are playing Lotto, and among a very large group, observing one person with this unlikely characteristic is not so surprising.

The final instance in the previous example is known as *data dredging* or *data snooping*, where an unexpected phenomenon is observed in a sample. This can be legitimately used to motivate further research, but the original sample should not be used further.

We now step away from these difficulties, assume that we do have "good data", and focus on techniques with which we will analyse them. While these will often be difficult to master, the issues we are now assuming away are much more thorny!

Once the sample data have been acquired then we look at the data and try to organise it in order to answer questions such as: do we have lots of large values and only a few small? Do we have a fairly even numbers of large and small values? Are there any values which are much more common than others?, etc. To answer these questions we look at how the data are *distributed* or the shape of the data. There are many techniques available and we will consider some of the most commonly used in the next sections.

2.3 Stemplots, dotplots and histograms

Stemplots, dotplots and histograms are three techniques which give the same basic summary of the data by looking at how they are distributed. They can be used for all quantitative data, although the histogram is specifically designed for use with continuous data. Implementing these techniques for sample data is discussed in the following sections.

2.3.1 Stemplots

A stemplot is a picture of how the data are distributed. They are sometimes known as stem and leaf plots. They are very useful with small data sets and professional statisticians use them all the time in the initial stages of a problem in order to get a feel for what the data are like. It is easiest to see how they work through an example.

Example 2.8

Fifteen psychiatric nurses in an acute ward were given a test to measure their level of stress (in order to see if they were approaching unsafe levels). A high score in this test indicates a high level of stress and the maximum possible score is 100. The data follows:

58 68 73 73 44 61 67 41 37 66 04 72 71 44 39

To draw a stemplot of these data:
(1) Separate each data value into a *stem* and a *leaf*. The leaf is the single digit at the right-hand end of the value. The stem shows the remaining digits, and has no set

length. Write with a vertical line separating the stem and the leaf. So the first value 58 will be written 5|8 where 8 is the leaf.

(2) Scan the data, identifying the smallest and largest values. Here 37 is the lowest and 74 the highest. These extreme values give you your smallest and largest stems. List these vertically to the left of a vertical line.

```
3 |
4 |
5 |
6 |
7 |
```

(3) Add the leaves to the right of the vertical line in the appropriate place. Looking at the first three data values, 8 goes after the stem of 5, 8 after the stem of 6 and 3 after the stem of 7, etc. Since we are using the stemplot to look at the shape of the data, it is important that the leaves get an equal space, and that they are vertically aligned, as shown.

```
STEM | LEAF
  3  | 7 9
  4  | 4 1 4
  5  | 8
  6  | 8 1 7 6 4
  7  | 3 3 2 4
```

This is an *unordered* plot.

(4) Redo, ordering the leaves from smallest to largest for each stem with smallest next to the vertical and working outward (to the right). This gives the *ordered* plot.

```
3 | 7 9
4 | 1 4 4
5 | 8
6 | 1 4 6 7 8
7 | 2 3 3 4
```

(5) Put in a *key* so that it is clear that 5|8 = 58 and not 5.8 or 0.58.
Note: any decimal points or commas in a data value are not recorded as this information is all held in the key.

```
3 | 7 9
4 | 1 4 4              key
5 | 8                  5 | 8 = 58
6 | 1 4 6 7 8
7 | 2 3 3 4
```

The stemplot gives us an immediate feel for the shape of the data, without any distractions such as commas, decimal points, or extraneous significant figures. The centre of the data is apparent (around 58 to 60), as are any extreme or outlying data, and the largest and smallest values. The most striking feature of this stemplot is that the data seem to split into two groups each with a peak (modes).

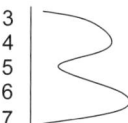

Such a graph is called *bimodal*. Perhaps this indicates that we have two groups of nurses, the group that is doing OK and the group that is at risk of burnout.

The stemplot is an excellent way to sort data. It helps because of the way we first allocate the data to a number of categories (according to the stem) leaving us with smaller lists of numbers (the leaves) to sort. Stemplots can also be used to compare two data sets by drawing a *back-to-back* plot of the data.

Example 2.9

A further group of 13 nurses from a medical ward score the following on the stress test:

$$45 \quad 50 \quad 48 \quad 38 \quad 47 \quad 35 \quad 62 \quad 39 \quad 52 \quad 71 \quad 39 \quad 51 \quad 48$$

We can attach this to our previous stemplot like this

```
     Medical        Psychiatric
     9 9 8 5  | 3 | 7 9
     8 8 7 5  | 4 | 1 4 4              key
       2 1 0  | 5 | 8                  3 | 7 = 37
           2  | 6 | 1 4 6 7 8
           1  | 7 | 2 3 3 4
```

Note that we must now carefully label which plot belongs to which data set. Also note that in the left-hand plot the values of the leaves increase from right to left. *Both plots have their smallest digits closest to the stem.* From this back-to-back plot we can clearly see that the medical ward nurses are less stressed than the psychiatric nurses with the values skewed toward the lower end of the scale.

With larger data sets you often get very long strings of data. For convenience we can split each stem into two, one carrying leaves of 0 to 4 and the other carrying leaves of value 5 to 9. Such a stemplot is called a *split-stem* stemplot.

Example 2.10

$$2 \mid 0\,1\,1\,2\,3\,3\,4\,5\,5\,6\,7\,8\,9$$

Becomes

```
     2 | 0 1 1 2 3 3 4
     2 | 5 5 6 7 8 9
```

Smaller numbers such as 5.3, 4.7 ... and 0.58, 0.64 are written as

```
     4 | 7           5 | 8
     5 | 3           6 | 4
```

with a key 4|7 = 4.7 for the first, and 5|8 = 0.58 for the second. Without the key, the two stemplots look very much the same, and we cannot infer what the original data were. If your data consist of larger numbers such as 585, 603, 597 your stemplot would look like this:

$$
\begin{array}{c|c}
58 & 5 \\
59 & 7 \\
60 & 3 \\
\end{array}
\qquad \text{key} \\
\qquad 58|5 = 585
$$

Note that there is always just one digit in the leaf and that is the rightmost one. If the range of the data is too great, and the data have many significant figures, then a stemplot can be impractical as we need too many leaves. A dotplot may give a better representation of the data.

2.3.2 Dotplots

Dotplots can be used with large or small data sets. You draw a horizontal line scaled appropriately for your data and individual pieces of data are represented by dots above this line. Duplications of individual data values are stacked on top of each other. The data from the 13 medical ward nurses above would be represented on a dotplot as follows:

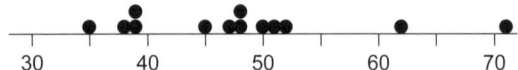

2.3.3 Histograms

When we have a large data set that consists of quantitative (measurement) data then one of the graphs most commonly used to describe and summarise the data is the histogram. The shape information it provides is not unlike the stemplot or dotplot, but it is much more useful for large data sets because we no longer retain the individual values, and these do not clutter the graph.

To draw a histogram the range of values in the data is found and then divided into subgroups also known as classes. The number of values in each subgroup is counted and the count or per cent of the data in each class is displayed in a bar. These counts are called *frequencies*.

Frequency = number of observations of a particular type, f
Relative frequency = frequency ÷ n
= frequency ÷ $\sum f$
= proportion
A table of frequencies is called a frequency table or *frequency distribution*.

2.3 Stemplots, dotplots and histograms

Historically, the area of a histogram bar is equal to the number (frequency) of data values in the subgroup. More common these days is the practice of giving all the subgroups the same range and then we only need to make the bar heights proportional to the counts. In this book we will use the convention that the subgroups must be of equal width. The easiest way to see how this works is through an example.

Example 2.11

The following is a set of data from fifty 10-year-old children from the three lowest socio-economic groups on a test to measure literacy level (maximum score = 100). It is only possible to score in integers so the data are discrete.

```
43  59  68  54  60  51  61  78  55  65  58  66  49  61  73  60  69
51  69  50  65  55  79  63  42  66  46  61  72  59  66  56  69  42
64  50  60  51  68  57  77  62  74  58  68  59  67  43  60  70
```

This is not a very large data set by statistical standards but already you can see how difficult it is to get a feel for the data by eye. To construct a histogram of this data we start by drawing up a frequency table. This has four steps:

(1) Find the maximum and the minimum of the data set. Here they are 79 and 42 respectively. The *range* of the data is the difference between the maximum and the minimum. Here the range $= 79 - 42 = 37 \approx 40$. [This curly equals sign $\approx$ means "is approximately equal to".]
(2) Divide this range into classes of equal size. You have to choose how many classes and there is no absolute right way to do this but a useful rule of thumb is to choose between 5 and 12 classes. If you choose fewer than 5 or more than 12, you will often obscure the shape of the histogram because you divided the data too coarsely or too finely. In this case $40 = 8 \times 5$ so choosing 8 groups each of width 5 seems a reasonable idea. They should cover the entire range of the data, and not overlap.
(3) Using a tally if you like, *count* the number of pieces of data in each group and record in a table. Using the data set above and following these steps results in the following frequency table:

Class	Tally	Frequency f_i
40 - 44	||||	4
45 - 49	||	2
50 - 54	⩗|	6
55 - 59	⩗ ||||	9
60 - 64	⩗ ⩗	10
65 - 69	⩗ ⩗ ||	12
70 - 74	||||	4
75 - 79	|||	3
		$\sum f_i = 50 = n$

The end points of range of values that specify a class, e.g. 40 and 44 in the class 40 - 44 are called *class limits*. f_i = frequency in the ith class so $f_1 = 4$ (the frequency in the first class) and $f_5 = 10$. The middle column is usually not shown in the finished frequency table.

(4) Rewrite the table using *class boundaries* instead of class limits so that there are no gaps between classes and so that there is no ambiguity about where any score should

be counted, i.e. choose boundaries so that no actual score lies on a boundary. There are several ways to do this but the simplest is the following:

Class boundaries	Frequency f_i	Relative frequency p_i	
39.5 - (44.5)	4	0.08	$(\frac{4}{50})$
44.5 - (49.5)	2	0.04	$(\frac{2}{50})$
49.5 - (54.5)	6	0.12	.
54.5 - (59.5)	9	0.18	.
59.5 - (64.5)	10	0.20	.
64.5 - (69.5)	12	0.24	
69.5 - (74.5)	4	0.08	
74.5 - (79.5)	3	0.06	
	$\sum f_i = 50 = n$	$\sum p_i = 1$	

We use 39.5 - (44.5) to represent values greater than or equal to 39.5, and less than 44.5., i.e. $39.5 \leqslant x < 44.5$. Note that relative frequencies are always less than 1 as they represent fractions of a whole and they always add to 1.

Now we are in a position to draw the histogram. Start with a vertical and horizontal axis. The vertical axis represents the frequencies (or relative frequencies) and the horizontal axis represents the values of the data. We draw one vertical bar for each group and the height of the bar is the frequency (or relative frequency). There are no gaps between bars unless there is a class which is empty, i.e. has no scores in it.

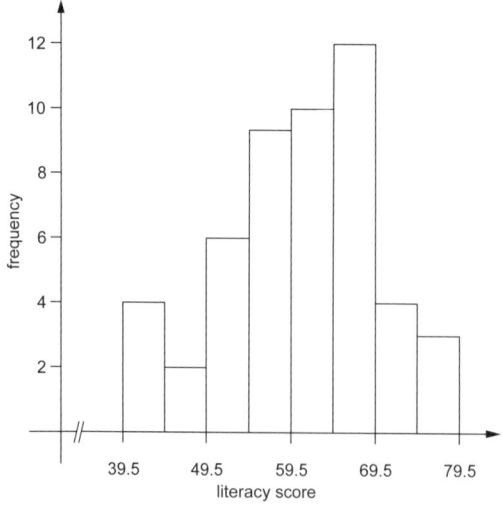

The symbol // through the horizontal axis in the histogram above indicates that the horizontal axis is not starting from 0. If we don't break the axis, we can end up with a lot of wasted space, as shown in the following representation of the bottom part of the histogram. The vertical axis should never be broken.

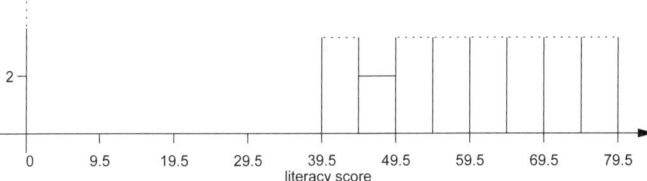

A good histogram should have an explanatory title and the scales of each axis should be clear. Both axes should be labelled.

You can draw a frequency histogram and a relative frequency histogram on the same graph simply by adding another vertical axis on the right for relative frequency as shown in Figure 2.3.

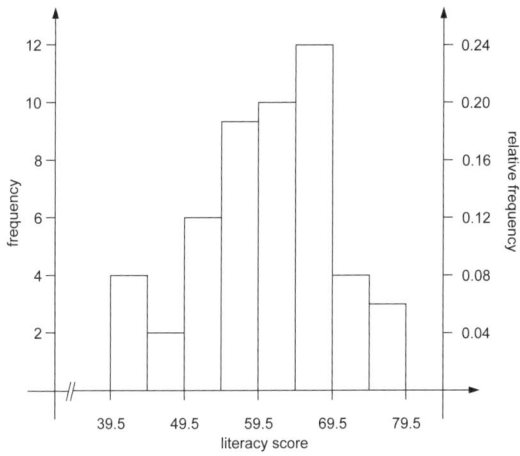

Figure 2.3 Histogram with a relative frequency axis

Whenever we wish to compare two groups of different sizes, relative frequency histograms are preferable, e.g. comparing the scores of 50 low socioeconomic children with the scores of 200 middle decile children. If we compare using frequency histograms, the one for the larger sample will dwarf the smaller one, whereas the heights of both relative frequency histograms will add to one.

Histograms may have several peaks (or modes). Multiple modes often imply the existence of several distinct groups in the data.

In the social sciences, especially psychology, data are often ordinal, for example coming from a questionnaire where people have to respond by checking the category that best describes their level of agreement with a statement from

usually coded as 4, 3, 2, and 1. Common practice in such a situation is to regard the data as continuous but rounded so that a histogram can be used.

It can be useful to have a single line that approximates the histogram. This helps to show the overall shape of it and any trends. Such a line is the *frequency polygon* and is used a lot in the social sciences, geology and physical geography. It is constructed by marking the midpoints of the tops of the histogram bars and joining these with straight lines. At each end the line goes to the axis point half-a-class-width away. This would be the midpoint of the next class, with zero frequency. At the left-hand end if the first group of the histogram starts at zero you have the option of starting at zero instead. Frequency polygons are shown with and without the histogram in Figure 2.4. These are very useful for comparing the shapes of the distributions of two or more data sets.

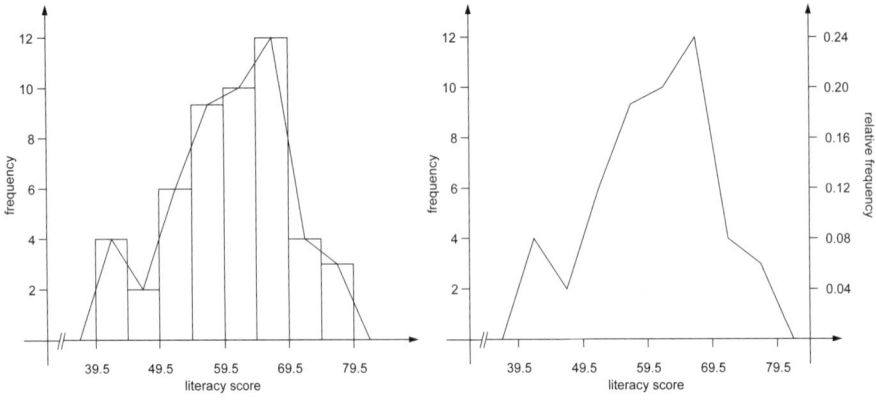

Figure 2.4   Frequency polygons

2.4   Bar charts

Data that are not measurements but counts of individuals within categories are displayed in a *bar chart*. For categorical data this is the chart of choice. We arrange the categories along the horizontal axis of the plot, and place a bar for each category with height corresponding to the frequency or relative frequency in this category. Unlike in the histogram, we leave a space between the bars, to reflect the fact that the categories are separate from one another, and to highlight the fact that we are dealing with categorical rather than measurement data.

If the categorical variable is ordinal, then the categories should be in order on the horizontal axis. Nominal variables do not have a built-in order, so we may order them in a variety of ways. Common ordering schemes are: alphabetical, or in descending or ascending order according to observed frequency.

Example 2.12

Here are three bar charts:

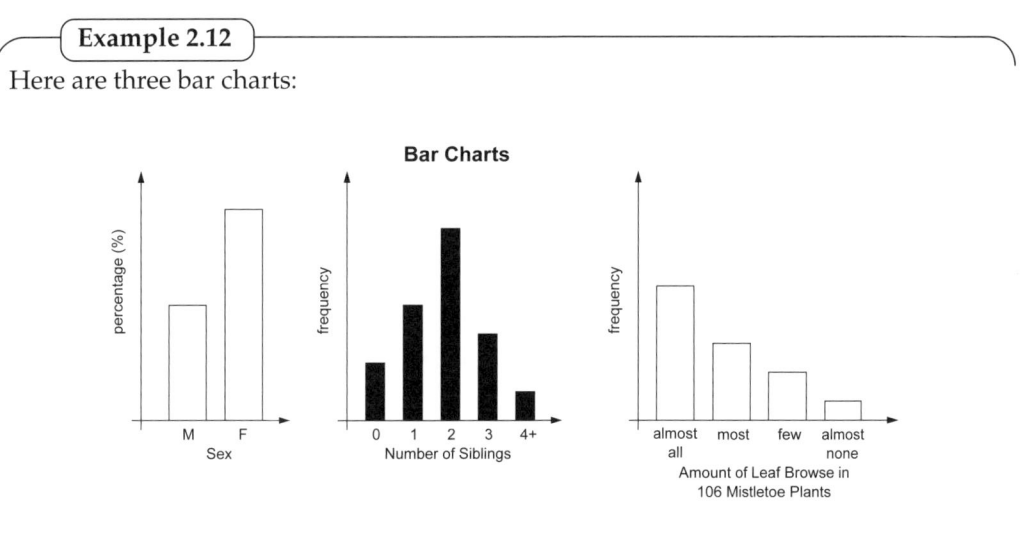

There are variations such as stacked bar charts and side-by-side bar charts in more complicated situations.

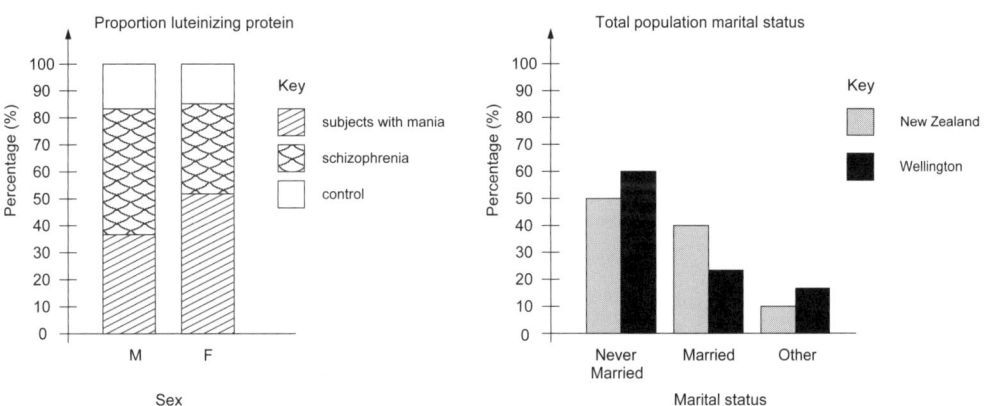

Figure 2.5   Stacked and side-by-side bar charts

Another graph used to look at behaviour between groups or over time is the *line chart*, where we omit the bars, and join the tops of the bars by a line. This form emphasises changes from one category to the next. If the categories are time, then the plot is called a *time series plot* (see Chapter 5).

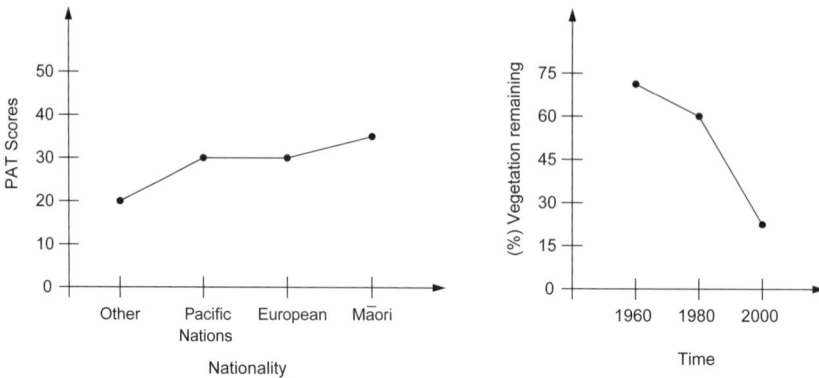

Figure 2.6 Line charts

2.5 Summary

Graphical tools are an important first port of call for a statistical analysis. Visualisation of the data allows us to clearly identify some of the basic features of the data.

When we are comparing two groups, we need to draw plots that are on the same scales, so that the comparison is fair and accurate.

All of these graphs help us get a feel for the shape of a data set but we usually need more than that in order to get a clear idea of the data. We need to find numbers that describe characteristics of the data such as where the middle of a data set is, what its maximum and minimum values are, what range of values is spanned by the data set, whether most values are close to the middle or very spread out and so on. These properties of the data will be described in the following chapter.

Exercises

Classifying variables

For the following questions, use the framework in Figure 2.2. The numbers after the questions refer to the exercise in which they appear.

2.1 Classify the following variables:
  (a) size of natural gas reserves (2.8)
  (b) music genre (2.29)
  (c) prevalence of smoking (3.11)
  (d) population size (3.12)
  (e) number of drinks (3.36)
  (f) spill size (medium/large) (5.13)
  (g) size of loss inflicted (7.12)
  (h) age group (12.35)

2.2 Classify the following variables:
  (a) number of drugs used (2.31)
  (b) volume of spirit-based drinks available (5.18)
  (c) depression score (5.22)
  (d) number of active litterers (6.12)
  (e) time spent ruminating (10.6)
  (f) frequency of child memory contributions (10.9)

(g) level of competence of ecological adaptation (10.10)
(h) level of cannabis use (12.9)

2.3 Classify the following variables:
(a) male kea weights (2.9)
(b) kea gender (2.9)
(c) oil spill size (4.16)
(d) distance to closest large landmass (4.21)
(e) number of conifer species (4.21)
(f) soil acidity level (4.25)
(g) accuracy of tail loss (9.44)
(h) water quality (12.28)

2.4 Classify the following variables:
(a) purpose of cellphone usage (2.27)
(b) real returns on land (5.9)
(c) overnight cash rate (5.3)
(d) household rent (5.4)
(e) insulation payback period (9.10)
(f) country (12.11)
(g) number of firms delisted (12.11)
(h) level of formal education (12.12)

Sampling

2.5 Using the random number table on page 324, describe how to collect a simple random sample of $n = 12$ individuals from a population of size:
(a) 12,034;
(b) 5000;
(c) 40.

2.6 Read the introduction to the exercises listed below, and found later in the book. For each, describe which aspects of the sampling technique you would need clarified before generalising to the population (which you should identify).
(a) Cellphone usage (2.27)
(b) Fishing activities (2.34)
(c) Female labour force (4.5)
(d) Consumer rights awareness (9.26)
(e) Travel mode to work (9.36)
(f) Sex before marriage (9.41)
(g) The "Oracle of Oberhausen" (9.37)
(h) Anxiety levels (10.33)

2.7 A July 2010 stuff.co.nz poll asked readers: "should police have greater access to guns?" Before placing my vote, I read there had been 6461 "yes" votes and 1851 "no" votes. Comment on this sampling method. What might be learned from these data?

Stemplots and dotplots

2.8 The following proved natural gas reserves (in trillion cubic metres), recoverable "in the future from known reservoirs under existing economic and operating conditions", are given in the June 2009 *Statistical Review of World Energy*.

Country	Reserve
Australia	2.51
Bangladesh	0.37
Brunei	0.35
China	2.46
India	1.09
Indonesia	3.18
Malaysia	2.39
Myanmar	0.49
Pakistan	0.85
Papua New Guinea	0.44
Thailand	0.30
Vietnam	0.56
Other Asia Pacific	0.39

(a) Draw a stemplot of these reserve data.
(b) Comment on any features.
(Data from bp.com)

2.9 The following table contains the weight (in grams) of a random sample of 15 fledgling male kea.

860	860	880	880	800
840	900	880	800	880
840	840	830	780	940

Draw a stemplot of these weights, and comment on any notable features. *(Thanks to Clio Reid for the data)*

2.10 The annual population growth rates (in %) in 1996 and 2006 for selected countries are:

Country	1996	2006
Australia	1.3	1.5
Canada	1.1	0.9
Denmark	0.6	0.3
England and Wales	0.3	0.6
France	0.3	0.7
Japan	0.2	0.0
Netherlands	0.4	0.1
New Zealand	1.6	1.2
Norway	0.5	0.7
Scotland	−0.2	0.4
Sweden	0.1	0.7
United States	1.2	1.0

(a) Calculate the change in growth rate over the decade, i.e. subtract the 1996 rate from the 2006 rate.

(b) Draw a stemplot of these changes.

2.11 The Victoria University of Wellington 2010 Accommodation Guide lists the following rates (in $) per week for university halls of residence. Where a range is provided, both ends of the range are included in the sample:

```
160  289  290  200  165  130
218  205  205  230  140  160
296  290  304  180  170
198  224  230  140  180
```

(a) Draw a stemplot of the data.
(b) Now draw a dotplot of the data.
(c) Which of the two graphs do you prefer?
(d) Comment on the features of the data.
(*Data from* vuw.ac.nz/accommodation)

2.12 A random sample of fried chips in Auckland from ten outlets gives the following percentage content of total fat:

```
8.7   9.0   9.6   8.4   9.4
7.9  11.2  10.3  15.1  16.5
```

Draw a stemplot of the data.

2.13 Following Exercise 2.11, the walking times (in minutes) from the halls to the Kelburn Campus are as follows:

```
40   4  15   5  15   2  15   2
 5   2   5  15  15   5  10
```

Draw a stemplot of these times, and comment on any notable features.
(*Data from* vuw.ac.nz/accommodation)

2.14 The following table gives the percentage of seats in parliament held by women for the countries ranked 1-20 on the Human Development Index, and for those ranked 101-120 (out of 182 countries in the 2009 report). Data from three countries were not available.

Country	%	Country	%
Norway	36	Paraguay	14
Australia	30	Sri Lanka	6
Iceland	33	Gabon	17
Canada	25	Algeria	6
Ireland	15	Philippines	20
Netherlands	39	El Salvador	19
Sweden	47	Syria	12
France	20	Fiji	-
Switzerland	27	Turkmenistan	-
Japan	12	Palestine	-
Luxembourg	23	Indonesia	12
Finland	42	Honduras	23
USA	17	Bolivia	15
Austria	27	Guyana	30
Spain	34	Mongolia	4
Denmark	38	Viet Nam	26
Belgium	36	Moldova	22
Italy	20	Eq. Guinea	6
Liechtenstein	24	Uzbekistan	16
New Zealand	34	Kyrgyzstan	26

Draw a back-to-back stemplot of these samples. Comment on any features.
(*Data from* hdr.undp.org)

2.15 Two Rongotai College math classes were asked to estimate the price of a frozen chicken at the supermarket. The table below summarises their responses:

Year 10 class							
10	12	12	6	3.4	6	12	8	10
5	10	10	12.8	15	8.5	13	16	11
15	10	7.8	13	6	15	16		

Year 11 class							
7.5	5.9	8	4.9	15	6.9	9.6	10	16
9	6	12.5	14	10	5	8.9	8	10
10	7.8	7.9	9.6	8.9	9			

Draw a back-to-back stemplot of these data, and summarise any patterns.
(*Thanks to Ross Findlay, 10DM & 11MY for the data*)

2.16 Invasive wasps may compete with native ants for food and nest sites. The following table gives the number of ants trapped by pitfall trapping over a four-day period at each site.

Invaded sites							
1	0	5	6	3	1	0	4
13	12	2	4	1	0	0	0
1	3	13	6	7	10	23	4
6	55	11	9	2	4	4	10
5	2	61	12	5	4	3	58

Uninvaded sites							
3	2	1	4	3	5	5	2
4	12	10	4	3	11	2	0
5	3	4	1	0	8	3	2
0	24	7	6	3	0	1	8
2	0	0	3	3	1	0	0

Draw a back-to-back stemplot of these data, and comment on any prominent features.
See also Exercises 10.29 and 10.4.
(*Thanks to Catherine Duthie, VUW, for the data*)

Frequency tables and histograms

2.17 John has 52 Led Zeppelin tracks on his MP3 player. Their lengths (in minutes) are shown in the following histogram:

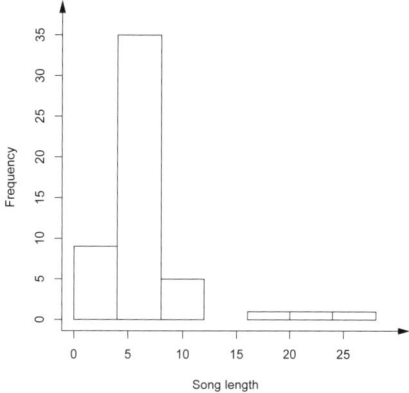

(a) Form a frequency table on the basis of this histogram, and calculate the relative frequencies.
(b) Copy the histogram and add a frequency polygon to it, paying special attention to the extremes of the sample.
(c) If these song lengths are to be compared with those for another artist with 37 tracks, which would you use, frequency or relative frequency on the vertical axis? Why?

2.18 A sample of 70 New Zealand firms yields the following weighted average cost of capital estimates (WACC, in %).

Range	Frequency
6-(8)	24
8-(10)	27
10-(12)	11
12-(14)	4
14-(16)	2
16-(18)	1
18-(20)	0
20-(22)	1

(a) Why is the histogram an appropriate display for these data?
(b) Draw a histogram of the WACC sample.
(*From pwc.com/nz/en/cost-of-capital*)

2.19 The following frequency table summarises the shell weights (in grams) of

Mytilus californianus at Cambria on the coast of California, USA.

Range	Frequency
36-(39)	6
39-(42)	19
42-(45)	22
45-(48)	21
48-(51)	17
51-(54)	7
54-(57)	2

(a) Why is a histogram the appropriate graph with which to represent these data?
(b) Draw a histogram of the shell weights.
See also Exercise 10.19.
(*Thanks to Nicole Phillips for the data*)

2.20 An online survey to investigate participation levels and types in online communities gathered the ages of respondents. These ages are summarised with frequencies f in the following table:

Age	f	Age	f
21-25	5	46-50	11
26-30	12	51-55	6
31-35	17	56-60	3
36-40	15	61-65	2
41-45	11		

(a) Draw a histogram of the data and comment on its features.
(b) Why can't you draw a stemplot of these data?
See also Exercise 3.25.
(*Data from Morse, Who's contributing? Do personality traits influence the level and type of participation in online community? VUW thesis, 2009*)

2.21 The reign lengths of Chinese emperors, from Qin Shihuangdi (221-210 BC) to Puyi (1908-1911) are summarised in the following table:

Reign	Freq	Reign	Freq
0-(5)	55	35-(40)	2
5-(10)	25	40-(45)	3
10-(15)	23	45-(50)	3
15-(20)	21	50-(55)	1
20-(25)	13	55-(60)	0
25-(30)	5	60-(65)	2
30-(35)	8		

(a) Draw a histogram of these data.
(b) Summarise the features of the reign lengths.
See also Exercise 3.31.
(*Thanks to Estate Khmaladze, John Haywood & Ray Brownrigg for the data*)

2.22 The following table summarises the percentage of total dietary fat consumption made up of animal fats for 173 countries in the period 2003-2005. New Zealand's figure is 70%.

Share	f	Share	f
0-(10)	6	50-(60)	30
10-(20)	16	60-(70)	22
20-(30)	30	70-(80)	10
30-(40)	22	80-(90)	0
40-(50)	36	90-(100)	1

Draw a histogram of these data.
See also Exercise 3.30.
(*Data from* fao.org)

2.23 A sample of 13-year-old boys measured social support scores, and these are summarised in the following table.

Score	Frequency
10-(15)	25
15-(20)	71
20-(25)	47
25-(30)	22
30-(35)	6
35-(40)	1

(a) What is the sample size?
(b) Calculate relative frequencies for each score range.
(c) Draw a relative frequency histogram for the data.

(d) Add a relative frequency polygon to your plot.
(Thanks to Paul Jose for the data)

2.24 The following table summarises the year of release of the 250 top films according to imdb.com's user base as at June 2010.

Decade	Freq	Decade	Freq
1920s	5	1970s	24
1930s	11	1980s	28
1940s	20	1990s	39
1950s	34	2000s	61
1960s	25	2010s	3

(a) Excluding the three films released in 2010 (*Toy Story 3*, *Kick-Ass*, and *How to Train Your Dragon*), draw a histogram of these data.
(b) Why does it make sense to omit the three films released in 2010?
(Data from imdb.com*)*

2.25 Researchers examining predation of the Pacific Sand Dollar *Dendraster excentricus* measured the size of a large number of intact shells, and broken shells. The following stemplot summarises the diameters (in mm) of a sample of broken shells:

```
1 | 4
1 | 789999
2 | 0124
2 | 55666777779
3 | 000112222444
3 | 55688
4 |
4 |
5 |            Key: 1 | 4 = 14 mm
5 |
6 | 1
```

(a) Draw up a frequency table for these data, using 5 mm bins.
(b) Draw a histogram for the data.
(c) Add a frequency polygon to your histogram.
(Thanks to Jeff Radford for the data)

Bar graphs

2.26 In 2005/06, 68142 New Zealanders were referred to the mental health service, while in 2006/07 there were a total of 68145 referrals. The sources, and their relative sizes (in %) are given in the table below, by year.

	05/06	06/07
General practitioner	28.5	29.5
Self or relative referral	20.6	18.8
Hospital (non-psychiatric)	8.7	11.8
Police	5.5	6.0
Justice	5.0	4.8
Adult community mental health service	4.5	4.2
Psychiatric inpatient	3.6	3.6
Accident and emergency	3.3	3.3
Education sector	1.9	1.6
All other sources	18.4	16.2

(a) Show these proportions using a bar graph, with a different shading pattern for each year.
(b) Summarise any obvious features in the data.
(c) Why is it better to have all the data on a single bar graph, rather than two separate ones side by side?
(Data from moh.govt.nz*)*

2.27 A survey on children's ownership and use of cellphones identified 27% of the 600 6- to 13-year-old New Zealanders had their own cellphone. Phone usage is summarised in the following table:

Purpose	%
Play games	82
Send or receive txts	81
Make calls	47
Take pictures	43
Send or receive pxts	25
Listen to music	15
Browse the internet	14
Listen to radio	8
Use chatrooms	3
Call, txt, or pxt strangers	2

(a) Display these data in a bar graph.
(b) How do we know the respondents are asked for *all* purposes for which they use their phone?
(*Data from* bsa.govt.nz)

2.28 A random sample of New Zealanders was asked if they think that "NASA faked the first moon landings for publicity". The responses, by political preferences are as follows:

	No	Yes	%(yes)
Labour	804	279	25.8
National	995	322	24.4
Greens	337	109	24.4
NZ First	34	14	29.2
Act	149	8	5.1
United Future	16	1	5.9
Maori Party	17	9	34.6

(a) Why is a bar graph of the number of "yes" responses less informative than a bar graph of the percentage of "yes" answers?
(b) Classify the "political preference" variable. Could it be considered ordinal?
(c) Draw a bar graph of the percentage of "yes" answers and comment on its features.

(*Special thanks to Marc Wilson for the data*)

2.29 The tracks on John's MP3 player are listed in the following genres.

Genre	Frequency
Alternative & Punk	212
Blues	24
Classical	35
Electronica/Dance	45
Folk	8
Hip Hop/Rap	287
Metal	270
Pop	102
Reggae	49
Rock	561
World	11

(a) How many songs are on John's MP3 player?
(b) Calculate relative frequencies for these genres.
(c) Draw an appropriate graph displaying these data with genres in alphabetical order.
(d) Redraw your graph with the genres sorted from largest to smallest.
(e) Which graph do you think is most effective?
(f) What property of the genres allows us to present our preferred graph?

2.30 The Australian ant *Monomorium sydneyense* was recently discovered in Tauranga. To investigate the effects of its colonies on native ant communities, peanut butter was placed as a lure in the vicinity of *M. sydneyense* nests. The frequencies of the following different species of ants seen foraging on the peanut butter over various sites and a period of hours were recorded.

Species	Count
I. anceps	43
M. sydneyense	46
M. antarcticum	15
P. vaga	25
P. rugosula	76
T. bicarinatum	9

Draw a bar graph of the data.
(*Thanks to Lloyd Stringer for the data*)

2.31 The New Zealand Arrestee Drug Abuse Monitoring programme investigated drug and alcohol use in a sample of 2206 detainees in 2005. 965 met the inclusion criteria. The table below gives the number of drugs used (m), from alcohol, amphetamines, cannabis, cocaine, ecstasy, hallucinogens, heroin, methadone, methamphetamines and tranquilisers among the sample.

m	%	m	%	m	%
0	1	4	13	8	6
1	5	5	10	9	4
2	27	6	8	10	3
3	16	7	8		

(a) Draw a bar graph of these data.
(b) Why is a bar graph more appropriate than a histogram for the number of drugs used?

See also Exercises 3.8 and 4.40.

(*Data from* police.govt.nz)

2.32 In 2005, approximately 10,000 New Zealanders responded to *Sunday Star Times*' "Great Morality Debate". Among the questions was "do you think it is wrong for two consenting adults of the same sex to have sexual relations?" 42% of respondents answered in the affirmative. The following table disaggregates the "yes" response by intended political party vote:

Party	%
ACT	33
Christian Heritage	98
Destiny	96
Greens	4
Labour	9
Māori Party	20
National	56
NZ First	54
United Future	90

(a) Draw a barplot of these data.
(b) Is "intended recipient of party vote?" a nominal or ordinal variable? What does politicalcompass.org/nz2008 suggest?
(c) Redraw your bargraph with the parties ordered as follows: Greens, Māori Party, NZ First, Labour, United Future, National, ACT, Christian Heritage, Destiny. Is there any political basis for this ordering?

2.33 The following table gives the regular time interval (Δ, in minutes) for Wellington City buses during the day, by route number (R). 0 denotes the cable car.

R	Δ	R	Δ	R	Δ
0	10	8	30	18	30
1	12	9	30	20	30
2	15	10	30	21	60
3	10	11	15	22	30
5	20	14	30	44	30
7	20	17	20	47	60

(a) Draw a stemplot of these data.
(b) Draw a histogram of the times.
(c) Draw a bargraph of the times.
(d) What are the relative merits of the three plots?

(*Data from* metlink.org.nz)

2.34 Fishing activities often capture species other than those targeted (bycatch). Some shrimp fisheries have bycatch rates as high as 20:1. The following table lists bycatch of large birds and mammals, recorded during one year of the New Zealand Ministry of Fisheries observer programme which monitors a random sample of fishing activities.

Bycatch	Frequency
Albatross	113
Dolphins	21
Petrels	97
NZ Fur Seals	151
Sea lions	11
Shearwaters	89
Whales	1

(a) Draw a bar chart of these data.
(b) What other information would you need before drawing conclusions about the levels of bycatch in New Zealand fisheries?

(*Data from* fish.govt.nz)

2.35 The twelve-month prevalence of anxiety disorders (in %) by gender in the New Zealand population is as follows:

Disorder	M	F
Panic disorder	1.3	2.0
Agoraphobia without panic	0.4	0.8
Specific phobia	4.3	10.1
Social phobia	4.5	5.6
Generalised anxiety	1.4	2.6
Post-traumatic stress	1.6	4.2
Obsessive-compulsive	0.7	0.5
Any anxiety disorder	10.7	18.6

(a) Represent these data in an appropriate bar chart.
(b) Summarise features in these data.
(*Data from Te Rau Hinengaro: The New Zealand Mental Health Survey, 2006*)

2.36 The Dunedin Multidisciplinary Health and Development Study is a long-running cohort study of 1037 children born in Dunedin in 1972-1973. It is a very important source of information on the health, development and behaviour of young people. At age 26, 331 of the 498 males in the study visited their GP at least once, compared with 429 of the 479 females. The number of visits for the year for each group is shown in the following table:

Frequency	Male	Female
Never	167	50
Once	136	66
Twice	74	104
3-5 times	81	149
6-11 times	27	72
> 11 times	13	38

(a) Calculate relative frequencies for the males and females.
(b) Draw a single bar graph comparing the two groups, with frequency of visit on the horizontal axis, and relative frequencies (from a) on the vertical axis.
(c) Why is it important to use relative frequency rather than frequency in b?
(*Figures published in the New Zealand Medical Journal, 2006*)

2.37 A sample of 2804 non-institutionalised individuals assessed prevalence of obsessive-compulsive symptom dimensions, with results:

Symptom	%
Any symptom dimension	13.0
Contamination / Cleaning	1.8
Harm / Checking	7.8
Symmetry / Ordering	3.1
Hoarding	2.6
Sexual / Religious	0.7
Moral issues	1.4
Somatic obsessions	4.6

(a) Draw a bar graph of these data.
(b) Do you think multiple symptoms are common among the sample? Why, or why not?
(*Data from Fullana et. al., Obsessive-compulsive symptom dimensions in the general population: results from an epidemiological study in six European countries, Journal of Affective Disorders, 2010*)

Chapter 3

Summary statistics

3.1 The centre of a data set

Often it is useful to specify what a typical or "middle" value of some data is. For a symmetric data set (one with a symmetrically shaped graph) this is easy (because the common measures agree what the middle of the data actually is) but it is hard to see what would be reasonable with a data set like the one in Figure 3.1.

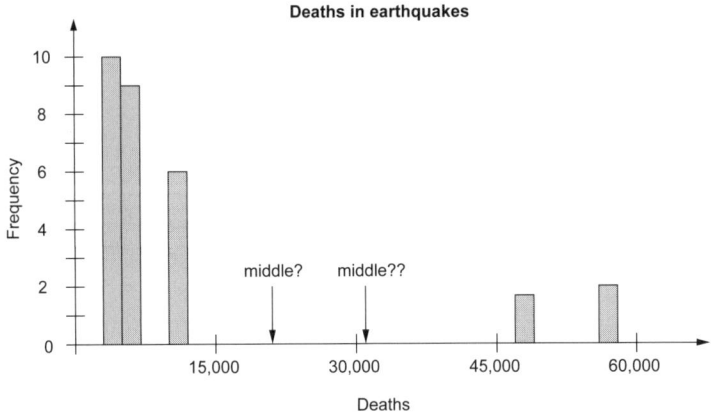

Figure 3.1   An asymmetric data set

Because of such situations we have several different measures of the "middle", which are used with different types of data. The mean, median, geometric mean and harmonic mean are just a few. The two most common are the *mean* and the *median*.

3.1.1 Mean

The mean or common arithmetic average of a collection of data is worked out in the same way whether you are talking about a population or a sample but they are given different labels so that readers can know whether they are looking at sample data or a whole population. In both cases, all the data are added up and then divided by the number of pieces of data.

- The mean of a *population* of size N with individual values x_i is:

$$\mu = \frac{1}{N} \sum_{i=1}^{N} x_i.$$

- The mean of a *sample* of values of size n is:

$$\bar{x} = \frac{1}{n} \sum_{i=1}^{n} x_i.$$

The population mean is a property or *parameter* of the whole population, and it is a single value. Greek letters are typically used to indicate population parameters. Here μ = "mu", the Greek letter "m", is used. The sample mean $\bar{x}$ varies from sample to sample, since if we choose a different n observations, the sample mean will be different.

Example 3.1

The following are the times (in minutes) taken to complete a puzzle for a sample of nine 70-year-olds who are suspected of having lost some cognitive function.

$$2 \quad 5 \quad 5 \quad 8 \quad 8 \quad 8 \quad 11 \quad 11 \quad 14$$

This small set of data is obviously not the whole population of 70-year-olds, i.e. it is a sample. The sample mean time it took this group to do the puzzle was:

$$\bar{x} = \frac{1}{9}(2 + 5 + 5 + 8 + 8 + 8 + 11 + 11 + 14) = 8 \text{ minutes.}$$

The best way of evaluating a sample mean is to use a calculator with statistical functions. Doing this is discussed in Appendix B.

While the sample mean is based on a relatively simple formula, there is always the possibility that we might enter the data incorrectly into our calculator. The mean should always lie in the middle of the data, and we should always check that this is so. In Example 3.1, we see the mean lies exactly in the middle of the data, and so we can be confident we have not made any calculation errors.

3.1.2 Median

Another measure of the middle of a distribution is the *median*. This is the value that half the observations lie above and half lie below. Consequently, we have two alternative, but equivalent, definitions of the median.

> 50% of all values in a data set are less than or equal to the value called the *median*.

> 50% of all values in a data set are greater than or equal to the value called the *median*.

With data that have an odd number of observations the median is simply the middle observation after the data are ranked in order from smallest to largest. Since the median is the middle number in the sample, the first thing we have to do when we are calculating a median is sort the data. A stemplot is often a useful way of doing this.

Example 3.2

Consider the following data set:

Original Observations	89	73	62	90	86	95	78
Ordered	62	73	78	86	89	90	95
Place	1st	2nd	3rd	4th	5th	6th	7th

(4th = median)

Here $n = 7$ and the median is the value 86 as this has three observations below it and three above it.

When the data set has an even number of members, it is less clear exactly which observation is the middle one. In this case, the median lies halfway between the two values in the middle of the sample. We find this middle value by averaging the two adjacent observations, as shown in the following example.

Example 3.3

With the following eight data the median lies between 78 (the 4th observation) and 86 (the 5th observation).

Ordered	62	73	73	78	86	89	90	95
Place	1	2	3	4	5	6	7	8

median = 82

The median is then $(78 + 86)/2 = 82$. Note that four values are below 82 and four above.

If the data have a symmetric distribution, then the mean and the median of the data set will be the same. In an asymmetric distribution (usually called a skew distribution) the mean will be pulled away from the median in the direction of any outlying values.

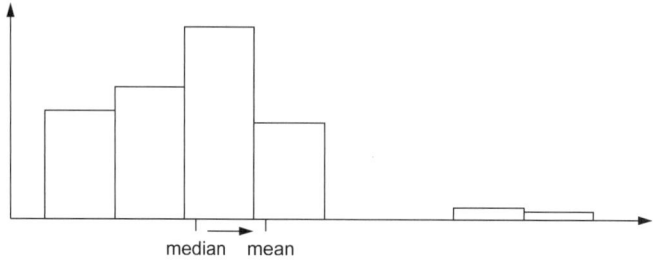

If you suspect any very large or very small values to be in error, then it is wise to use the median as it is *robust* to the effect of single values, i.e. only changes to the middle values in a sample can have any effect on the median. An advantage of the mean, when the data do not include outliers, is that it contains more information than the median as it uses all the data values and not just the middle-ranked value(s).

3.1.3 Percentiles

The pth percentile of a data set is the observation which has $p\%$ of observations less than or equal to it. We have already met the 50th percentile, which has the special name of median. Two other commonly used percentiles are the upper quartile (UQ), and the lower quartile (LQ).

> The *lower quartile* (LQ) is the 25th percentile of a data set, i.e. 25% of the observations are below it.

> The *upper quartile* (UQ) is the 75th percentile of a data set, i.e. 75% of observations are less than it.

To find the quartiles, as well as the median, and any other percentile, we use the following procedure. It is not the only way of finding percentiles, but it is one which is easy to use.

To find the pth percentile of a data set we use the following procedure:

- write $p\%$ as a decimal, i.e. the 35th percentile gives $p = 0.35$
- order the data from smallest to largest
- calculate np, e.g. if $n = 50$ and $p = 0.35$, $np = 17.5$
    - if np is a whole number, go halfway between that number and the next to find the required percentile, e.g. if $np = 14$, go halfway between the 14th and 15th scores
    - if np is not a whole number *round up* to the next whole number, and choose that value, e.g. if $np = 17.5$, take the 18th value.

Example 3.4

Suppose we have $n = 10$ observations, which we have already ordered from smallest to largest:

5   7   9   9   10   11   16   16   21   25

We will find the median, lower quartile, upper quartile and 30th percentile using the method above.

The median is the 50th percentile and has $p = 0.5$. Calculating $np = 10 \times 0.5 = 5$ which is a whole number, so the median is halfway between the 5th and 6th observations, i.e. median $= (10 + 11)/2 = 10.5$.

The lower quartile (LQ) is the 25th percentile and has $p = 0.25$. Consequently $np = 10 \times 0.25 = 2.5$, which is not a whole number so we round up and take the 3rd observation, i.e. LQ $= 9$.

The upper quartile (UQ) is the 75th percentile and has $p = 0.75$. This gives $np = 10 \times 0.75 = 7.5$ so we take the 8th value, i.e. UQ $= 16$. An alternative is to note that the UQ is the LQ for the reverse ordered data set. The LQ was the 3rd value from the bottom, so for the UQ, we take the 3rd value *from the top* which again is 16.

The 30th percentile has $p = 0.3$, so $np = 10 \times 0.3 = 3$, a whole number and we take the value halfway between the 3rd and 4th values, which gives us $(9 + 9)/2 = 9$.

A description of the centre of a collection of data is usually not enough on its own to describe it. We usually need some idea of the spread or variability of the data in order to know if the measure of centre we have used is typical of the data and whether data values are mostly similar to the central value or whether they vary greatly. We will look at several measures of variability such as the range, interquartile range, standard deviation and variance, and also at the related issues of outlier detection. In the process we meet another useful graph.

3.2 Spread or variability

Many statisticians would argue that statistics is primarily about the study of variation and in particular, of how far away data are from the centre of their distribution. Two data sets can have the same mean or median but their histograms or frequency polygons can demonstrate vastly different distributions of data as we see in Figure 3.2.

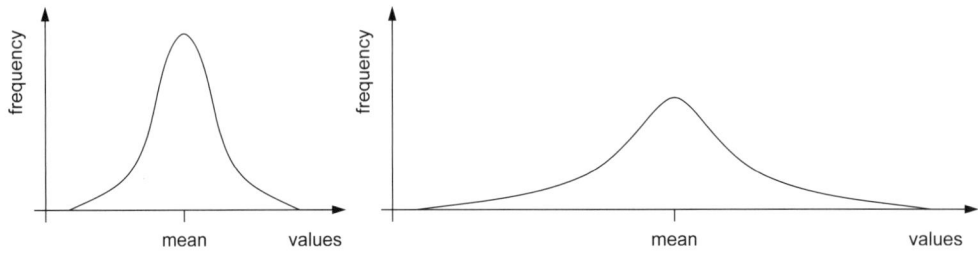

Figure 3.2   Spread

3.2.1 Range

The simplest measure of spread is the range.

> The *range* of a data set = largest value − smallest value.

The range, while useful because it is very quick and easy to compute, has limited value because it is very sensitive to outliers; one extreme value can change it markedly. This is not a good property as it means that one error, if it is a very small or very large data value, will distort the range hugely. Another reason it is not enough on its own as a measure of spread is that two data sets can have the same mean and the same range and yet still be very different.

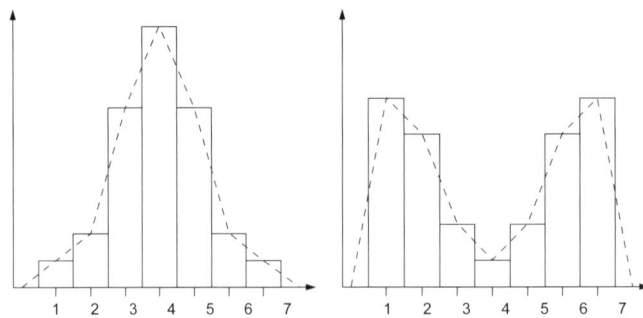

Figure 3.3   Two samples with the same range

Two data sets with the same range, but quite different shape properties are shown in Figure 3.3. Both of the samples have a mean of 4 and a range of 7 but the first has the bulk of the observations between 3 and 5 while the second has the bulk of the observations between 1 and 2 and between 6 and 7. The bulk of the observations are in the centre for the first and in the tails for the second. Alternative measures of spread will reflect this difference.

3.2.2 Interquartile range, IQR

The distance between the lower and upper quartiles gives another measure of spread, the range covered by the middle 50% of the data. This difference is called the *interquartile range* (IQR), and is the range of the middle 50% of the data.

> The interquartile range (IQR) is the difference between the upper and lower quartiles
>
> $$IQR = UQ - LQ.$$

Example 3.5

In Example 3.4, we saw UQ = 16 and the LQ = 9, so IQR = 16 − 9 = 7.

Like the median, the IQR is not often changed by one or two rogue values and is robust to outliers.

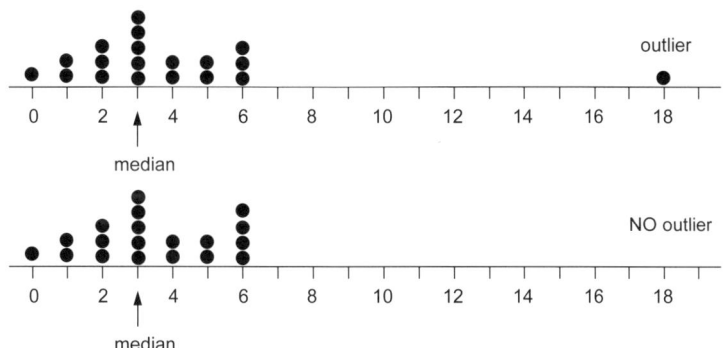

Figure 3.4 Same IQR, but not range

Figure 3.4 shows two samples which are identical except for one value. In the top sample, there is an observation 18, whereas in the second, this is replaced by a value 6. Both samples have the same medians (3), upper quartiles (5) and interquartile ranges (5 − 2 = 3). However, the ranges (18 − 0 = 18 vs 6 − 0 = 6) of these samples are quite different due to the outlier in the top sample.

3.2.3 Variance and standard deviation

The range and IQR are very crude measures of spread in the sense that they do not directly use a great deal of the sample data. They also do not indicate how much of the data is close to the extremes. For this reason the standard deviation and variance are the most common measures of spread used. They are based on all the data, unlike the range and IQR. They should only be used when it is appropriate to use the *mean* as your measure of the centre of the data.

> The population variance of a population of size N is given by
> $$\sigma^2 = \frac{1}{N} \sum_{i=1}^{N} (x_i - \mu)^2.$$
> The sample variance of a sample of size n is given by
> $$s^2 = \frac{1}{n-1} \sum_{i=1}^{n} (x_i - \bar{x})^2.$$

The population variance is denoted σ^2 = "sigma squared". Sigma is the Greek letter "s".

These formulae have other versions which can be easier to use:

$$\begin{aligned} s^2 &= \frac{1}{n-1} \sum (x_i - \bar{x})^2 \\ &= \frac{1}{n-1} \left(\sum x_i^2 - n\bar{x}^2 \right) \\ \text{or} \quad &= \frac{1}{n-1} \left(\sum x_i^2 - \frac{1}{n} \left(\sum x_i \right)^2 \right). \end{aligned}$$

The variance is useful for mathematical reasons; however, it doesn't make a lot of sense as a number. The primary reason is that it is not on the scale of the data, e.g. if our length data are measured in metres (m) the variance is measured in square metres (m^2), a unit of area! To correct for this, we typically report standard deviations, which are found by square-rooting the variance.

> For either population or sample data,
> $$\text{standard deviation} = \sqrt{\text{variance}}.$$

The population standard deviation is
$$\sigma = \sqrt{\sigma^2}.$$
The sample standard deviation is
$$s = \sqrt{s^2}.$$

It can be difficult to see what the standard deviation and variance represent from the computational formulae. Basically, the standard deviation is a measure of how far the pieces of data are from the mean *on average*. The differences $x_i - \bar{x}$ tell us how far away from the mean the data are, and $\frac{1}{n-1}\sum$ is an averaging process (albeit one with $n-1$ in the denominator, not n). The average of the $x_i - \bar{x}$ is always zero, since the mean lies in the middle of the data. To get around this, we square before averaging. The squaring also serves to remove the sign from the difference, i.e. an observation two below the mean contributes $(-2)^2 = 4$ and an observation two above the mean contributes $2^2 = 4$ and so it only matters that the observation is two away from the mean. The square-root we perform for the standard deviation undoes the square and makes the standard deviation have the same units as the data.

The variance and standard deviation $= 0$ only when there is no spread at all, i.e. when all the data are the same. Otherwise s and $s^2 > 0$ (as are σ and σ^2).

Example 3.6

Consider again the following nine pieces of data, whose mean we found in Example 3.1:

$$2 \quad 5 \quad 5 \quad 8 \quad 8 \quad 8 \quad 11 \quad 11 \quad 14$$

A dotplot of this data is

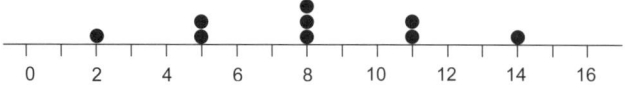

with $\bar{x} = 8$ and $n = 9$. In order to find the variance or standard deviation, we need to evaluate $\sum x$ and $\sum x^2$ for this data. We might calculate these by using a table:

i	1	2	3	4	5	6	7	8	9	Sum
x_i	2	5	5	8	8	8	11	11	14	72
x_i^2	4	25	25	64	64	64	121	121	196	684

and we then evaluate the formula. The variance is

$$s^2 = \frac{1}{n-1}\left(\sum x^2 - \frac{1}{n}\left(\sum x\right)^2\right) = \frac{1}{8}\left(684 - \frac{1}{9}(72)^2\right) = \frac{1}{8} \times 108 = 13.5.$$

The standard deviation is $s = \sqrt{s^2} = \sqrt{13.5} = 3.674$ (3dp). You should ensure you can get this number using the statistical mode on your calculator. Doing this is discussed in Appendix B.

3.3 Sample statistics for repeated or grouped data

In Examples 3.1 and 3.6 we had a sample in which there were many repeated observations, i.e. the observation 5 was present twice, 8 was present three times, and 11 twice. In terms of frequencies, the observation $x_1 = 2$ has frequency $f_1 = 1$, $x_2 = 5$ has frequency $f_2 = 2$, $x_3 = 8$ has frequency $f_3 = 3$, and so on. We can simplify the calculation of the mean and standard deviation for these data using the following formula.

> If your data have values x_i with corresponding frequencies f_i, then
>
> $$\bar{x} = \frac{1}{n}\sum f_i x_i \quad \text{and} \quad s^2 = \frac{1}{n-1}\left(\sum f_i x_i^2 - \frac{1}{n}\left(\sum f_i x_i\right)^2\right).$$

Example 3.7

Consider again the data in Examples 3.1 and 3.6. We can represent them in a frequency table, to which we add the entries $f_i x_i$ and $f_i x_i^2$.

x_i	f_i	$f_i x_i$	$f_i x_i^2$
2	1	2	4
5	2	10	50
8	3	24	192
11	2	22	242
14	1	14	196
Sum		72	684

We note that $\sum f_i x_i = 72$ is the same as the sum of the data, and $\sum f_i x_i^2 = 684$ is the same as the sum of the data squared. The sample mean and variance are

$$\bar{x} = \frac{1}{n}\sum f_i x_i = \frac{1}{9} \times 72 = 8$$

and

$$s^2 = \frac{1}{n-1}\left(\sum f_i x_i^2 - \frac{1}{n}\left(\sum f_i x_i\right)^2\right) = \frac{1}{8}\left(684 - \frac{1}{9}(72)^2\right) = \frac{1}{8} \times 108 = 13.5.$$

3.3 Sample statistics for repeated or grouped data

The numerical aspects of these formulae are exactly the same as the earlier examples, just the formulae are slightly different.

The formulae for finding the mean and variance from repeated data are also very useful for grouped data. When the data is presented in a frequency table you can get an *approximation* to the mean by assuming that all values in a group are close to or equal to the middle of the group, i.e. the midpoint of the group boundaries.

> If your data is grouped, with frequency f_i in the class with midpoint x'_i, then
>
> $$\bar{x} \approx \frac{1}{n}\sum f_i x'_i \quad \text{and} \quad s^2 \approx \frac{1}{n-1}\left(\sum f_i x'^2_i - \frac{1}{n}\left(\sum f_i x'_i\right)^2\right).$$
>
> These estimates are *approximations* for the true sample mean and standard deviation.

The actual observations in the formula for repeated data have been replaced by the estimated observations x'_i. If we do not have the actual data, we replace each observation in the sample by the midpoint of the class it falls in. We find the midpoint by averaging the class boundaries. The mean and standard deviation are found using the same formulae as before, but now the estimates are only approximate, and we use the "approximately equal to" sign $\approx$ to show this.

Example 3.8

Q: Estimate the sample mean and standard deviation for the data from Example 2.11 based on the frequency table:

Class boundaries	Frequency f_i
39.5 - 44.5	4
44.5 - 49.5	2
49.5 - 54.5	6
54.5 - 59.5	9
59.5 - 64.5	10
64.5 - 69.5	12
69.5 - 74.5	4
74.5 - 79.5	3

A: Here, we pretend we do not have the actual data (which can be found in Example 2.11). In the first class, we have four observations and we approximate each of these by the class midpoint $x'_1 = \frac{1}{2}(39.5 + 44.5) = 42$. (Note that the actual observations were $42, 42, 43, 43$ with an average of 42.5.) In the second class, we have two observations, which we approximate by their class midpoint $x'_2 = \frac{1}{2}(44.5 + 49.5) = 47$. (These observations were actually 46 and 49 with an average of 47.5.) Continuing, we obtain the table:

Class boundaries	Class midpoint x'_i	Frequency f_i	$f_i x'_i$
39.5 - 44.5	42	4	168
44.5 - 49.5	47	2	94
49.5 - 54.5	52	6	312
54.5 - 59.5	57	9	513
59.5 - 64.5	62	10	620
64.5 - 69.5	67	12	804
69.5 - 74.5	72	4	288
74.5 - 79.5	77	3	231
		$\sum f_i = 50 = n$	$\sum f_i x'_i = 3030$

The sample mean is approximately $\bar{x} \approx \frac{1}{n} \sum f_i x'_i = \frac{1}{50} \times 3030 = 60.6$. Note that we still have 50 observations: four of 42, two of 47, six of 52, etc. If we were to do the standard deviation in the same way, we would end up with some pretty big numbers in the table. A better way is to use our calculator's statistical mode. Entering the midpoints, rather than the data, we find $\sum f_i x'^2_i = 187770$ and $s \approx \sqrt{84.7347} = 9.21$ (2dp).

The estimates are pretty close to the actual values $\bar{x} = 60.58$, and $s = 9.42$ (2dp). You should check these using the actual data given in Example 2.11.

3.4 Summarising the features of sample data

In this section we present a technique which allows us to summarise graphically a data set in such a way that we can easily read off its centre, its spread, its range, and whether or not any outliers are present. This graph is the *boxplot*.

3.4.1 Boxplots

While the median, and lower and upper quartiles are useful summary statistics for a data set, they do not give a strong sense of the *shape* of the data, nor do they give any indication of the length of the tails of the distribution, i.e. how far below and above the quartiles the data stretch. We can improve the summary by incorporating the maximum and minimum values of the data. From these five values we can then draw a *boxplot* to display the data.

Boxplots are an excellent way to summarise a data set and they also give some information about its shape. They are sometimes called box-and-whisker plots.

3.4 Summarising the features of sample data

> To draw a boxplot of a set of data:
>
> 1. Order the data from smallest to largest.
>
> 2. Draw a scale that spans the range of the data (either vertical or horizontal).
>
> 3. Mark on the scale the five values: minimum, maximum, LQ, UQ, and median.
>
> 4. Draw a box with ends at LQ and UQ parallel to the scale, any width you like.
>
> 5. Mark a line through the box at the median.
>
> 6. Draw lines (whiskers) out from the end of the box to the maximum and minimum values.

Example 3.9

For the 10 measurements in Example 3.4, we draw a box which extends from the LQ of 9 to the UQ of 16, with a vertical bar shown at the median 10.5. The whiskers extend from the ends of box, to the maximum of 25 and the minimum of 5, as shown.

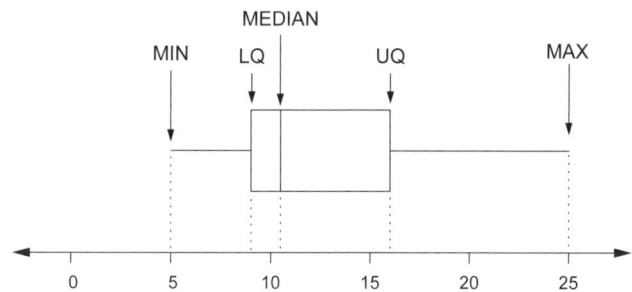

Boxplots show the centre of the data (using the median), the spread (through the IQR and range), they indicate the minimum and maximum of the data, and also reveal whether or not the data are distributed symmetrically about the median. If the data *are* distributed symmetrically then the two parts of the box will be roughly equal in size as will the two whiskers.

We can use side-by-side boxplots to compare the distributions of two sets of data, as shown in Figure 3.5. There we see two boxplots with the same range $(20 - 3 = 17$ and $22 - 5 = 17)$, different centres (12.5 and 17) with the first more symmetric than the second which has a very short upper tail. Placing the boxplots next to each other highlights both the differences and similarities between the two data sets.

44 Chapter 3 Summary statistics

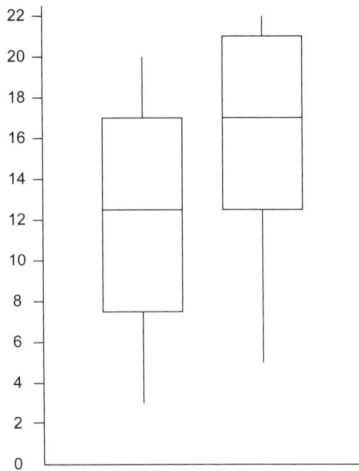

Figure 3.5   Side-by-side boxplots

3.4.2 Outliers

So far we have used the word *outlier* in the general sense. It describes an odd value that seems to be lying all by itself away from the main body of data. To make identification easier, there is a technical definition of what constitutes an outlier. Having identified any outliers in a sample, these need to be checked to see if they are due to errors of measurement or recording. If not, they may be genuine data points which we need to investigate further to see what has made them so different from the rest of the data.

> An outlier is an observation
> - above UQ + 1.5 × IQR; or
> - below LQ − 1.5 × IQR.

To identify outliers, we

- work out the UQ, LQ and IQR
- evaluate UQ + 1.5 × IQR and LQ − 1.5 × IQR.
- Data values greater than the first or less than the second of these two values are outliers.

Example 3.10

In Example 3.4 we had LQ = 9 and UQ = 16 giving IQR = 7 and 1.5 × IQR = 1.5 ×

$7 = 10.5$. The upper cut-off is $UQ + 1.5 \times IQR = 16 + 10.5 = 26.5$ and the lower cut-off is $LQ - 1.5 \times IQR = 9 - 10.5 = -1.5$. No values are less than -1.5 or greater than 26.5 (the minimum is 5 and the maximum 25), so there are no outliers in this data set.

3.4.3 Modified boxplots

We use a modified boxplot to clearly signal the presence of outliers in a sample, and also to give a clearer picture of the shape of the remainder of the data, i.e. the sample with any outliers removed.

If there are no outliers in the data, the modified boxplot is identical to the (unmodified) boxplot.

> If there are outliers in the data, then when you draw the modified boxplot the whiskers extend out only to the smallest and largest observations that are *not outliers* and the outliers have to be plotted as *separate points*.

A modified boxplot is shown in Figure 3.6.

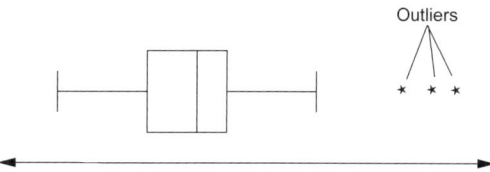

Figure 3.6   A modified boxplot

Example 3.11

Here we draw boxplots for the nurse data used in Examples 2.8 and 2.9. The back-to-back stemplot from Example 2.9 was

```
     Medical       Psychiatric
     9 9 8 5  | 3 | 7 9
     8 8 7 5  | 4 | 1 4 4              key
         2 1 0| 5 | 8                  3|7 = 37
             2| 6 | 1 4 6 7 8
             1| 7 | 2 3 3 4
```

and this gives us the sorted data we need for the five values used in each boxplot.

For the medical nurses, $n = 13$, and the psychiatric nurses number $n = 15$. The following tables outline the calculations we require:

46 Chapter 3 Summary statistics

	Medical			Psychiatric		
Statistic	np	Score	Value	np	Score	Value
Minimum		1st	35		1st	37
LQ	3.25	4th	39	3.75	4th	44
Median	6.5	7th	48	7.5	8th	64
UQ	9.75	10th	51	11.25	12th	72
Maximum		13th	71		15th	74
IQR			12			28
LQ − 1.5 IQR			21			2
UQ + 1.5 IQR			69			114

Only one value is outside the limits LQ − 1.5 IQR, UQ + 1.5 IQR for the two samples and this is 71 for the medical sample. This is the only outlier. The modified boxplots are:

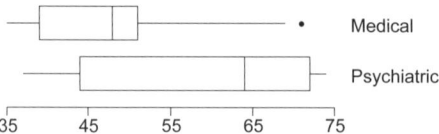

Note that the right whisker for the medical nurse boxplot extends only to the value 62.

While the focus in this chapter has been on calculator use, a spreadsheet is a useful tool for producing these summary statistics. This tool is available for any modern computer. Appendix C gives instructions for finding common summary statistics using a spreadsheet.

Exercises

Summary statistics & boxplots

3.1 Using the data in Exercise 2.13:
(a) Draw a modified boxplot of the walking times.
(b) Calculate the mean and standard deviation of the sample without the outlier present. Compare these to the mean and standard deviation of the entire sample.

3.2 Following Exercise 2.15:
(a) Calculate the mean and standard deviation for the Year 10 boys' prices.
(b) Repeat for the Year 11 boys' prices.

(c) Comment on your statistics.

3.3 Draw a modified boxplot of the sand dollar diameters in Exercise 2.25.

3.4 Following Exercise 2.14, calculate the mean and standard deviation of the two sets of percentage of seats held by women. Comment on any patterns in the estimates.

3.5 The *Permanent Forest Sink Initiative* allows landowners to claim "sink credits" which can be used to offset emissions or traded. The following table gives carbon sequestration figures for various common forest types in

New Zealand (in tonnes of CO_2 per hectare), i.e. the quantity of carbon dioxide which can be absorbed by the plants.

Vegetation type	Stock
Pasture	11
Woody scrub	128
Manuka-dominated	238
Mixed native forest	525
Radiata pine (low)	550
Manuka/Kanuka	554
South Island indigenous shrub	598
Radiata pine (high)	814
Mountain beech	938
Lowland podocarp/broadleaf	1238

Calculate a mean and standard deviation for these figures.

(Data from maf.govt.nz/forestry/pfsi)

3.6 A random sample of black, non-waterproof "volumising" mascaras gives the following prices (in $ per millilitre or gram):

6.99	4.29	5.78	8.77	7.33
1.27	2.56	2.11	2.31	3.17
3.57	7.87	3.40		

(a) Calculate the mean and standard deviation of the prices.
(b) Draw a modified boxplot of the sample.

3.7 The New Zealand Speed-cubing Championships were held at Te Papa in July 2010. During this event, Australian Felix Zemdegs set a world record for solving the 4×4 cube and narrowly missed the world record for the average time for solving five 5×5 cubes. The organisers calculate the average using a *trimmed mean*, for which the minimum and maximum times are discarded, and the mean of the remaining three times is calculated. The world record averages (Ave, in seconds) prior to this event are given in the table, along with the times (in seconds) they are based on:

	Cube			
Rnd	2×2	3×3	4×4	5×5
1	2.22	8.91	38.09	82.44
2	2.19	8.83	37.83	70.68
3	2.93	10.91	55.53	73.06
4	3.28	9.90	48.46	72.55
5	1.03	8.69	39.47	74.46
Ave	2.45	9.21	42.01	73.36

(a) For each of the cubes, calculate the sample mean times, i.e. using all five observations.
(b) For each of the cubes, calculate the sample median times, i.e. using all five observations.
(c) Comment on the organisers' use of the trimmed mean. What do you think the advantages and disadvantages of this measure are over the mean and median calculated here?

(Data from worldcubeassociation.org)

3.8 The New Zealand Arrestee Drug Abuse Monitoring programme investigated drug and alcohol use in a sample of 2206 detainees in 2005. Use the frequency table of drug use in Exercise 2.31 to answer the following questions.
(a) What is the modal number of drugs used among this sample?
(b) What is the average number of drugs used?
(c) What is the standard deviation of the number of drugs used?
See also Exercise 4.40.

(Data from police.govt.nz)

3.9 The number of coins and bank notes (in thousands) in circulation in New Zealand as at 31 December 2009 is given in the following table:

Coin	Number	Note	Number
10c	150,507	$5	22,029
20c	136,553	$10	19,704
50c	62,659	$20	71,092
$1	77,944	$50	21,304
$2	70,436	$100	14,761

(a) What is the average frequency of coin in circulation?
(b) What is the average frequency of note in circulation?
(c) What is the average *value* of coin in circulation?
(d) What is the average *value* of note in circulation?

(Langwasser, Recent trends and developments in currency – 2009, RBNZ Bulletin, 2010)

3.10 Using the data in Exercise 2.16, draw side-by-side boxplots of the two samples. Do the data seem to support the suggestion that ants will be less common at invaded sites? See also Exercise 10.4.

3.11 The prevalence (in %) of tobacco smoking among populations of the Western Pacific Region as at December 2009 is:

Country	%	Country	%
Australia	22	Nauru	50
Cambodia	28	New Zealand	21
China	32	Palau	24
Cook Islands	38	Philippines	33
Fiji	13	South Korea	30
Japan	28	Samoa	41
Lao	40	Singapore	22
Malaysia	28	Tonga	39
Marshall Is.	21	Tuvalu	37
Micronesia	24	Vanuatu	29
Mongolia	26	Viet Nam	23

(a) Draw a stemplot of these rates.
(b) Calculate the mean and standard deviation of the rates.
(c) Draw a boxplot of the data. Can any of the rates be considered outliers? If so, which countries do these belong to?

(*Data from* wpro.who.int)

3.12 The populations in 2006 for selected countries (in millions) were:

Country	Population (m)
Australia	20.7
Canada	32.6
Denmark	5.4
England and Wales	53.7
France	63.2
Japan	127.8
Netherlands	16.3
New Zealand	4.2
Norway	4.6
Scotland	5.1
Sweden	9.1
United States	298.6

Draw a modified boxplot of these data.

3.13 Natural gas reserves for various countries in the Asia-Pacific region were given in Exercise 2.8.
(a) Draw a boxplot of the data. Are there any outliers?
(b) Calculate the sample mean and standard deviation of these data.
(c) Compare the sample mean and sample median. What property of the data is influencing the difference between these figures?

(*Data from* bp.com)

3.14 Following from Exercise 2.36:
(a) What prevents you from estimating the average number of GP visits for the males and females?
(b) For the males and females who visit the GP 0-11 times (inclusive), approximate the mean and standard deviation of the number of GP visits.
(c) If you had included the subjects who had been to their GP at least once a month, would the averages have been lower, the same, or

higher? How about the standard deviations?

3.15 The rainfall in selected Gilbert Islands and Tuvalu, in inches per year for 1963 and 1973 was:

Island	1963	1973
Makin	124.6	140.5
Butaritari	114.9	122.8
Marakei	57.5	30.2
Abaiang	70.8	102.8
Tarawa	45.3	135.9
Maiana	37.5	123.4
Abemama	21.5	99.2
Kuria	21.3	113.9
Aranuka	18.5	118.6
Nonuti	14.6	151.8
Tabiteuea	15.3	115.1
Beru	14.7	117.2
Nikunau	19.8	70.7
Arorae	14.7	122.7
Nanumanga	45.8	93.1
Nuitao	45.8	116.1
Nui	73.2	122.1
Vaitupu	78.8	116.7
Niulakita	147.9	139.4

(a) Calculate the mean and standard deviation of the rainfall in each of the years.
(b) Draw side-by-side boxplots of the two samples. Comment.
(c) Identify any outliers in the two samples.
(*Data from Morrison, Teaching Exploratory Data Analysis, New Zealand Journal of Geography, 1983*)

3.16 In February 2010, 64 intrepid cyclists started the inaugural Kiwi Brevet, an 1100km dirt cycle tour around the top of the South Island. The times of the 55 finishers are given in the following stemplot, with key 4 | 1 = 4.1 days:

```
4 | 1233334445559
5 | 000222333334
6 | 0011122333444444459
7 | 0000045555
```

(a) Calculate the mean and standard deviation of the finishing times.
(b) Identify any outliers in the sample.
(*Data from* kiwibrevet.blogspot.com)

3.17 Following from Exercise 3.16:
(a) Calculate the average speed for each the 55 riders (in kilometres per day) by dividing the distance (1100 km) by the time (in days).
(b) Calculate the sample mean of the speeds in (a).
(c) Using the average time in Exercise 3.16(a), show that 1100 km divided by this average does not equal the average in (b). (This phenomenon is a special case of *Jensen's inequality*.)

3.18 The following sample gives the stated percentages of peanuts (minimum % in some cases) for a sample of peanut butter brands in the New Zealand market:

```
96  97  85  63  85  63  99
94  96  95  94  92  92  64
89  88  89  89  63
```

(a) Draw a stemplot of these percentages.
(b) Now draw a boxplot of the peanut contents.
(c) Which of the stemplot and boxplot do you think is the more effective way of displaying these data?
(d) There was a group of low-fat peanut butters in the sample. Is this group evident in either plot?

3.19 The following stemplot gives the ages of ascent of 51 Roman emperors, from Nerva (AD 96) to Theodossius (AD 379):

```
0 | 4 8 9
1 | 2 2 4 7 8
2 | 0 1 1 1 2 8
3 | 0 0 1 2 2 3 5 5 7 9 9
4 | 1 2 2 3 3 4 4 5 5 7
5 | 0 1 1 4 5 6 8 9
6 | 0 2 5 6 7
7 | 3 5 9       Key: 0|4 = 4 years
```

Calculate the mean and standard deviation of the ages of ascent of these emperors.
(Data from Khmaladze, Brownrigg & Haywood, Brittle power: On Roman Emperors and exponential lengths of rule. Statistics and Probability Letters, 2007)

3.20 The Human Development Index (HDI) combines many variables to produce a single measure of each country's level of development. One variable is the year in which a woman first became speaker or presiding officer of parliament or one of its houses. Of the 25 most highly ranked countries as of the 2009 report, Austria met this requirement in 1927, Denmark in 1950, and New Zealand in 2005 (with Margaret Wilson). France, Liechtenstein, Singapore and Hong Kong have not met this requirement. The remaining data are given in the following stemplot (with key: 197 | 2 = 1972):

```
197 | 2 2 4 7 9
198 | 7 2 9
199 | 3 8 1 3 1 9 2
200 | 4 4 5 7
```

(a) What is wrong with the stemplot above? Redraw it and correct the omission.
(b) Find the median year for this group of countries.
(c) Taking care to account for the four countries that have not met the requirement, confirm whether or not Austria's and Denmark's years are outliers.

(d) Draw a modified boxplot for this sample. How might you handle the missing data?
(Data from hdr.undp.org)

3.21 The following table gives the energy intensity for various industries in New Zealand in 2007. Energy intensity is the sector's energy input, divided by its contribution to GDP (in megajoules per dollar):

Sector	Intensity
Wood, pulp, paper	23.83
Basic metal products	20.32
Non-metallic minerals	14.95
Dairy products	11.05
Chemicals	5.46
Forestry, logging	4.32
Dairy farming	4.23
Mining, exploration	4.13
Other agriculture, fishing	3.28
Meat processing	3.19
Other food processing	2.43
Publishing, printing	2.34
Textiles	2.33
Non-specified manufacturing	1.75
Construction	0.67

(a) Draw a modified boxplot of these data.
(b) Calculate the sample mean intensity and compare it to the sample median.
(c) The overall intensity is 5.44. Why does this differ from both the sample mean and median?
(Data from eeca.govt.nz)

3.22 The 2009 All Blacks squad has heights given in the following stemplot, with key 17 | 0 = 170 cm

```
       Backs       |    | Forwards
                98 | 17 |
   4443322221100   | 18 | 1 3 3 4 4
                65 | 18 | 5 6 6 7 7
                   | 19 | 0 0 3 3
                   | 19 | 6 6 6 6 9
                   | 20 | 0 0 2 2
```

Draw modified boxplots for the two groups within the team. Comment on any differences.
See also Exercise 10.11

(*Data from* allblacks.com)

Repeated or grouped data

3.23 *Hwa-byung* (HB, with literal translation "anger disorder" or "fire disease") is an example of culturally-bound psychological disorder. Hwa-byung is specific to Korea, and includes symptoms of subjective anger, expressed anger, a sensation of heat and feelings of hate. A random sample of Korean psychiatric patients was screened for HB and a variety of other psychiatric conditions. The number of conditions (in addition to HB for the HB group) is presented below.

	0	1	2	3
With HB	47	108	28	0
Without HB	10	74	11	2

Calculate the mean and standard deviation (each to 3dp) of the number of conditions (in addition to HB, if suffered) for the two groups.
See also Exercise 10.2.
(*From Min & Suh, The anger syndrome hwa-byung and its comorbidity, Journal of Affective Disorders, 2010*)

3.24 The table below summarises the number of plays of each of the tracks on John's MP3 player.

Plays	Freq	Plays	Freq
0	8	7	6
1	450	8	2
2	614	9	1
3	312	10	0
4	123	11	0
5	57	12	1
6	29	13	1

(a) What is the most common number of times the tracks have been played? What name is given to this number?
(b) Calculate the mean and standard deviation of the number of plays.

3.25 Ages of participants in online communities were summarised in Exercise 2.20.
(a) Assuming the 21-25 group contains individuals who report ages of 21, 22, 23, 24 and 25, what is the appropriate class midpoint?
(b) Approximate the mean and standard deviation of the ages.
(c) If these were not ages, but weights in kilograms, how would your answers to (a) and (b) differ?

3.26 Following Exercise 2.18:
(a) What are representative values for each of the WACC groups?
(b) Estimate the sample mean and standard deviation for the WACC figures.
(c) From the frequency table alone, are you able to classify the largest WACC figure as an outlier?

3.27 Following Exercise 3.22, make a frequency table of the heights of the *backs*. Using this repeated data, calculate the mean and standard deviation of the heights.

3.28 The 2007/08 Active NZ Survey collated information on club and centre membership using a sample of 4443 New Zealanders. The age profile of those surveyed is, and their membership rates were, as follows:

Age	n	%
16-24	523	51.6
25-34	713	32.3
35-49	1288	30.6
50-64	948	30.3
≥ 65	971	34.5

Assuming a maximum age of 94, estimate the average age of adult club members (over the age of 16).
(*Data from* sparc.org.nz)

3.29 The following are the closing prices of the NZX50 Index constituents on Friday 2 July, 2010.

0.38	0.39	0.48	0.53	0.65	0.66
0.69	0.70	0.86	0.88	0.90	0.91
1.05	1.14	1.20	1.24	1.46	1.59
1.64	1.76	1.84	1.87	1.87	1.99
2.03	2.15	2.15	2.33	2.35	2.43
2.74	2.84	2.88	3.05	3.06	3.40
3.70	3.85	4.02	4.65	5.00	5.73
6.09	6.11	6.28	6.60	7.20	7.66
25.55	25.59				

(a) Draw a modified boxplot of the sample.
(b) Calculate the sample median and interquartile range of the sample, first without the two largest observations, and then with.
(c) Calculate the sample mean and standard deviation of the sample, first without the two largest observations, and then with.
(d) Comment on the effect of the two largest values on these summary statistics. (*Data from* nzx.com)

3.30 The percentage of total dietary fat consumption made up of animal fats for 173 countries in the period 2003-2005 was given in Exercise 2.22.
(a) What is an appropriate representative value for each of the histogram classes?
(b) Estimate the mean and standard deviation of the percentages.
(c) Compare these to the true sample mean and standard deviation: $\bar{x} = 41.4$ and $s = 18.1$, both to 1dp. Why do they differ?
(*Data from* fao.org)

3.31 The reign lengths of Chinese emperors, from Qin Shihuangdi (221-210 BC) to Puyi (1908-1911) are summarised in a table in Exercise 2.21.
(a) Estimate the mean and standard deviation of the reign lengths.
(b) Compare these to the true sample mean and standard deviation: $\bar{x} = 13.2$ and $s = 12.7$, both to 1dp. Why do they differ?

3.32 Estimate the sample mean and standard deviation of the mussel weights in Exercise 2.19.

3.33 The 2006 Census collected data on how people travelled to work on census day. The following table summarises the relative frequency of distance travelled (in kilometres) by main mode of transport: bicycle (Bike), public transport (PT), company car (CC), private car (PC), walking or jogging (WJ) and other (Oth):

km	Bike	PT	CC	PC	WJ	Oth
0-1	3	1	2	1	9	3
1-2	21	7	12	12	44	14
2-5	45	31	25	29	29	28
6-10	20	28	21	24	5	17
11-20	7	21	21	20	3	16
21-50	3	9	14	11	5	15
51-100	1	1	3	2	2	2

(a) Estimate the mean and standard deviation of the distances travelled for each mode of transport.
(b) What additional information would you need to estimate the average distance travelled for all workers? (*Data from* stats.govt.nz)

3.34 Estimate the mean and standard deviation of the social support scores in Exercise 2.23.

3.35 In a survey of 1000 New Zealanders, 15% believed they had been scammed or tricked out of money. The following table summarises the size of the loss inflicted on these people.

Size	Rel. freq.
Less than $1000	60%
$1000 to $4999	19%
$5000 to $9999	6%
$10,000 to $19,999	2%
$20,000 or more	13%

(a) Comment on the use of a midpoint of $40,000 for the largest scams.
(b) Estimate the average scam size among those who believe they have been scammed.
(c) Estimate the average scam size among all New Zealanders surveyed.
(Data from consumeraffairs.govt.nz)

3.36 A study into binge drinking by university students in New Zealand surveyed recent drinking habits. 249 respondents had not had an alcoholic drink in the last four weeks. However, 502 males and 693 females had consumed alcohol in this period, and their numbers of drinking sessions are summarised in the table below:

Sessions	Males	Females
1-7	41.4	51.2
8-14	36.5	34.5
15-21	19.1	13.4
22-28	3.0	0.9

(a) What is a representative number of drinking sessions for each of the groupings in the table?
(b) Approximate the sample mean and standard deviation of the number of drinking sessions for these students using the information above.
(c) The students who had no alcohol have been excluded from the table. If they had been included, what effect do you think they would have had on the mean and standard deviation?

See also Exercises 10.7 and 12.27.
(Data from Kypri, Langley, McGee, Saunders & Williams, High prevalence, persistent hazardous drinking among New Zealand tertiary students, Alcohol & Alcoholism, 2002)

3.37 Using the data in Exercise 2.25:
(a) Calculate the sample mean and standard deviation of the shell diameters using the stemplot.
(b) Approximate the sample mean and standard deviation using the frequency table or histogram.
(c) Why do your answers in (a) and (b) differ?
(d) Repeat excluding the outlier from the sample. Which of the mean and standard deviation does it affect most?

Chapter 4

Describing bivariate relationships

Bivariate statistics occur when each member of a sample has two measurements made on it instead of just one. For example, we might measure: the age and weight of a person, the soil acidity and basal area of a plant, the bail decision and the ethnicity of an accused person, the amplitude of the shock wave and the distance of a sensor from an earthquake epicentre, or the advertising expenditure of a firm and its sales figures. This sort of double measurement results in a data pair for each sample member (X_i, Y_i), where X_i is the measurement of characteristic X on subject i and Y_i is the measurement of characteristic Y on subject i.

Throughout this chapter, we examine the relationship between tree size and age, for data from a book by Chapman and Demeritt, *Elements of Forest Mensuration* published in 1936. The data are of the diameter at breast height of chestnut oak trees for a sample of 14 such trees grown in poor soil, and are given in Table 4.1. The X variable is the age of the tree, and the Y variable is its diameter at breast height. We have 14 data pairs, from our sample of $n = 14$ trees.

Tree	Age (years)	DBH (inches)	Tree	Age (years)	DBH (inches)
1	4	0.8	8	23	4.7
2	8	1.0	9	28	6.0
3	8	3.0	10	30	6.0
4	10	3.5	11	33	8.0
5	13	3.5	12	35	7.0
6	16	4.5	13	38	7.0
7	20	5.5	14	42	7.5

Table 4.1 Oak tree data (DBH = diameter at breast height)

Simply looking at the data in tabular form does not tell us much about it, even though we may get the sense that as the age of the tree increases so does its diameter at breast height. This relationship is not perfect though, since the 42-year-old tree has a smaller diameter than the 33-year-old tree.

If we have a sensible reason to suspect that one of the variables is affecting the other, then this variable is usually the one labelled X and is called the *explanatory* variable (note it is not always the case that one of X or Y explains or causes the other). Y is the *dependent* or response variable. In the oak tree data, it is clear that the diameter of the tree cannot cause its age; however, its age is likely to be an important determinant of its diameter. Consequently, the tree's age is labelled X, and its diameter Y.

We can display ungrouped bivariate data on a scatterplot (or scattergram).

4.1 Scatterplots

The scatterplot is a simple graph that allows us to visualise the relationship between two variables. It consists of two axes, and points plotted on these axes. The position of the point allows us to read off its value on the X variable and its value on the Y variable. An example with two points is shown in Figure 4.1.

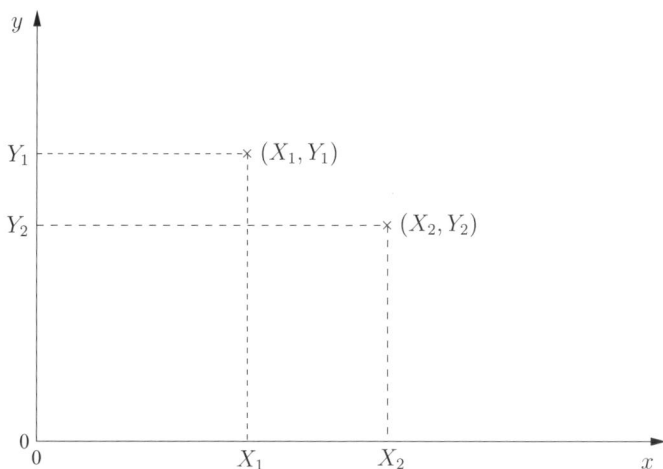

Figure 4.1   Two points on a scatterplot

When we plot the variables, the explanatory variable (if we have specified one) goes on the x-axis (horizontal), and the dependent variable goes on the y-axis (vertical). We arrange the axes so that the points take up the majority of the plot area, breaking the axes if necessary. This is demonstrated in Figure 4.2, where much better use of the plot area allows us to see the points in more detail.

56 Chapter 4 Describing bivariate relationships

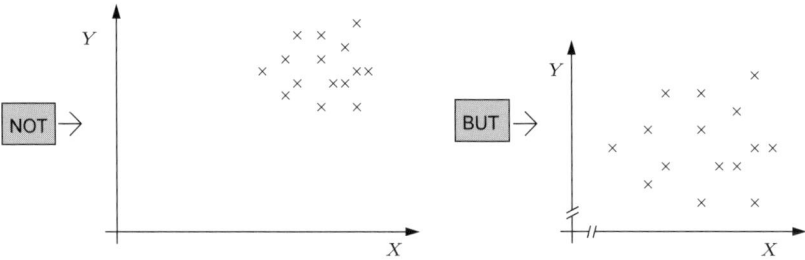

Figure 4.2 Breaking the axes

We now draw a scatterplot of the oak tree data.

Example 4.1

The following plot is of the oak tree data in Table 4.1.

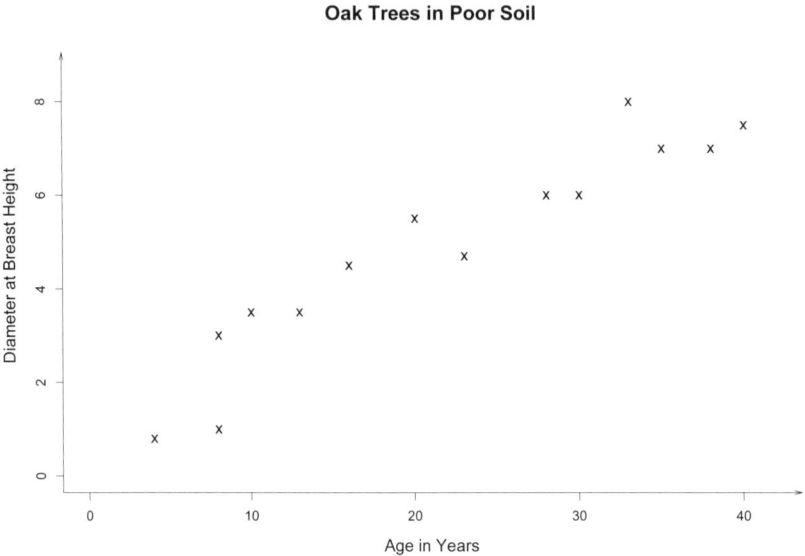

Note that we label both the x- and y-axes, and provide a title for the plot. In this case, it is reasonable to leave $(0,0)$ in the bottom-left corner, and so we do not break either of the axes.

The oak tree sample has no repeated points. If your sample does have data pairs that are the same, you need to make sure that is clear on your plot either by placing (n) next to the point where n is the number of repeats, or by *jittering* (moving slightly away from the spot) all the points with the same coordinates. An example is shown in Figure 4.3.

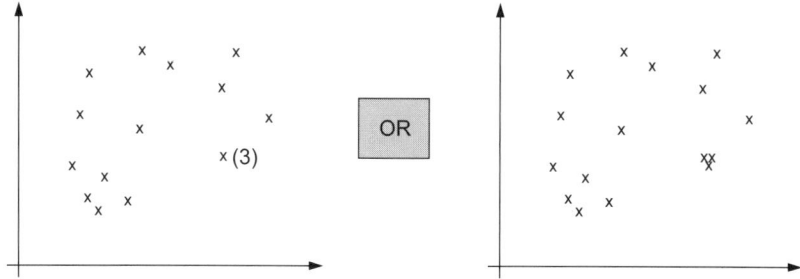

Figure 4.3 Treatment of repeated points in a scatterplot

The points in the plot of the oak tree data look to lie roughly in a straight line. As the ages of the trees increase, their diameter *tends* to increase. As with other graphical representations of data, we have associated summary statistics for describing bivariate data. The main way of summarising a relationship between two variables is by using a *correlation coefficient*. There are several types of correlation, and these are covered in Section 4.2. After considering correlations, Section 4.3 introduces *regression*, a technique used to *describe* the relationship between the two variables.

4.2 Correlation

Having collected some bivariate data and drawn the scatterplot of these data, it is natural to ask "what sort of relationship exists between these two variables, and how *strong* is this relationship?" A measure that sums up this property is called a *correlation coefficient* and this gives a measure of the *strength* of a relationship between an X and a Y variable. There are many correlation coefficients, each appropriate for particular types of data: Pearson's correlation coefficient, Spearman's correlation coefficient, Kendall's tau, the point-biserial correlation coefficient, etc. Questions about the strength of what looks like a *straight-line* relationship where both variables are *continuous* are answered by computing Pearson's correlation coefficient, examined in Section 4.2.1. We also look at a correlation based on ranks, Spearman's correlation, in Section 4.2.2.

4.2.1 Pearson's linear correlation coefficient

Pearson's linear correlation coefficient is the most commonly used correlation.

> When books refer to *the* correlation coefficient, Pearson's is usually the one they mean

It measures the strength of a straight line, or *linear* relationship, between two *measurements*. It doesn't apply for data that are not measurements, or variables that do not have a linear relationship.

> The best computational formula for r is
>
> $$r = \frac{\sum XY - \frac{1}{n}\sum X \sum Y}{\sqrt{\left(\sum X^2 - \frac{1}{n}(\sum X)^2\right)\left(\sum Y^2 - \frac{1}{n}(\sum Y)^2\right)}}$$
>
> This is commonly simplified to
>
> $$r = \frac{S_{XY}}{\sqrt{S_{XX}S_{YY}}}$$
>
> where
>
> $$S_{XY} = \sum XY - \frac{1}{n}\sum X \sum Y$$
> $$S_{XX} = \sum X^2 - \frac{1}{n}\left(\sum X\right)^2$$
> $$S_{YY} = \sum Y^2 - \frac{1}{n}\left(\sum Y\right)^2$$
>
> and n is the *number of points*.

It is usual to use lower-case letters x and y when we are dealing with actual data. To compute Pearson's correlation coefficient we might set up a table like this:

x_i	y_i	x_i^2	y_i^2	$x_i y_i$
4	0.8	16	0.64	3.2
8	1.0	64	1.00	8.0
⋮				⋮

from which we obtain the sums $\sum x_i$, $\sum y_i$, $\sum x_i^2$, $\sum y_i^2$ and $\sum x_i y_i$. [Details are given on summation and order of operations in Appendix A.] The table reminds us that to evaluate $\sum x_i^2$ we square the x_i before we add them, likewise with $\sum x_i y_i$ we multiply the pairs before we add them. Since no limits are given for the sums, we sum over all the available data, which consists of n pairs.

The simplest way to make any sense of this formula is to see it in use.

Example 4.2

We calculate the correlation coefficient for the oak tree data in Table 4.1. We should enter the data into our calculator, following the technique in Appendix B. This gives the following statistics:

$$\sum x = 308 \quad \sum y = 68 \quad \sum x^2 = 8804 \quad \sum y^2 = 397.98 \quad \sum xy = 1843.8$$

with $n = 14$ pairs, and

$$S_{XX} = \sum x^2 - \tfrac{1}{n}\left(\sum x\right)^2 = 8804 - \tfrac{1}{14}(308)^2 = 2028$$
$$S_{YY} = \sum y^2 - \tfrac{1}{n}\left(\sum y\right)^2 = 397.98 - \tfrac{1}{14}(68)^2 = 67.69429$$
$$S_{XY} = \sum xy - \tfrac{1}{n}\sum x \sum y = 1843.8 - \tfrac{1}{14}(308)(68) = 347.8.$$

Pearson's correlation for these data is

$$r = \frac{S_{XY}}{\sqrt{S_{XX}S_{YY}}} = \frac{347.8}{\sqrt{2028 \times 67.69429}} = 0.9386 \text{ (4dp)}.$$

If your calculator has a bivariate mode, you may be able to get the correlation directly. This is because the calculator evaluates the same formula that we have evaluated.

What does the correlation tell us? Some of the basic features of a correlation follow.

> A correlation coefficient r has the following properties:
> - r is always between -1 and 1.
> - $r < 0$ indicates a negative linear relationship between X and Y, and $r = -1$ is a *perfect* negative *linear* relationship.
> - $r = 0$ indicates no *linear* relationship between X and Y.
> - $r > 0$ indicates a positive linear relationship between X and Y, and $r = 1$ is a *perfect* positive *linear* relationship.

Examples of positive and negative relationships are shown in the scatterplots in Figure 4.4. In the left plot, we see a *positive* relationship, where as X increases, Y tends to increase. Matching the behaviour of the points, r is also positive. The opposite is shown in the right scatterplot with a *negative* relationship. When X increases, Y tends to decrease, and the correlation is negative. The sign of r matches the nature of the relationship, so the correlation $r = 0.9386$ indicates a positive relationship between the oak diameters and their age, supporting our comments after Example 4.1.

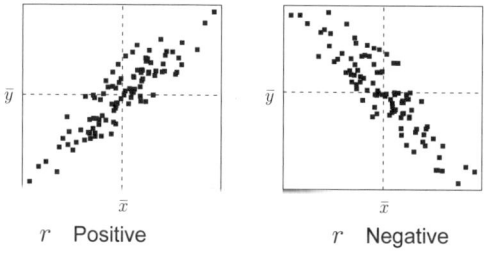

Figure 4.4 Using a correlation to summarise the sign of a relationship

When reporting a correlation, you should not only mention its *sign*, i.e. whether the relationship is positive or negative, but also its *size*, i.e. how *strong* the relationship is. We use the terms "weak", "moderate", and "strong" to describe the strength of the relationship; however, these terms are inexact, and people do not always agree where the cut-offs should be. We prefer the following rules of thumb:

> $|r|$ is the absolute value of r and we ignore the sign.
>
> - $|r| < 0.3$ indicates a weak linear relationship.
> - $0.3 < |r| < 0.7$ indicates a moderate linear relationship.
> - $|r| > 0.7$ indicates a strong linear relationship.
> - $|r| = 1$ indicates a perfect linear relationship.
>
> Here $|r|$ denotes the absolute value of r, which means we ignore the sign, and just look at the size of r.

Scatterplots corresponding to the above classification are shown in Figure 4.5. We can use the sample means $\bar{x}$ and $\bar{y}$ to divide the area of each scatterplot into four regions. The plot in the top row shows an example of two variables with zero correlation, i.e. $r = 0$. A feature of this plot is that there is random scatter of points within the plot area and we cannot identify any pattern in these data. A vertical line is added at $\bar{x}$ and a horizontal line at $\bar{y}$ and we can see that the number of points in each area of the plot is roughly the same.

The second row of Figure 4.5 contains plots where X and Y have negative correlation. A feature of these plots is that there are more points in the top left and bottom right areas of the plot. As r gets closer to -1, we see that the points become more like a straight line with negative slope. We say that the relationship between the two variables is getting stronger, and this is consistent with the rules above for interpreting $|r|$: $r = -0.3$ is a weak relationship, $r = -0.7$ is a strong relationship, and $r = -0.95$ is a very strong relationship.

The opposite is seen in the third row of Figure 4.5, where the variables have positive correlation. Here the points mainly lie in the bottom left and top right areas of the plot. Again, as r gets closer to 1, the points become more like a straight line, this time with positive slope. Again, we would say that the relationship between the two variables is getting stronger, and this again is consistent with the rules above for interpreting $|r|$: $r = 0.3$ is a weak relationship, $r = 0.7$ is a strong relationship, and $r = 0.95$ is a very strong relationship. The only difference between these three plots and the ones above them, is that the relationship is positive, rather than negative.

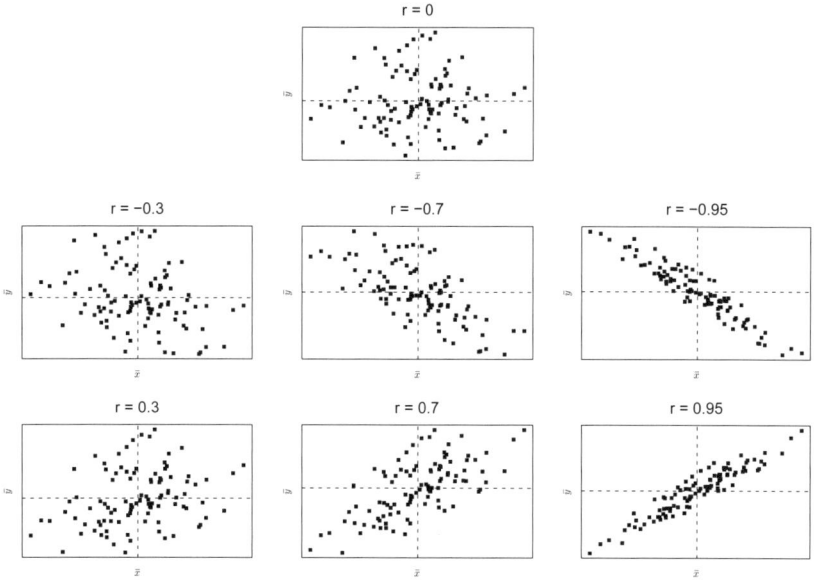

Figure 4.5 Scatterplots of X and Y with correlations equal to $r = 0$ (top row), $r = -0.3, -0.7,$ and -0.95 (second row) and $r = 0.3, 0.7,$ and 0.95 (third row)

If the points lie exactly on a straight line, $r = \pm 1$, with the sign depending on the slope of the line. This indicates a *perfect* relationship, which is positive if $r = 1$ and negative if $r = -1$. Note that the weaker the relationship, i.e. $r = 0, = -0.3$ and $r = 0.3$, the more circular the scatter is, while the strong relationships look more like a straight line.

A low value for r does not necessarily mean that the X and Y variables are not related – they may be related but not in a linear fashion. An example of this is the relationship between stress and performance shown in Figure 4.6. There we see too laid-back and unstressed won't give your best performance; a little more stress and performance goes up; but after a certain point any more stress causes performance to start to go down.

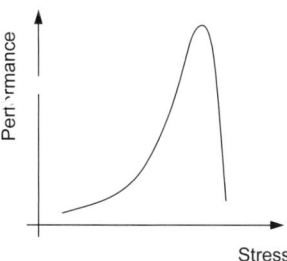

Figure 4.6 A non-linear relationship

We reiterate that while we *can* calculate r for any pair of variables, it is a measure of the strength of the *linear* relationship between X and Y, so it only makes sense to calculate it for variables where a linear relationship is reasonable. Further, both X and Y variables should be quantitative variables. If the relationship between X and Y is not linear, or one or both of them are not quantitative variables, we should not calculate r.

> When calculating r, both variables must be quantitative, and there should be a linear relationship between them.

In many cases, two variables do not have a linear relationship between them (as identified by a scatterplot), or they are not both measurement data. We now consider calculation of an alternative correlation coefficient which may be useful for such variables.

4.2.2 Spearman's rank-order correlation coefficient

If any of the following situations occur:

- your data are not measurements but rankings, e.g. a rating (Likert) scale such as

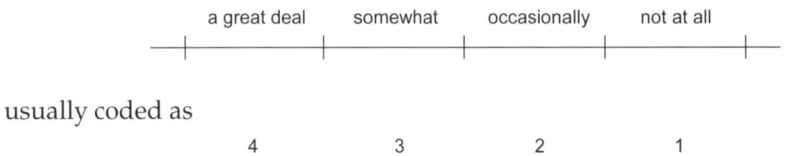

usually coded as

    4   3   2   1

where the numerical scores are an ordinal variable and indicate a definite order, i.e. here a "4" is more positive than a "3", although the size of the numbers is not important (we could just as well code as $40, 30, 20, 10$ or $2.5, 2, 1.5, 1$.);

- the data are measurements but you are only interested in rankings, e.g. if your test mark is 53 it is important for you to know whether this is near the top or the middle or the bottom;

- your data are measurements but your scatterplot clearly shows a relationship that is not a straight-line relationship;

- your sample includes outliers either in the X variable or the Y variable,

then you should not use Pearson's correlation and should instead consider one of the rank-order correlation coefficients. We will only deal with one in this book but there are a number of them. We will use *Spearman's rank-order correlation coefficient*. To distinguish this from r, we use r_s for Spearman's correlation. This correlation is based on ranks, and as such, we require that both X and Y be ordinal variables.

The rank for score X_i tells us the position of X_i in the sample. The smallest observation gets rank 1, the next rank 2, and so on, up to the largest, which gets rank n. When we have allocated ranks to a sample of size n, we can check that we have done this correctly by adding them up and noting that the sum should equal $n(n+1)/2$. [This follows from the mathematical result that $1 + 2 + \cdots + n = n(n+1)/2$.]

4.2 Correlation

> Spearman's rank-order correlation coefficient, r_s
>
> $$r_s = 1 - \frac{6 \sum d_i^2}{n^3 - n}$$
>
> $d_i = \text{rank } x_i - \text{rank } y_i.$

To compute this correlation coefficient use the following method:

1. Order the x_i and y_i values separately from lowest to highest (a stemplot might be useful for this).

2. Compute $d_i = \text{rank } x_i - \text{rank } y_i$ for each data pair.

3. Evaluate the formula for r_s.

Example 4.3

Consider the following table of data on ten subjects who each had their age and their reaction time to a stimulus (in seconds) recorded. The data and their scatterplot are as follows:

Subject	Age	Time
1	18	0.70
2	20	0.85
3	23	1.05
4	19	0.95
5	27	1.20
6	31	1.25
7	21	1.00
8	26	1.10
9	24	1.30
10	22	1.15

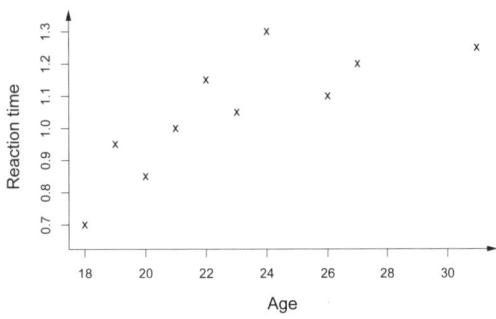

While the data are measurements, the relationship between the two variables appears non-linear (as x increases beyond a point, y increases less), and the 31-year-old subject is considerably older than the others in the sample (a possible outlier). Consequently, we will calculate Spearman's correlation.

We see that the lowest age is 18, so this gets rank "1". The next age is 19, and this gets rank "2", down to the highest age, which is 31, and gets a rank of "10". The smallest reaction time is 0.7, and this will get rank "1", 0.85 will get rank "2" and so on down to 1.3 which gets rank "10". Adding these to the table, and differencing the ranks $d_i = \text{rank } x_i - \text{rank } y_i$ we have:

Subject	Age, x_i	Reaction time, y_i	rank x_i $r(x_i)$	rank y_i $r(y_i)$	difference, d_i $= r(x_i) - r(y_i)$	d_i^2
1	18	0.70	1	1	0	0
2	20	0.85	3	2	1	1
3	23	1.05	6	5	1	1
4	19	0.95	2	3	−1	1
5	27	1.20	9	8	1	1
6	31	1.25	10	9	1	1
7	21	1.00	4	4	0	0
8	26	1.10	8	6	2	4
9	24	1.30	7	10	−3	9
10	22	1.15	5	7	−2	4
					$\sum d_i = 0$	$\sum d_i^2 = 22$

The sum of squared differences $\sum d_i^2$ is best evaluated using your calculator's statistical functions. We can enter the differences d_i, check $\sum d_i = 0$, and obtain $\sum d_i^2$ for use in the formula for Spearman's correlation.

We now evaluate the correlation

$$r_s = 1 - \frac{6 \sum d_i^2}{n^3 - n} = 1 - \frac{6 \times 22}{1000 - 10} = 1 - \frac{132}{990} = 1 - 0.1333 = 0.8666 \text{ (4dp)}$$

which indicates a strong positive (rank-order) relationship between age and reaction time as seen in the scatterplot.

We interpret Pearson's and Spearman's correlations in the same way, always reporting *sign*, i.e. whether the relationship is positive or negative, and *size*, i.e. how strong the relationship is. In Pearson's case, the correlation is a measure of the strength of the *linear* relationship, and both variables should be quantitative. In Spearman's case, we do not restrict the way Y increases with X, and allow the variables to be ordinal only.

r_s has the same basic properties as r, and always lies between −1 and 1 for instance. If there were a perfect relationship between age and reaction time, i.e. the older you are the slower your reaction time, then for each subject the x and y rankings would be exactly the same, and all the differences and their squares would be 0, with $r_s = 1$.

In the previous example, suppose that subjects 1 and 4 were both 18 instead of 18 and 19. How then would we rank their ages? The two ages cannot have different ranks, since the scores are the same. So the two 18-year-old subjects take positions 1 and 2 and since we have no reason to favour one subject over the other we give both these subjects the average of the places they have taken, i.e. they both get a rank equal to the average of 1 and 2 which is ($\frac{1+2}{2}$) = 1.5. Using this technique ensures that again the ranks add to $n(n+1)/2$.

Example 4.4

Nine subjects completed a survey in which they were asked to rate various statements on a seven-point Likert scale. Although the relationship between the ratings is approximately linear, these are not measurements, so Pearson's is invalid. Responses for two of these statements are as follows (notice the treatment of the repeated observation):

Subject	Rating 1	Rating 2
1	2	6
2	3	6
3	3	7
4	4	4
5	4	5
6	5	2
7	5	3
8	5	3
9	6	2

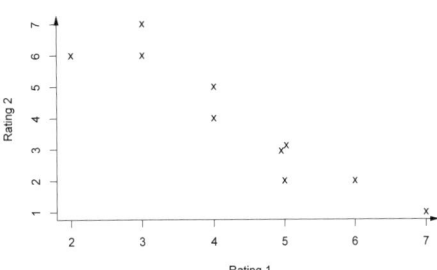

After ordering the first ratings from lowest to highest, we rank them as follows:

Rating 1	2	3	3	4	4	5	5	5	6
Counter	1	2	3	4	5	6	7	8	9
Rank	1	$\frac{2+3}{2} = 2.5$		$\frac{4+5}{2} = 4.5$		$\frac{6+7+8}{3} = 7$			9

and similarly for the second ratings:

Rating 2	2	2	3	3	4	5	6	6	7
Counter	1	2	3	4	5	6	7	8	9
Rank	$\frac{1+2}{2} = 21.5$		$\frac{3+4}{2} = 3.5$		5	6	$\frac{7+8}{2} = 7.5$		9

Using these ranks, we complete the table:

Subject	Rating 1 x_i	Rating 2 y_i	rank x_i $r(x_i)$	rank y_i $r(y_i)$	difference, d_i $= r(x_i) - r(y_i)$	d_i^2
1	2	6	1.0	7.5	−6.5	42.25
2	3	6	2.5	7.5	−5.0	25.00
3	3	7	2.5	9.0	−6.5	42.25
4	4	4	4.5	5.0	−0.5	0.25
5	4	5	4.5	6.0	−1.5	2.25
6	5	2	7.0	1.5	5.5	30.25
7	5	3	7.0	3.5	3.5	12.25
8	5	3	7.0	3.5	3.5	12.25
9	6	2	9.0	1.5	7.5	56.25

$\sum d_i = 0$, indicating we've correctly allocated ranks, and $\sum d_i^2 = 223$. The correlation is

$$r_s = 1 - \frac{6 \times 223}{729 - 9} = 1 - \frac{1338}{720} = 1 - 1.8583 = -0.8583 \text{ (4dp)}.$$

Correlation coefficients measure the strength of a relationship between two variables, and answer the question "how strongly are these variables related (if at all)?" If there *is* a linear relationship between the variables, regression, the topic which follows, enables us to take a big step forward and *predict* other values from existing data.

4.3 Regression

Linear regression is a technique that can be used to estimate a linear relationship between two variables X and Y. In this situation we believe that the variable Y is dependent on the explanatory variable X and that the size of X will have a direct bearing on the size of Y. We require that both X and Y be *quantitative* variables, with a linear relationship between them. While we *can* do regression for any paired variables, unless X and Y are linearly related, the results will not make much sense.

The first stage in a regression analysis is always to draw a scatterplot of the data. This helps us establish whether or not the linear relationship is a sensible one. Having drawn the points, we now seek the "best" line through these points. We will learn how to find this line, and how we can use it to *describe* the relationship between the two variables.

4.3.1 Choosing the regression line

As with the oak tree example, bivariate data from a sample rarely lie precisely in a straight line. Because of this, there is some flexibility as to how we choose a line through the points. Figure 4.7 shows sample data with three possible lines describing the relationship, all of which are reasonable. There are many more possibilities than the three lines shown.

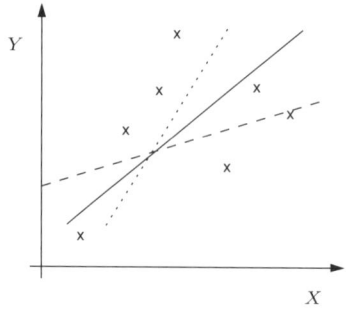

Figure 4.7   Drawing lines through points

Whichever of the lines we finally choose to represent the data, the line can be represented mathematically as

$$Y = a + bX = \text{intercept} + \text{slope} \times X.$$

If the line slopes upward (from left to right), then the number b is positive, and if it slopes downward, then b is negative. The intercept tells us what Y is when X is zero. If the y-axis is drawn at $X = 0$ (i.e. with no breaks), then the line cuts the y-axis exactly at $Y = a$.

If we can draw many different lines through a collection of points, how do we choose the one that summarises the Y variable best, given the X variable? We do this by looking at the *vertical* distances of the points away from the approximating line. These distances are denoted $e_1, e_2, \ldots, e_n$ where e_i is the vertical distance of the ith sample point (X_i, Y_i) from the line.

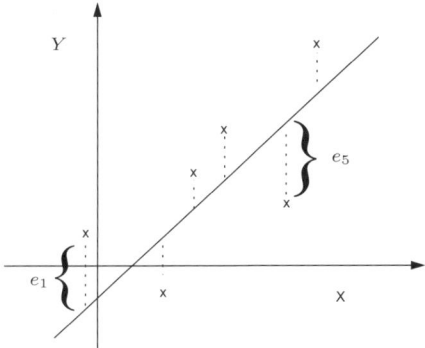

Figure 4.8   Residuals

We want to choose the line that has the points packed as tightly as possible in the vertical direction around that line. The distances e_i measure closeness, and we square them to remove the sign (just like in the sample variance, where we average $(X_i - \bar{X})^2$).

> We find the particular line for which $\sum e_i^2 = e_1^2 + e_2^2 + \cdots + e_n^2$ is as small as possible.

The sum $\sum e_i^2$ is sometimes called the *residual sum of squares*, and denoted RSS. The line that makes RSS smaller than any other line is called the *least squares regression line*, the *line of best fit*, or simply the *regression line*. Of all possible straight lines you could draw through the data, this line will be on *average* the closest to the scatterplot points (measured vertically). This particular choice of line is discussed in more detail in Section 4.4.2 (this material is optional).

4.3.2 Estimating the regression line

We now consider how the regression line is determined. We begin by presenting a *model* for how we believe the data actually behave. Under the premise of this model, we believe that the Y data depend on the X data in a linear fashion, but that we are prevented from observing this linear relationship exactly, by random errors.

> The regression model is
>
> $$Y_i = \alpha + \beta X_i + \epsilon_i$$
>
> where the ϵ_i are random errors with mean zero and constant variance σ^2, and α and β are the parameters (i.e. the intercept and slope) of the linear relationship.

On average, an observation Y_i is equal to the line $\alpha + \beta X_i$, since the average of the ϵ_i is zero. We cannot observe α, β and the variance of the errors σ^2 as they are properties of the whole population so in practice we *estimate* them from sample data. The sample data are of the form seen in Table 4.1, where each pair is represented as a row in the table.

We call our estimates of α and β, a and b respectively, and the line we draw on our scatterplot is

$$\hat{Y} = a + bX.$$

This line will go through the point $(\bar{X}, \bar{Y})$, the averages of the sample data. For the actual observed values of X_i, the "Y hat" values are given by $\hat{Y}_i = a + bX_i$ and are called the *fitted values*. $\hat{Y}$ is used so the reader knows we are talking about a line calculated from sample values, and it is only an estimate of the "true" line, with true slope equal to β and true intercept equal to α.

Calculating a and b

Finding the formulae for a and b can be done using calculus. We will not concern ourselves with *how* the formulae are obtained, but we do need these formulae so that we can calculate the line.

> The slope coefficient b is given by
>
> $$b = \frac{\sum(X_i - \bar{X})(Y_i - \bar{Y})}{\sum(X_i - \bar{X})^2}.$$

> The intercept coefficient a is given by
>
> $$a = \bar{Y} - b\bar{X}.$$

As with the sample variance formula, we can rewrite the formulae for a and b in ways that are easier to calculate:

> Calculating the regression line $\hat{Y}_i = a + bX_i$ is best done using
>
> $$b = \frac{\sum XY - \frac{1}{n}\sum X \sum Y}{\sum X^2 - \frac{1}{n}(\sum X)^2} \quad \text{and} \quad a = \frac{1}{n}\sum Y - b\frac{1}{n}\sum X$$
>
> often expressed in shorthand as $b = \frac{S_{XY}}{S_{XX}}$ and $a = \bar{Y} - b\bar{X}$.

Since a depends on the value of b, we always have to calculate b first (unless we use our calculator, which does both at once).

> The *fitted values* in the regression are $\hat{Y}_i = a + bX_i$. These are the values of the line at the observed values of X_i.

The estimate b is the *slope* of the regression line. This tells us the average change in the dependent variable Y, for change of one unit in the explanatory variable X. The estimate a is the *intercept* of the regression line. This tells us what we would expect Y to be if $X = 0$. For many regressions, this quantity will not be useful, since $X = 0$ may be impossible.

> The vertical distances from the points to the line are called the *residuals*. They are
> $$e_i = Y_i - \hat{Y}_i$$
> where Y_i is the observed value and $\hat{Y}_i$ is the fitted value.

We previously made the residual sum of squares $\sum e_i^2$ as small as possible. This resulted in the choice of a and b for the regression line. Residuals can be positive or negative, depending on whether the point is above or below the line. Points above the line have positive residuals, and those below have negative residuals. Occasionally, a point may lie exactly on the line, in which case its residual will be exactly zero. Because of the way we choose a and b, the residuals $e_i = Y_i - \hat{Y}_i$ always add to zero. This forces the line to always pass *through the middle of the data*.

For a given point (X_i, Y_i) we recap, and note:

$$
\begin{aligned}
Y_i &= \text{observed value} \\
\hat{Y}_i = a + bX_i &= \text{fitted value} \\
&= \text{estimated value given by line} \\
e_i = Y_i - \hat{Y}_i &= \text{residual} \\
&= \text{observed} - \text{estimated}.
\end{aligned}
$$

In this section we have used capital letters X and Y to denote the notional sample data. In the following examples, we have actual sample data, and so we revert to lower-case x and y.

Example 4.5

We now find the regression line for the oak tree data in Table 4.1. We should enter the data into our calculator, following the technique in Appendix B. As in Example 4.2, we have

$$\sum x = 308 \quad \sum y = 68 \quad \sum x^2 = 8804 \quad \sum y^2 = 397.98 \quad \sum xy = 1843.8$$

with $n = 14$, the number of *pairs* of data we have. Also,

$$S_{XX} = \sum x^2 - \tfrac{1}{n}\left(\sum x\right)^2 = 8804 - \tfrac{1}{14}(308)^2 = 2028$$
$$S_{XY} = \sum xy - \tfrac{1}{n}\sum x \sum y = 1843.8 - \tfrac{1}{14}(308)(68) = 347.8.$$

These, and the sample means $\bar{x} = \tfrac{1}{n}\sum x = 22$ and $\bar{y} = \tfrac{1}{n}\sum y = 4.8571$ (4dp), are used to find the regression line slope and intercept terms

$$b = \frac{S_{XY}}{S_{XX}} = \frac{347.8}{2028} = 0.1715 \quad \text{and} \quad a = \bar{y} - b\bar{x} = 4.8571 - 0.1715 \times 22 = 1.0841$$

giving the regression line $\hat{y} = 1.0841 + 0.1715x$.

Note that b is positive as we would expect from the shape of the plot in Example 4.1, and the positive correlation in Example 4.2. You should get into the habit of always checking results against any plot you have to ensure they are consistent with each other. If they are not, then you must have made a computational (or plotting) error.

To add the regression line to a plot of the data, we need to find at least two points that lie on the line and join them. The point $(\bar{X}, \bar{Y})$ will always lie on the line so you could plot that. To find another point on the line you can take any convenient value of X, and by substituting that value into $\hat{Y} = a + bX$, find the matching $\hat{Y}$ value, or, more easily, input the X value into your calculator, and use it to display $\hat{Y}$ directly. It is best to use points that are far away from one another, so using the largest and smallest X values can be a good choice.

Example 4.6

For the oak tree data, the youngest tree has $x = 4$, and a fitted value of $\hat{y} = 1.0841 + 0.1715 \times 4 = 1.77$ (2dp). The oldest has $x = 42$, and a fitted value of $\hat{y} = 1.0841 + 0.1715 \times 42 = 8.29$ (2dp). Adding these two points to the plot, and joining them, we add the regression line to the data.

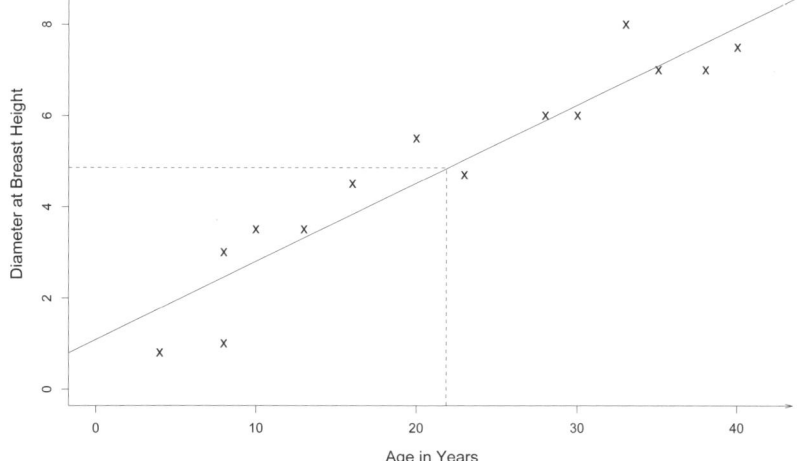

Notice the line passes nicely through the middle of the points. As a check, we can also see that the line passes through the point $(\bar{X}, \bar{Y})$, as indicated by the dotted lines.

4.3.3 Prediction

The regression line can be interpreted as follows: "given X_i, what do we expect Y_i to be?" The answer is the fitted value $\hat{Y}_i$. We can calculate $\hat{Y}_i$ corresponding to X_i data we have collected, and from the fitted values and the Y_i data, calculate residuals.

We can also calculate $\hat{Y}_i$ when we do not have corresponding Y_i values. This is called *prediction*. We use the estimated line, as well as an X value to obtain a prediction $\hat{Y}$. In practice, Y might be a variable of interest, but one which is difficult to measure, e.g. leaf area. The X variable might be a good predictor of X in the sense that having fitted the regression line to some data, the SSE is small, and further, this X variable might be relatively easy to measure, e.g. leaf length.

Example 4.7

Suppose we have recorded body weight, X, and blood pressure, Y, for twenty 48-year-old Samoan women. We have $n = 20$ pairs of data (x_i, y_i) where x_i is the weight of the ith woman and y_i is her blood pressure. The correlation coefficient answers the question "how strong is any linear relationship between X and Y?" Since the data are measurements, and if the relationship is a linear one, then simple regression answers the question "how can we predict the blood pressure of *another* woman whose weight we know, i.e. can we predict another point on the scatterplot if we know x but not y?" We can use the fitted regression line to do this.

When we form predictions, we must ensure that the X variable for the prediction lies within the range of the data we have used to estimate the regression line. The reason for this is that we cannot be sure that the linear relationship holds outside the range of the data we have.

Example 4.8

If another tree not in the original oak tree sample was 25 years old ($x = 25$), we can predict its diameter at breast height by substituting $x = 25$ into the equation

$$\hat{y} = a + bx = 1.0842 + 0.1715 \times 25 = 5.3716$$

and so we would predict a diameter of roughly 5.37 inches.

Example 4.9

Suppose we have data on the length of time spent studying for a test, X, and the test scores achieved, Y, with data and possible relationships shown in the following plot.

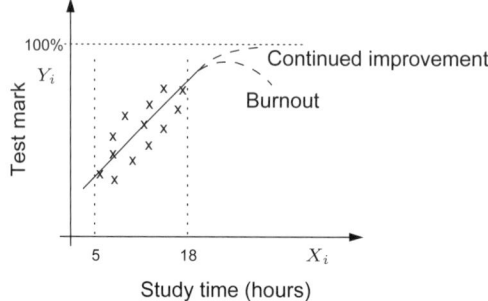

The students in our sample all studied between 5 and 18 hours for the test. Then we can only predict test scores for study times between 5 and 18 hours. This is because we cannot know what studying for 25 hours may lead to. It could yield an exceptional score or possibly lead to burnout and a poor score. In general, inferences are NOT valid outside the original data range. This is because a linear relationship may no longer hold. This is obviously the case in this example, because scores in the test can never increase indefinitely, i.e. we can't do better than 100%!

It is not surprising that there are links between Pearson's correlation and the regression line, given the similarity of the formulae for r and b. The link is called the *coefficient of determination*.

> The connection between the correlation coefficient, r, and the regression line is that $r^2 =$ the proportion of variation in the Y values that is explained by the regression line, i.e. that is explained by the X values. Here r^2 is the *coefficient of determination*.

Since $-1 \leqslant r \leqslant 1$, it follows that $0 \leqslant r^2 \leqslant 1$, so the coefficient of determination, commonly referred to as "R-squared", is a proportion.

4.3.4 Assumptions

In order for a regression analysis to be valid, we need to make certain assumptions. The majority of these can be checked by looking carefully at the sample data.

We can fit a least squares regression line to *any* bivariate data but to be confident of the validity of any predictions we make we need three things to be true of the population from which the sample was drawn.

1. The population must have *a straight-line relationship between the X and Y variables*. That is, there is a relationship $Y_i = \alpha + \beta X_i + \epsilon_i$, where α and β are unknown parameters of the population and ϵ_i is the true error which is also unknown and estimated by the residual, e_i.

2. The errors $\epsilon_1, \ldots, \epsilon_n$ must be *independent*, which is ensured if we have used random sampling.

3. The errors must have *zero mean* and a *constant variance*. The zero mean assumption ensures that the line passes though the middle of the data, and that the sum of the residual distances above the line is equal to the sum of those below.

How can we check whether these assumptions hold for a particular data set? We do it by looking at the residuals, $e_i = Y_i - \hat{Y}_i$. We calculate the residual for each data point and use these as a new data set. At this stage, we are particularly interested in a single plot: a scatterplot with the X_i values of the data on the horizontal axis and with the residuals e_i on the vertical axis. This allows us to look for constant variance of the residuals, and also of the validity of the linear model.

Non-constant variance will often be indicated by a funnel shape in the scatterplot of e_i against X_i. An example of this is seen in Figure 4.9. Alternatively, other shapes may be evident. If the constant variance assumption is valid, the points should lie above and below the x-axis in relatively even bands.

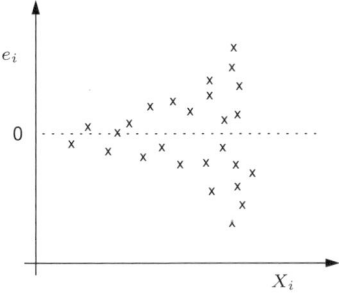

Figure 4.9 Non-constant variance in regression residuals

If there is a noticeable curve in the scatterplot, it indicates that the linearity assumption is unlikely to be satisfied. An example of this is seen in Figure 4.10, which indicates that a *curvilinear* relationship between X and Y is more appropriate than a linear one.

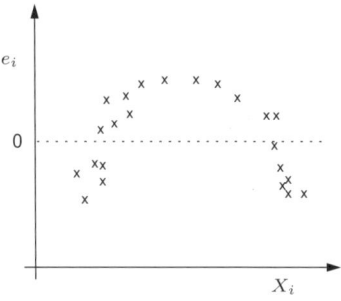

Figure 4.10   Non-linearity in regression residuals

Example 4.10

Below are the residuals for the oak tree data of Table 4.1. The first of these is calculated as follows:
$$\hat{y}_1 = a + bx_1 = 1.0842 - 0.1715 \times 4 = 1.77$$
giving $e_1 = y_1 - \hat{y}_1 = 0.8 - 1.77 = -0.97$. The negative residual indicates that the first point is below the line. The residuals add to -0.01 which is just due to rounding the fitted values (and the residuals) to two decimal places. This sort of rounding error is quite acceptable and doesn't indicate any mistake.

x_i	y_i	$\hat{y}_i$	e_i
4	0.8	1.77	−0.97
8	1.0	2.46	−1.46
8	3.0	2.46	0.54
10	3.5	2.80	0.70
13	3.5	3.31	0.19
16	4.5	3.83	0.67
20	5.5	4.51	0.99
23	4.7	5.03	−0.33
28	6.0	5.89	0.11
30	6.0	6.23	−0.23
33	8.0	6.74	1.26
35	7.0	7.09	−0.09
38	7.0	7.60	−0.60
42	7.5	8.29	−0.79
		Sum	−0.01

The following scatterplot is of the residuals against the ages of the trees. This plot shows that the residuals are fairly evenly distributed above and below the centre line with no obvious funnelling, which supports the assumption of constant variance. However, there is a point out on its own (indicated in the plot as a possible outlier) that needs to be investigated. Also, as indicated, we see a possible curve in the data, so there is perhaps some cause to doubt the suitability of the linear relationship.

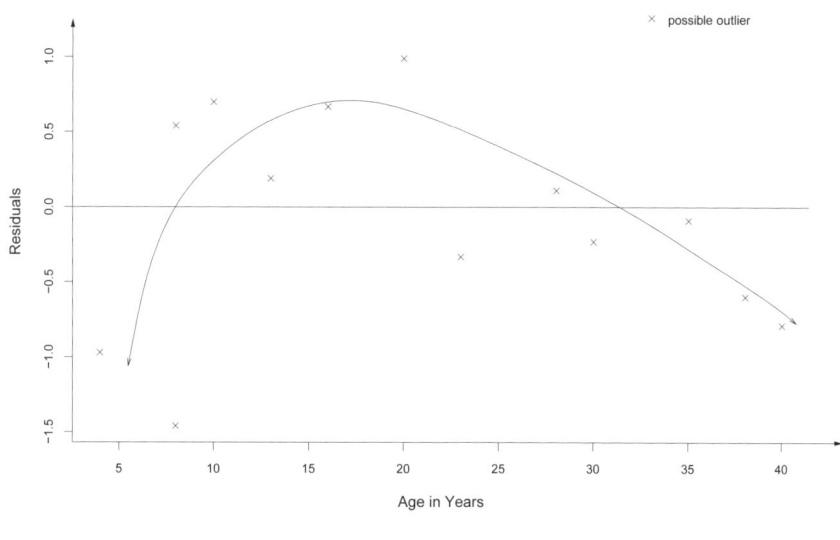

As with univariate summary statistics, correlation and regression are usefully done in a spreadsheet. Instructions for calculating Pearson's and Spearman's correlations and the regression line slope and intercept in a spreadsheet are given in Appendix C.

4.4 Optional technical material

4.4.1 Pearson's linear correlation coefficient

An alternative formula for Pearson's correlation is

$$r = \frac{\sum_{i=1}^{n}(X_i - \bar{X})(Y_i - \bar{Y})}{\sqrt{\sum_{i=1}^{n}(X_i - \bar{X})^2 \sum_{i=1}^{n}(Y_i - \bar{Y})^2}}$$

where $\bar{X}$ and $\bar{Y}$ are the sample means of the X and Y data respectively. This formula is similar to the one we used to explain the sample variance, $s^2 = \frac{1}{n-1}\sum_{i=1}^{n}(X_i - \bar{X})^2$. The denominator (bottom line) of r features the same sum of squares terms as the sample variance, but this time for both X and Y. These terms scale the numerator of r to ensure $-1 \leq r \leq 1$. They also ensure r is dimensionless, meaning r has no units – it is just a number. This is unlike the mean and standard deviation of X, which both have the same units as X.

The numerator (top line) of r is a sum of a series of multiplications. We add the terms $(X_i - \bar{X})(Y_i - \bar{Y})$, which for observation X_i is the deviation from its mean $\bar{X}$, times the deviation for the paired observation Y_i from its mean $\bar{Y}$. We can divide the scatterplot into four regions as shown in Figure 4.4.1.

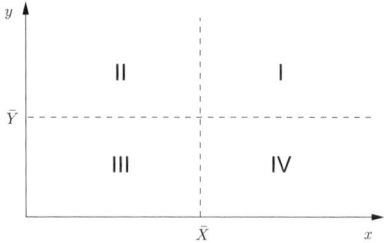

Figure 4.11 The quadrants in a scatterplot

Referring to Figure 4.4.1:

- in quadrant I, $X_i - \bar{X} > 0$ and $Y_i - \bar{Y} > 0$, so $(X_i - \bar{X})(Y_i - \bar{Y}) > 0$
- in quadrant II, $X_i - \bar{X} < 0$ and $Y_i - \bar{Y} > 0$, so $(X_i - \bar{X})(Y_i - \bar{Y}) < 0$
- in quadrant III, $X_i - \bar{X} < 0$ and $Y_i - \bar{Y} < 0$, so $(X_i - \bar{X})(Y_i - \bar{Y}) > 0$
- in quadrant IV, $X_i - \bar{X} > 0$ and $Y_i - \bar{Y} < 0$, so $(X_i - \bar{X})(Y_i - \bar{Y}) < 0$.

If there is a negative relationship, most of the points lie in quadrants II and IV, so most of the terms $(X_i - \bar{X})(Y_i - \bar{Y})$ will be negative and r will be negative. Conversely, if there is a positive relationship, most of the points lie in quadrants I and III, so most of the terms $(X_i - \bar{X})(Y_i - \bar{Y})$ will be positive and r will be positive. If there is no relationship between X and Y, we'd expect the positive and negative terms in quadrants I and II to cancel out, and likewise for quadrants III and IV. As a result, r should be close to 0.

It also turns out that if the points lie exactly on a straight line, $r = \pm 1$, with the sign depending on the slope of the line.

4.4.2 The method of least squares

A problem immediately arises: unless $|r| = 1$, the points do not lie on a line, in which case, how do we choose the "best" line? Let's look at the situation a bit more closely through an example. Suppose we had collected the following data:

X	1	2	3	4
Y	1.5	3.4	3.7	5.4

These points are shown in Figure 4.12 along with three lines, marked A, B and C. Line A has the equation $y = 0.5 + 1.2x$, line B has the equation $y = 1 + x$ and line C has the equation $y = 3.5$. It is pretty obvious from the scatterplot that the third line isn't a very good description of the data: although the middle points are close to the line, the extreme points are not at all close. On the other hand, lines A and B both pass through the data, and are reasonably close to every point. We see that line A is closer to the extreme points than line B, but line B is closer to the middle points than line A. So which is better? And is either of them "best"?

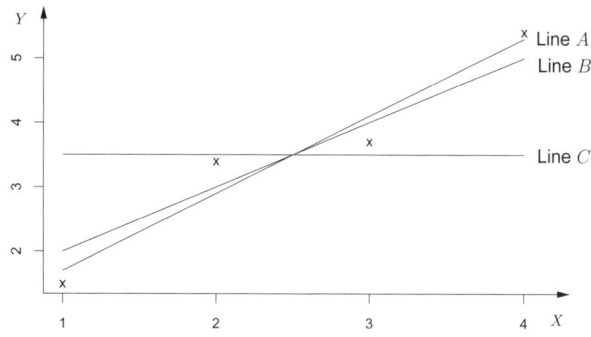

Figure 4.12 Three potential linear relationships between X and Y

For each of the points in Figure 4.12, we can calculate how far away the point is to each of the lines. We measure this distance in terms of vertical height, i.e. we fix the x value, and measure how close the line is to the y value. These distances are shown in Table 4.2. We notice that for all three lines, the vertical distances from the line add to zero. This is a good thing, as it indicates that on average, the lines are in the middle of the data.

		Errors			Squared errors		
X	Y	A	B	C	A	B	C
1	1.5	0.2	0.5	−2.0	0.04	0.25	4.00
2	3.4	−0.5	−0.4	−0.1	0.25	0.16	0.01
3	3.7	0.4	0.3	0.2	0.16	0.09	0.04
4	5.4	−0.2	−0.4	1.9	0.04	0.16	3.61
	Sums	0.0	0.0	0.0	0.49	0.66	7.66

Table 4.2 Errors, and squared errors for the lines (A, B and C) in Figure 4.12

When we calculated the sample standard deviation, we used *squared deviations*. The deviations were squared so that negative errors were treated the same way as positive errors, and so that larger errors had a bigger penalty than smaller errors. The squared deviations are reported in Table 4.2. These are added up to form the sum of squared errors (SSE) for each line. Line C, which we have already decided is not a good summary of the data, has an SSE that is 7.66. In contrast, the other two lines have SSEs equal to 0.49 for A, and 0.66 for B. We notice that these are considerably smaller than 7.66, confirming our impression that lines A and B describe the relationship between X and Y much better than line C. Line A has a smaller SSE than line B, indicating that on average, line A is closer to the observations than line B.

On the basis of squared errors, line A describes the data better than either of the other two lines. In fact, line A is determined by a method called *ordinary least squares* regression (OLS), and is the best line under the criterion that the SSE should be as small as possible. No other line through the four points in Figure 4.12 has a smaller SSE than line A.

Exercises

Scatterplots and correlation

4.1 List these six correlation coefficients in order

$$-0.75, -0.25, 0, 0.1, 0.3, 0.9$$

starting from that which indicates the weakest linear relationship to that which indicates the strongest linear relationship.

4.2 Consider a population that has a correlation of approximately zero between two characteristics X and Y.
(a) Sketch a scatterplot of the population.
(b) On the same sketch identify a sample of that population that will have a correlation between X and Y of approximately one.

4.3 The following statements all contain errors. Identify and explain the error(s) in each case.
(a) The correlation between birth weight and weight at one year old is very strong, at 1.05.
(b) The correlation between length of prison sentence and ethnicity is approximately 0.6.
(c) Using a random sample of retirees, there's a weak negative correlation between years of education and unemployment of -0.86.
(d) There is a fairly strong correlation of $r = 0.7$ between men's and women's education levels in New Zealand.
(e) There's a strong correlation between car colour and popularity.

4.4 The data below give scores for 20 individuals on a "traditional religious belief" scale (Rel), and on a "fear of death" scale. Traditional religious belief is calculated as the mean of the four items (belief in: God, Satan, Heaven and Hell, and an immortal soul), with higher scores meaning more traditional religious belief. Fear of death is measured using Hoelters' 1970 Fear of Death scale and is the mean of approximate 40 items like "I am frightened by the thought that I could still be conscious when I die" and "People should have autopsies to ensure that they are really dead", etc. Scores are between 1 and 7, and higher scores mean greater death anxiety. Traditional religious belief is calculated as the mean of the four items (belief in: God, Satan, Heaven and Hell, and an immortal soul) in the traditional religious belief subscale of Tobacyk's paranormal belief scale. Again, higher scores mean more traditional religious belief (1=minimum, 7=maximum).

Rel	Fear	Rel	Fear
1.25	2.95	2.25	4.60
2.75	3.15	4.75	4.62
3.75	3.58	5.00	4.75
5.50	3.83	4.25	4.75
5.50	3.83	5.25	4.75
5.50	3.83	5.25	4.85
2.00	3.92	5.75	4.90
7.00	4.10	5.00	5.25
4.75	4.15	4.00	5.78
5.25	4.35	4.50	5.95

(a) Plot these points in a scatterplot.
(b) Comment on the conjecture that the unreligious and deeply religious are least anxious about death.
(c) Would Pearson's or Spearman's correlations be useful statistics to support the claim in (b)? Why, or why not?

(*Special thanks to Marc Wilson for the data*)

4.5 The following table gives the percentage of professional and technical workers who are female (x) and the percentage of legislators, senior officials and managers who are female (y) for countries ranked 1-20 on the Human Development Index (out of 182 countries in the 2009 report). Data from two countries were not available.

Country	x	y	Country	x	y
Norway	51	31	Luxembourg		
Australia	57	37	Finland	55	29
Iceland	56	30	USA	56	43
Canada	56	37	Austria	48	27
Ireland	53	31	Spain	49	32
Netherlands	50	28	Denmark	52	28
Sweden	51	32	Belgium	49	32
France	48	38	Italy	57	34
Switzerland	46	30	Liechtenstein		
Japan	46	9	New Zealand	54	40

(a) Draw a scatterplot of the data, and comment on any obvious features.
(b) Excluding the data for Japan, calculate the correlation between x and y.
(c) What effect do you think including the Japanese observation will have on your correlation?
(d) Check your claim in (c) by recalculating the correlation with the Japanese observation included.

(*Data from* hdr.undp.org)

4.6 The following table gives forecast (FC, in degrees Celsius) and actual (°C, in degrees Celsius) maximum and minimum temperatures for 20 February 2010, for towns and cities around New Zealand.

	Min		Max	
	FC	°C	FC	°C
Kaitaia	17	15	25	25
Whangarei	17	15	25	26
Auckland	16	15	24	24
Tauranga	16	14	24	24
Hamilton	11	9	25	24
Rotorua	12	10	24	23
Gisborne	14	12	26	28
Napier	14	11	25	25
Masterton	11	9	24	25
Palmerston North	14	11	23	22
Paraparaumu	14	17	20	21
Wellington	15	16	23	23
Nelson	14	12	26	25
Blenheim	11	8	27	26
Kaikoura	14	10	20	22
Christchurch	12	10	23	21
Timaru	12	9	25	24
Dunedin	13	13	21	20
Oamaru	13	9	24	20
Hokitika	13	10	19	19
Alexandra	12	11	25	29
Queenstown	12	9	21	25
Invercargill	13	9	19	18

(a) Calculate the correlation between the forecast and actual minimum temperatures.
(b) Calculate the correlation between the forecast and actual maximum temperatures.
(c) On the basis of (a) and (b) which of the two temperatures would appear easier to forecast?

See also Exercises 10.27 and 13.10.

(*Thanks to Paula Acethorp and the Meteorological Service of New Zealand Limited* metservice.com *for the data*)

4.7 Students from ten nations are asked to rate their own cultural characteristics on a variety of dimensions. The average ratings for two variables are given in the table below.

Country	Embeddedness	Intellectual autonomy
Argentina	4.70	4.96
Brazil	4.51	4.37
Germany	3.72	3.66
India	4.72	4.35
New Zealand	3.91	4.49
PRC (China)	4.91	4.47
Peru	5.15	5.07
Taiwan	4.80	4.59
UK	4.72	4.53
US	4.37	4.46

(a) Draw a scatterplot of these average ratings.
(b) Calculate Spearman's correlation coefficient for these data.

(Data from Fischer, Congruence and functions of personal and cultural values: Do my values reflect my culture's values? Personality and Social Psycholology Bulletin, 2006)

4.8 Rumination (contemplation or reflection) may become persistent and recurrent worrying. The following table contains a random sample of rumination and depression scores for 15 girls (13 years old) from a larger study of rumination in adolescents.

ID	Rumination	Depression
1	19	12
2	24	9
3	23	7
4	32	14
5	14	15
6	11	12
7	24	18
8	25	12
9	33	19
10	28	3
11	22	8
12	36	37
13	23	31
14	30	20
15	16	6

(a) Draw a scatterplot of these data.

(b) Under what conditions is Pearson's correlation suitable for these data?
(c) Calculate the correlation between the scores.
(d) Does your correlation allow you to conclude rumination causes depression in these young women?
See also Exercise 10.6.

(Many thanks to Paul Jose for access to raw data, from Jose & Brown, When does the gender difference in rumination begin? Gender and age differences in the use of rumination by adolescents, Journal of Youth and Adolescence, 2008)

4.9 Weta Digital owns one of the world's largest computers. The primary use of the world's 500 largest supercomputers as at May 2010 is tabulated below (categories with fewer than five machines are excluded), with number of machines, and the relative share of computing power. The use of 151 of the 500 are undisclosed, and 29 others are in categories with fewer than 5 machines. The remaining 320 are below:

Use	Count	Share
Aerospace	5	0.96
Database	5	0.43
Defense	16	3.83
Energy	6	0.65
Finance	53	4.90
Geophysics	24	2.37
Info Processing	16	1.80
Info Servers	33	3.40
Logistics	29	3.05
Research	82	30.28
Retail	7	0.69
Semiconductor	15	1.33
Service	7	0.67
Software	7	0.77
Telecom	8	0.73
Web	7	0.74

(a) Draw a scatterplot of these data. On the basis of your plot, which is the most appropriate correlation coefficient to summarise the link between number of supercomputers and share of computing power?

(b) Calculate the correlation between the two variables, and summarise the relationship.

(*Data from* top500.org)

4.10 The Reserve Bank of New Zealand measures global activity using a subset of economies with important trade links to New Zealand. Worldwide growth is calculated as an export-weighted average of growth in trading partner economies (GDP-16, in %), while worldwide inflation is calculated as an import-weighted average of inflation in those economies (CPI-16, in %). The following table lists the average weights in these measures, by economy.

Economy	GDP-16	CPI-16
Australia	28.80	21.30
US	13.30	11.60
EU	10.60	14.40
Japan	9.90	9.40
China	8.80	16.20
UK	5.00	2.70
Korea	4.10	3.50
Indonesia	3.20	2.50
Singapore	2.60	5.40
Malaysia	2.50	4.00
Taiwan	2.30	2.20
Hong Kong	2.30	0.50
Thailand	2.00	3.10
Philippines	2.10	0.40
Canada	1.60	1.70
Viet Nam	1.00	1.00

(a) Draw a scatterplot of these weights, and comment on the relationship between the points.

(b) Choose an appropriate correlation coefficient between these variables, and calculate it.

(*Data from* Zhang, *The evolution of New Zealand's trade flows*, RBNZ, 2009)

4.11 The following table gives bill lengths and widths (both in millimetres) from a random sample of 15 fledgling male kea at Aoraki/Mount Cook.

Kea	Length	Width
1	48.1	13.4
2	46.5	12.9
3	47.0	14.3
4	47.9	13.9
5	48.9	12.7
6	44.5	13.8
7	46.8	14.2
8	50.0	14.0
9	48.6	14.0
10	47.8	13.6
11	46.0	13.6
12	47.5	13.4
13	45.6	13.9
14	47.3	13.1
15	49.9	14.8

(a) Plot these data in a scatterplot.
(b) Calculate Pearson's correlation coefficient between the two variables.
(c) Repeat using Spearman's correlation.
(d) Which of the two correlations do you think is more appropriate in this instance?

(*Thanks to Clio Reid for the data*)

4.12 Using the Gilbert Islands and Tuvalu rainfall data from Exercise 5.14:
(a) Draw a scatterplot of the data.
(b) Identify outliers in each of the samples, and mark these on the plot.
(c) Are the outliers still notable once the bivariate relationship has been taken into account?
(d) Does the scatterplot help identify any other interesting patterns in the data?

4.13 The Global Peace Index is calculated for countries using a number of variables, including access to weapons, levels of organised conflict and relationships with neighbouring countries. The twelve most peaceful countries by the 2010 scores are in the table below, with their 2007 (global) ranks and scores (Iceland and Luxembourg were not rated in 2007):

	2010 Rank	2010 Score	2007 Rank	2007 Score
New Zealand	1	1.188	2	1.363
Iceland	2	1.212		
Japan	3	1.247	5	1.413
Austria	4	1.290	10	1.483
Norway	5	1.322	1	1.357
Ireland	6	1.377	4	1.396
Denmark	7	1.341	3	1.377
Luxembourg	8	1.341		
Finland	9	1.352	6	1.447
Sweden	10	1.354	7	1.478
Slovenia	11	1.358	15	1.539
Czech Rep.	12	1.360	13	1.524

Use the scores for the ten countries that were rated in 2010 and 2007.
(a) Draw a scatterplot of the data, with the 2007 scores on the x-axis.
(b) Why might Pearson's correlation not be appropriate for the scores?
(c) Why are the 2007 ranks not directly appropriate for calculating Spearman's correlation?
(d) Calculate Spearman's correlation for the scores of the ten countries.
(Data from visionofhumanity.org)

4.14 Attitudes towards homosexual sex for a sample of New Zealanders, by political preference, were given in Exercise 2.32.
(a) Political ideology is often plotted on two dimensions (e.g. left-right and libertarian-authoritarian as at politicalcompass.org/nz2008). Research a sensible ordering of the nine political parties given, providing references for your decision.
(b) Calculate the correlation between political ideology and attitude for this sample.
(c) Calculate Spearman's correlation using the ranking given in 2.32(c). Critique this rating if it differs from your own.

4.15 The following table lists the proportion (in %) of individuals in various Western Pacific nations who smoke tobacco products daily.

Country	Male	Female
Australia	19	16
Cambodia	42	5
China	57	3
Cook Islands	37	28
Fiji	18	2
Japan	39	10
Laos	57	12
Malaysia	42	2
Marshall Islands	31	4
Micronesia	25	14
Mongolia	43	6
Nauru	44	50
New Zealand	20	18
Palau	33	7
Philippines	43	9
South Korea	50	5
Samoa	56	18
Singapore	26	4
Tonga	60	11
Tuvalu	50	17
Vanuatu	46	5
Viet Nam	34	2

(a) Draw a scatterplot of these data.
(b) Add the line $y = x$ to your scatterplot.
(c) Comment on any interesting patterns in the data.
(Data from wpro.who.int)

4.16 The Deepwater Horizon spill in 2010 released close to one million tonnes (over seven million barrels) of oil into the ocean. The following table lists the spill sizes (in thousand tonnes) of some of the world's largest individual oil tanker spills, and their year of occurrence. The largest spill was 287,000 tonnes from the Atlantic Empress off the West Indies in 1979.

Year	Size	Year	Size	Year	Size
1967	119	1979	95	1991	260
1972	115	1980	100	1991	144
1975	88	1983	252	1992	74
1976	100	1985	70	1992	67
1977	95	1988	132	1993	85
1978	223	1989	80	1996	72
1979	287	1989	37	2002	63

(a) Draw a scatterplot of these data with year on the x-axis.
(b) Do there appear to be any outlying points in the sample?
(c) Excluding the four largest spills (those in excess of 150,000 tonnes), calculate a suitable correlation coefficient for the data.
(d) What does the sign of this correlation indicate? (*Data from* itopf.com)

4.17 ASB Bank Limited published the following Home Loan rates (in % p.a.) on 14 May 2010:

Term	Rate
Overnight	5.75
6 months	5.85
12 months	6.35
18 months	6.95
2 years	7.30
3 years	7.70
4 years	8.20
5 years	8.50

(a) Draw a scatterplot of these data, with term (in years) on the x-axis and rate on the y-axis.
(b) Calculate Pearson's correlation coefficients for these data.
(c) Calculate Spearman's correlation coefficients for these data.
(d) Which coefficient better reflects the relationship between the term and rate?

4.18 The 2007/08 Active NZ Survey identified the participation rates of New Zealanders in sport and recreation activities based on a sample of 4443 adults. Six of the top ten activities were common to men and women. The following table summarises the rates and ranks of these six activities:

	Men		Women	
	%	r	%	r
Walking	52.3	1	75.1	1
Gardening	37.3	2	48.7	2
Swimming	33.4	3	36.1	3
Cycling	28.0	5	27.0	4
Equipment-based	26.0	6	17.8	6
Jogging/Running	19.3	8	15.8	7

(a) Draw a scatterplot of the participation rates.
(b) Calculate and compare Pearson's and Spearman's correlation coefficients for these data.
(c) Summarise the relationship between the participation rates of men and women.
(*Data from* sparc.org.nz)

4.19 The Brooklyn wind turbine overlooks Wellington City and is the oldest commercial wind turbine in New Zealand. Its rotor has a diameter of 27 metres, and it has a power rating of 225 kilowatts (0.225 megawatts). The turbine was produced by the Danish firm Vestas. In the table below are the rotor diameters (d, in m) and power ratings (P, in MW) of Vestas wind turbines.

d	P	d	P
27	0.23	90	1.80
52	0.85	90	2.00
60	0.85	100	1.80
83	1.65	90	3.00
80	2.00	112	3.00

(a) Draw a scatterplot of these data with diameter on the x-axis.

(b) The power output is based on the area swept by the turbine, which is related to diameter squared. Is this non-linear relationship evident in the plot?
(c) Why is Spearman's the appropriate correlation coefficient for these data?
(d) Calculate Spearman's correlation.

(*Data from* vestas.com)

4.20 The following table gives bond ratings (a qualitative measure of riskiness) and excess yield (in % per annum) for a random sample of New Zealand corporate bonds. Theory would predict a negative relationship between the two. The ratings in the table are issued by Standard & Poor's, and are ordered in the table from most to least risky.

Bond	Rating	Excess yield
1	BB+	3.62
2	BBB+	0.72
3	BBB+	0.95
4	A-	1.75
5	A	0.90
6	A	0.92
7	A+	0.68
8	A+	0.92
9	AA-	1.72
10	AA-	0.66
11	AA-	0.85

(a) Noting the missing categories BBB- and BBB, draw a scatterplot of the data, with equal gaps between the rating categories.
(b) Is Pearson's correlation suitable for these data? Why, or why not?
(c) Calculate Spearman's correlation for the data, and comment on the theoretical expectation.

(*Thanks to Martin Lally for the data*)

4.21 A study to better understand spatial variation in coniferous tree species diversity surveyed islands in Barklay Sound, British Columbia, Canada. Isolation was measured as the distance to the closest large landmass (D below, in metres), and the islands were thoroughly surveyed for conifer trees. The following table gives the number of species on each of the 17 smallest islands in the sound (m):

Is	D	m	Is	D	m
1	8	1	10	30	1
2	8	2	11	40	1
3	9	4	12	60	3
4	10	3	13	70	3
5	10	4	14	70	1
6	10	3	15	75	1
7	15	6	16	75	2
8	20	1	17	90	1
9	20	4			

(a) Draw a scatterplot of the data with island isolation on the x-axis.
(b) Calculate Spearman's correlation between island isolation and the number of conifer species present on the island.
(c) Both variables are quantitative, nonetheless Pearson's correlation is unlikely to be suitable. Explain why this is the case.
(d) What other factors might affect the number of species on each island?

(*Data from Burns, Berg, Bialynicka-Birula, Kratchmer & Shortt, Tree diversity on islands: assembly rules, passive sampling and the theory of island biogeography, forthcoming in the Journal of Biogeography*)

4.22 The British National Corpus is a supposedly representative collection of British English texts, and contains about 100 million words of running text. Analysis of this text for the plural versions of a list of nouns ending in "f" gives the following frequencies (in %):
(In some cases, the noun is indistinguishable from the verb, e.g. dwarf,

which can be used as either a noun or a verb. In these case, a random sample of passages is analysed, and this sample gives an estimate of the number of noun occurrences.)

Noun	Count	% "-ves"
behalf	5	40.0
dwarf	326	82.8
half	406	2.7
hoof	256	19.1
loaf	663	99.2
scarf	186	3.2
wharf	61	16.4

(a) Plot these data in a scatterplot, with frequency on the horizontal axis.
(b) Based on your scatterplot, is a linear relationship appropriate between the two variables?
(c) Calculate Pearson's correlation for these data.
(d) Now calculate Spearman's correlation coefficient.
(e) Do the two correlation coefficients give the same suggestion about the relationship? Which, if either, do you think is more suitable?

See also Exercise 10.30.
(Data thanks to Laurie Bauer)

4.23 Introduction of Emissions Pricing in New Zealand is predicted to have a short-term negative impact on output, as firms' costs of production increase. The following table presents estimated "maximum value at stake" (MVAS, a measure of the short-run risk of the scheme) against estimated long-run change in gross output (%ΔGO) for various industries if an emissions price of $50 per tonne of CO_2 were introduced.

Industry	%ΔGO	MVAS
Food, beverage, tobacco	−0.15	3.84
Textiles, apparel	0.14	3.77
Wood, pulp, paper	−0.49	10.56
Printing, publishing	−0.14	0.69
Chemicals	−1.84	5.02
Non-metallic mineral	−2.19	14.11
Basic, fabricated metals	−3.26	13.31
Machinery, equipment	0.43	0.72
Other manufacuring	0.08	0.70

(a) Draw a scatterplot of these data.
(b) Calculate Pearson's correlation for sample.
(c) Calculate Spearman's correlation for the sample.
(d) Compare your two correlation estimates. Which do you think better summarises the relationship between the two variables? Why?
(Data from med.govt.nz)

4.24 The following table lists the national Agricultural Output Index (AOI, base year 1939) and the agricultural patent density for the Wellington region.

Year	AOI	Density
1880	300.83	0.00
1881	230.34	0.00
1882	230.49	1.54
1883	295.98	0.00
1884	279.24	2.80
1885	277.97	1.35
1886	195.81	1.26
1887	254.05	3.68
1888	288.79	9.72
1889	314.87	18.96
1890	326.39	13.91
1891	245.35	1.01
1892	310.44	4.91
1893	275.34	6.62
1894	238.20	4.61
1895	215.22	7.85

(a) Plot these data in a scatterplot, with density on the horizontal axis.
(b) Calculate Pearson's correlation for these data.

(c) Now calculate Spearman's correlation coefficient.
(d) Do the two correlation coefficients give the same suggestion about the relationship. Which, if either, do you think is more suitable?

(Thanks to Rebecca Craigie & Les Oxley for the data)

4.25 A project investigating the effects of intensified dairy farming on the nitrogen levels of groundwater systems seeks to ensure these practices can be sustainable. Physical and chemical properties of the vadose zone, the layer of the earth's crust between the root zone and the groundwater, are measured. The table gives acidity levels (pH) and nitrate concentrations (NO_3, in micrograms per gram of soil) at various sites in Canterbury.

Site	pH	NO_3	Site	pH	NO_3
1	6.84	0.73	9	6.95	0.10
2	6.99	1.07	10	6.94	0.47
3	6.58	0.82	11	7.00	0.27
5	6.81	0.91	12	7.11	0.00
6	6.91	0.66	13	7.15	0.10
7	6.97	0.50	14	7.16	0.22
8	6.87	0.53	15	6.80	0.25

(a) Draw a scatterplot of these data.
(b) Calculate an appropriate correlation coefficient for these data. Justify your choice.
(c) Summarise the relationship between pH and nitrate concentration.

(Thanks to Michelle Peterson, Plant and Food Research, for the data)

Regression lines

4.26 As cities grow, public transport becomes increasingly important for efficient movement to and from work. The following table gives the growth rates (in %) of train and bus usage in the Wellington region.

Year	Train	Bus	Year	Train	Bus
2001	4.7	0.4	2006	2.9	3.8
2002	5.5	1.8	2007	−4.1	−1.1
2003	2.2	−3.7	2008	3.9	7.6
2004	4.2	3.5	2009	0.3	−4.5
2005	4.8	5.7			

(a) Draw a scatterplot of these data.
(b) Comment on the suitability of Pearson's correlation and linear regression to summarise the relationship between the two variables.
(c) Estimate Pearson's correlation for the data.
(d) Estimate the linear regression line for the data.
(e) Summarise the relationship between these two variables.

(Thanks to Greater Wellington Regional Council for the data)

4.27 Kea (*Nestor notabilis*) are highly inquisitive alpine parrots. Behavioural ecologists wanted to find out if "bold" individuals were more likely to suffer from lead poisoning due to exploring items introduced by humans into their environment (such as huts, cars and rubbish bins). Kea were scored for behavioural diversity (BD). High BD scores indicate very 'outgoing', exploratory behaviour. Blood-lead concentrations (BLC, in $\mu g/L$) were also measured. The data for 18 kea studied in Aoraki / Mount Cook National Park are below:

Kea	BLC	BD	Kea	BLC	BD
1	3	2	10	38	3
2	7	0	11	42	7
3	19	3	12	42	0
4	24	0	13	43	1
5	25	5	14	43	13
6	29	4	15	44	13
7	30	0	16	48	3
8	33	10	17	52	11
9	37	9	18	56	10

(a) Draw a scatterplot of the data with behavioural diversity on the x-axis.
(b) Assuming behavioural diversity is a measurement, calculate Pearson's correlation.
(c) Estimate the regression line for these data, and add it to your plot.
(d) Predict the blood-lead concentration of a kea with behavioural diversity score of 6. Comment.
(e) Predict the blood-lead concentration of a kea with behavioural diversity score of 20. Comment.

(*Data from Reid, Exploration-avoidance and an anthropogenic toxin (lead Pb) in a wild parrot (kea: Nestor notabilis), VUW thesis, 2008*)

4.28 The following table is for domestic and international destinations which were part of Air New Zealand's May 2010 promotion. Distances (in kilometres, km), flight times (in hours, h) and costs (in $) are all ex-Auckland.

Destination	km	h	Cost
Kerikeri	197	0.67	39
Wellington	492	1.00	49
Christchurch	762	1.33	49
Nelson	507	1.42	89
Blenheim	516	1.33	100
Tonga	3391	4.22	169
Gold Coast	2227	3.67	178
Westport	608	1.83	178
Canberra	2301	2.87	339
Fiji	2155	2.63	354
Darwin	5139	6.38	676
Tahiti	4098	5.10	1055
Hawaii	7046	8.75	1624
Shanghai	9346	11.62	1856
Los Angeles	10478	13.02	1874
Vancouver	11331	14.08	1890
Hong Kong	9121	11.33	1917
San Francisco	10487	12.03	1949

In each of the following instances, draw a scatterplot and comment on suitability of linear regression, calculate Pearson's correlation coefficient, estimate the linear regression line, and interpret the intercept and slope estimates:
(a) flight duration against distance
(b) cost against distance
(c) cost against flight duration.
(*Data from* travelmath.com)

4.29 The following table gives changes in the unemployment rate (dU) and changes in GDP (dY) in New Zealand over the period 1989-2009.

Year	dU	dY
1989	0.0015	0.0190
1990	0.0018	0.0100
1991	0.0034	0.0140
1992	-0.0093	0.0220
1993	0.0126	-0.0050
1994	0.0616	-0.0080
1995	0.0493	-0.0190
1996	0.0424	-0.0140
1997	0.0360	0.0010
1998	0.0194	0.0060
1999	0.0053	0.0070
2000	0.0517	-0.0090
2001	0.0238	-0.0090
2002	0.0366	-0.0050
2003	0.0468	-0.0020
2004	0.0397	-0.0060
2005	0.0374	-0.0070
2006	0.0302	0.0000
2007	0.0193	-0.0010
2008	0.0325	-0.0010
2009	-0.0060	0.0080

(a) Draw a rough scatterplot of the data, with change in GDP on the x-axis, and comment on the suitability of the linear regression model.
(b) Regress change in the unemployment rate on change in GDP.
(c) If GDP increases by 1%, what would we expect to happen to the unemployment rate?

(*Thanks to Dean Hyslop for the data*)

For the following set of exercises, use the data in the specified question to estimate the regression line. The dependent variable is stated below.

4.30 Female legislators (4.5)

4.31 Depression scores (4.8)

4.32 CPI-16 weights (4.10)

4.33 Kea bill length (4.11)

4.34 Maximum observed temperatures (4.6)

4.35 Oil spill size (restrict to spills less than 200,000 tonnes) (4.16)

4.36 Long run change in gross output (4.23)

4.37 Nitrate concentration (4.25)

Fitted values and residuals

4.38 A study of sign language acquisition in children with Deaf parents studied young children between 8 and 36 months old. The following table summarises the average numbers of sign tokens used (expressive) and understood (receptive) by these children.

Age (months)	Expressive	Receptive
8-11	3.8	10.3
12-15	15.8	36.2
16-19	59.3	106.3
20-23	126.9	174.3
24-27	203.6	252.7
28-31	268.3	331.1
32-36	348.1	405.3

(a) Draw a scatterplot of the averages, with the number of expressive signs on the x-axis.
(b) Calculate Pearson's correlation for these data.
(c) Calculate Spearman's correlation for these data.
(d) Estimate a linear regression line for these data.
(e) Calculate the residuals from the regression and use these to make a recommendation about which correlation is most appropriate, and why.

(Data from Woolfe et al., *Early vocabulary development in deaf native signers: a British Sign Language adaptation of the communicative development inventories*, Journal of Child Psychology and Psychiatry, 2010)

4.39 The table below contains full retail prices (in May 2010, in $) for products where an identical product is available at exactly double the size.

Product	Single	Double
AA batteries, 2	5.49	9.99
Cheese, 500 g	10.58	14.65
Chewing gum, 14 pc	1.88	3.59
Chocolates, 225 g	8.49	15.99
Eggs, 1/2 dozen	2.05	4.15
Laundry powder, 500 g	4.89	6.99
Manuka honey, 250 g	4.99	9.98
Milk, 1L	2.35	4.29
Mouth wash, 500 mL	9.79	14.79
Olive oil, 500 mL	8.68	16.21
Olive oil spread, 500 g	4.79	8.59
Orange juice, 1 L	4.39	8.33
Plain biscuits, 250 g	2.85	6.74
Spread, 250 g	4.20	6.59
Sugar, 1.5 kg	3.39	6.70
Tea bags, 100	4.27	9.15
Tinned peaches, 410 g	2.19	4.54
Toilet paper, 4 rolls	3.89	6.99
Weetbix, 375 g	3.09	3.65

(a) Plot these prices in a scatterplot with the "single" prices on the x-axis. Does a linear relationship seem appropriate?
(b) Add to your plot the line $y = 2x$. Interpret this line.
(c) Add your estimated regression line to the plot. Interpret the coefficients.
(d) Calculate the residual for the plain biscuits and that for the cheese. Interpret these residuals.

4.40 Following from Exercise 2.31, participants in the New Zealand Arrestee Drug Abuse Monitoring survey were asked at what age they started using the drugs listed in the table below. The average age of first use among the users is tabulated by drug and gender.

	Male	Female
Alcohol	13	14
Cannabis	14	15
Cocaine	20	19
Heroin	19	21
Methadone	22	27
Methamphetamines	22	23
Ecstasy	21	20
Tranquillisers	19	17
Hallucinogens	17	18

(a) Plot a scatterplot of these data with males' average ages on the x-axis.

(b) Estimate the regression line, using the males' average ages as the independent variable.

(c) Calculate the residual for each drug. *(Data from police.govt.nz)*

4.41 ASB Bank Limited published the following term deposit rates (% p.a., for deposits over \$10,000) on 15 May 2010.

Term	Rate
12 months	5.10
18 months	5.60
2 years	5.50
3 years	6.00
4 years	6.25
5 years	6.75

(a) Draw a scatterplot of these rates against the home-loan rates from Exercise 4.17 for the same terms, and comment on the suitability of a linear relationship between these variables.

(b) Estimate the linear regression line for the data, using the deposit rates as the X variable.

(c) Calculate the residuals for the home loan rates, and confirm they sum to zero.

4.42 The following savings (in \$, per annum) are for 10 houses, and reflect the benefits of retro-fitting adhesive plastic double glazing. This film is disposed of and replaced once a year and the savings reflect this. The houses were randomly selected, and the costs of installation calculated using the cheapest prices for the available kits as at February 2009. Electricity savings were calculated using average retail electricity pricing and assumptions about improved insulation for a given ambient room temperature. The savings in the table below reflect average climate for Christchurch and Dunedin respectively.

House	Christchurch	Dunedin
1	246	336
2	56	118
3	147	228
4	9	70
5	83	137
7	114	198
8	118	190
9	61	103
10	123	192

(a) Plot a scatterplot of the savings, with Christchurch on the x-axis.

(b) Calculate the regression line for Houses 1-5 and 7-10. Interpret the estimated coefficients.

(c) Calculate the residual for each of the houses.

(d) Why might House 6, with observation (278,295) have been removed from the sample?

(Data from Smith, A cost benefit analysis of secondary glazing as a retrofit alternative for New Zealand homes, VUW thesis, 2009)

Chapter 5

Time series data

A special bivariate relationship that is often of interest is when the independent (x) variable is time. Many economic variables are of this type, but they occur in all areas. Time series data are frequently presented graphically – a common example is the trace of a seismograph. Many of us treasure graphs of our height and weight as recorded by our Plunket nurse for the first year or two of our lives.

Other examples of time series include: the population of New Zealand on census date, the number of New Zealanders incarcerated as of 31 December each year, New Zealand's estimated inflation rate, the maximum observed temperature at Wellington airport each calendar month, the weight of a tuatara from its "birth" measured each Wednesday morning through the first year of its life, the daily dosage of medication given to a mental health patient, or the level of the Karori Reservoir as measured by a floating pencil on rotating graph paper.

Notice that the data in the above examples are measured at different time intervals and probably will have different features.

- New Zealand's population is recorded every census date, every five years. The population has tended to grow in the past. In the future, changes will depend on birth and death rates, and immigration and emigration patterns.

- The number of people in prison is recorded every year. The actual numbers will depend on government policy, legislation, enforcement, the courts, and capacity within the penal system.

- New Zealand's inflation rate is reported by Statistics New Zealand every three months. The governor of the Reserve Bank of New Zealand is obliged to keep this variable between 1% and 3% at all times.

- The maximum temperature is measured every month. It will tend to be high in the summer months and low in the winter months. Global warming would imply that this will also tend to go up over time.

- The tuatara's weight is measured every week. Unless the tuatara gets sick we might expect its weight to go up with each measurement, but the increase may change through time depending on "growth spurts" and access to food supplies.

- The medication dosage is recorded every day. We might expect this to change very rarely. When it does change, it will go up if the patient's condition has worsened, or go down if the patient's condition has improved.

- The reservoir level is measured *continuously*. It will tend to go up after rain, and go down at other times.

The *frequency* with which these data are measured is the length of time between measurements. The interval of the census is five years and the measurement of the reservoir level was continuous. The other examples give data which are annual, quarterly, monthly, weekly, and daily. These are the most common sampling frequencies. The sampling frequency summarises the properties of the independent variable, time. We also usually report the *sample period* using the first time observation and the last, e.g. "we have annual data for 1980 to 2009 inclusive".

Observations of the dependent (y) variable are called *time series* data, or simply *a time series*. As with univariate and bivariate random samples, we use symbols as a shorthand for the data.

> A *time series* consists of T measurements from a population on a variable Y. The data can be listed
>
> $$Y_1, Y_2, \ldots, Y_T$$
>
> where T is the series length.

If we compare the labelling of the time series with how we might label a simple random sample $X_1, \ldots, X_n$ and a bivariate sample $(X_1, Y_1), \ldots, (X_n, Y_n)$ we notice several differences:

- we use t to index the values, not i, and the sample size is T, not n;

- the "x-variable" is the time index itself;

- the time series data are usually labelled X, and so the pairs are $(1, Y_1), \ldots, (T, Y_T)$.

This labelling convention is different from that used for random samples to remind us that time is involved, and that the data are collected in a certain order. When we collect a random sample, the order is not important, e.g. if we were writing down retail sales figures for New Zealand cities in May 2010, we could list them from largest sales to smallest, from largest city to smallest city, from northern-most city to southern-most, in alphabetical order, or use some other scheme. When we observe time series data, we always list them in time order with the earliest observation first, and the most recent observation last.

As in Chapter 4, where we analysed oak tree data, throughout this chapter we will use the rainfall data in Table 5.1 to illustrate techniques. It is from cliflo.niwa.co.nz and is the total rainfall for each month at The Chateau Earthquake and Weather Station (EWS) on the northern flanks of Mount Ruapehu. The regular nature of the observations permits the natural layout of the table, with complete years in the rows, and months in the columns.

	Jan	Feb	Mar	Apr	May	Jun	Jul	Aug	Sep	Oct	Nov	Dec
2007	154.6	44.0	281.8	127.6	148.2	335.0	316.8	309.6	132.0	368.4	195.0	229.0
2008	60.4	62.8	99.8	212.0	71.6	281.4	478.6	388.8	228.6	346.8	192.4	185.2
2009	128.8	282.6	73.4	147.2	223.0	115.2	246.0	228.4	192.0	230.2	180.6	197.0

Table 5.1   Total monthly rainfall (in mm) at The Chateau, Mount Ruapehu

As with most sample data, it is hard to discern much from the raw data, although we can see the smaller rainfall figures tend to be in the first few months of the year (January to March), while the largest are in the middle (July and August). While it may be useful to look at a stemplot, histogram or boxplot of these numbers, none of these take into account the natural order within the data provided by time.

5.1 Displaying time series data

A graph can be a useful way of identifying patterns in data, and time series data are displayed using a *time series plot*. This is a scatterplot, with time on the x-axis and the observations on the vertical y-axis. We plot the individual observations as points, and then join these points with *straight lines*.

> A time series plot is a scatterplot of Y_t against t with consecutive points joined by a straight line.

The straight lines allow the eye to focus on increases and decreases in the data. We join the first observation to the second with a straight line, the second to the third with a straight line, the third to the fourth, etc. As with all other plots, we must label the axes and provide a title. The x-axis label will typically be one of "Date", "Year", "Month" etc. If we choose "Time" as the label, it is useful to add the frequency, e.g. "Time (days)".

Example 5.1

The rainfall data in Table 5.1 consist of 36 observations: 154.6, 44.0, 281.8, 127.6, 148.2, 335.0, ..., 197.0. These numbers form the y-ordinates in the time series (scatter) plot. The x-ordinates are given by the dates, and they should be regularly spaced. There are several ways we could label them, e.g. "Jan 2007", "Feb 2007", ..., "Dec 2009" but the important thing is that they are equally spaced. Once we've identified the 36 points in the xy plane, we join these by straight lines, as shown in the following plot:

5.1 Displaying time series data

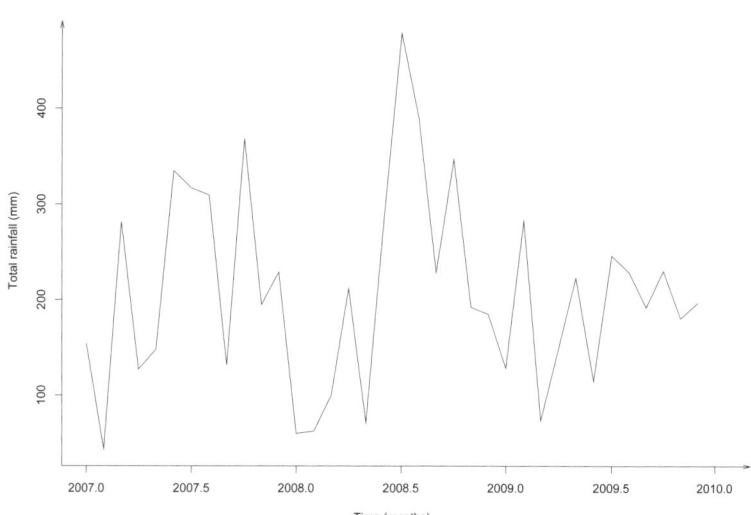

Note that we've chosen to label the axis 2007.0 through to 2010.0, corresponding to January 2007 through to January 2010. 2007.5 corresponds to July 2007.

The time series plot of the rainfall data is a little hard to interpret, since the numbers move around a fair bit. An alternative plot for data which may be influenced by the calendar (called *seasonal data*) is called a seasonal subseries plot. This consists of many small time series graphs – one for each "season".

Example 5.2

The following graph is a seasonal subseries plot for the rainfall data in Table 5.1.

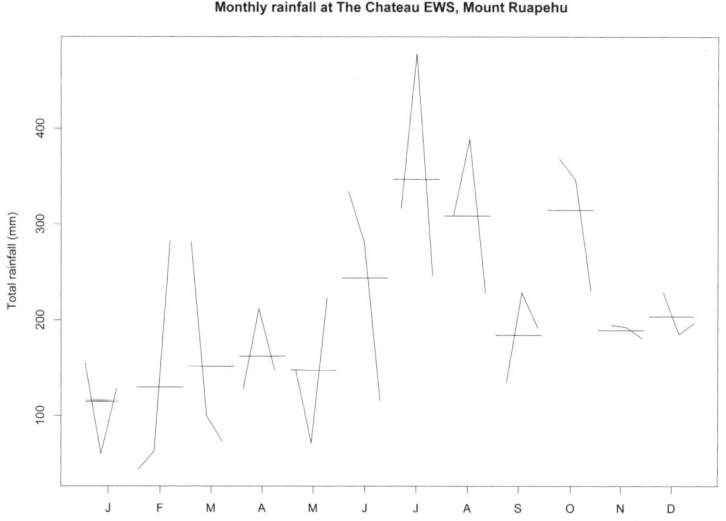

Each short time series in the graph is formed using a single column of the table, giving 12 series in total. Along the x-axis are the months, labelled "J", "F", etc. Above "J" are the three January observations, plotted as a time series. The horizontal bar is the median of those three values.

This plot better enables us to identify months with largest average rainfall, i.e. July, August and October, months with lowest average rainfall, i.e. January and February, months with high variation, i.e. February, March, June, July and August, and months with low variation, i.e. November and December. These features were difficult to identify in the time series plot of the data.

5.2 Time series components

One fundamental property of a time series is the sample period: how *often* observations are measured and *over what period of time*. In contrast, the features of the dependent variable are less easy to describe.

It can be useful to think of a time series as the combination of unobserved *components*, each of which focuses on one prominent aspect of the data. Even though we only get one observation at each point in time, we can think of the observation as being made up of these various parts. These components are not observed, and different people will allocate them in different ways.

We will focus on three components: the seasonal, the trend, and the irregular.

- The *seasonal* component is very often related to weather patterns, and repeats itself through time. Monthly and quarterly data are very likely to have a seasonal pattern for this reason, while annual data are very unlikely to have one. Daily data might have a seasonal pattern because of employment patterns within a week, e.g. number of hours' sleep each night might tend to be longer on Sunday through Thursday nights, and shorter on Friday night and Saturday night. Hourly data might be influenced by behaviour patterns due to daylight hours.

- The *trend* or *trend-cycle* describes longer-term variation, e.g. over a period of many years for annual data. Economic variables often have repeating periods of increase followed by periods of decrease, and this is called the *business cycle*. Unlike a seasonal pattern, this does not repeat itself over a fixed period of time. The lengths of business cycles have changed through history, whereas the seasons repeat every year.

- The *irregular* component is simply what we do not allocate to the trend or seasonal components. Usually the irregular is assumed to be random, and can be thought of as a regression residual.

> The common time series components are the *trend* or *trend-cycle*, the *seasonal* and the *irregular*.

The components themselves are indexed by time, and are often labelled T_t for the trend, S_t for the seasonal, and ϵ_t for the irregular. We typically will not know the trend, seasonal and irregular components and will only observe their combination, i.e. the time series data. Writing a time series as these three unobserved components is illustrated in the following example.

Example 5.3

An athlete in a major sporting event in April 2011 has running as part of her preparation. Her training plan is over a three-year period, and is based around 3-monthly blocks, within which she runs a set distance per week. Initially she builds up her training slowly, before increasing more rapidly. She then maintains a set pattern for a period, before reducing training volume leading up to the event. She trains most in spring and summer, less in autumn and least in winter. When she is unwell or busy at work she misses sessions, but this doesn't happen very often. Sometimes she runs further than she was programmed to do. Here are her statistics decomposed into three components:

t	y_t	T_t	S_t	ϵ_t	t	y_t	T_t	S_t	ϵ_t
Autumn, 2008	14.9	15	−1	0.9	Spring, 2009	37.1	35	2	0.1
Winter, 2008	14.1	17.5	−3	−0.4	Summer, 2010	37.5	35	2	0.5
Spring, 2008	21.8	20	2	−0.2	Autumn, 2010	33.5	35	−1	−0.5
Summer, 2009	27.3	25	2	0.3	Winter, 2010	31.7	35	−3	−0.3
Autumn, 2009	28.5	30	−1	−0.5	Spring, 2010	27.1	25	2	0.1
Winter, 2009	32.7	35	−3	0.7	Summer, 2011	16.1	15	2	−0.9

Notice the regular pattern in the seasonal component (−1, −3, 2, 2) which translates into 1 km/week below average in Autumn, 3 km/week below average in Winter, and 2 km/week above average in Spring and Summer. These add to 0 over a whole year. The actual distances run each quarter are found by adding up the three components, i.e. $y_t = T_t + S_t + \epsilon_t$. The time series plot below shows that the trend is the dominant component for these data.

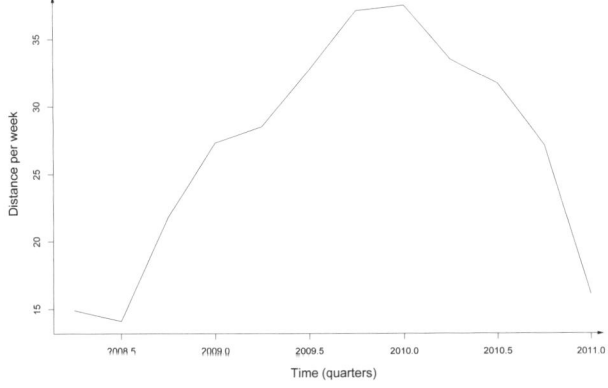

It can be useful to have trend, seasonal and irregular components in mind when we describe a time series plot, and we comment on each in turn if possible.

- The trend component describes the general movement of the data over the entire sample period, e.g. up initially, and then down.

- The seasonal component describes *regular* variation around this trend, e.g. high in summer months and low in winter months.

- The irregular component describes the rest. We may attribute unusual observations to the irregular, and we may even provide a suggested reason, e.g. visitor arrivals to New Zealand were unusually high in September 2011, probably due to the Rugby World Cup.

Not all time series will have all three components. The rainfall data in Table 5.1 might be decomposed into only two: a seasonal and an irregular. The seasonal component might be estimated by the subseries medians shown in Example 5.2, while the irregular might be the deviations from these medians. Since the monthly subseries are not consistently going up or down over the three-year period, we might choose not to estimate a trend.

There are many different ways of combining unobserved components. The most common are the *additive* and *multiplicative* models. The running data in Example 5.3 are an example of this, where the three components are added to give the data. The components are multiplied together to give the data in a multiplicative model. A multiplicative model is used when the variation in the seasonal *and* irregular components goes up when the trend goes up.

Example 5.4

The monthly short-term overseas visitor arrivals to New Zealand from January 1975 to December 2009 are shown (in thousands) in the following plot. We see that as the level of the data goes up, so too does the variation about this level. We might use a multiplicative model to represent this series.

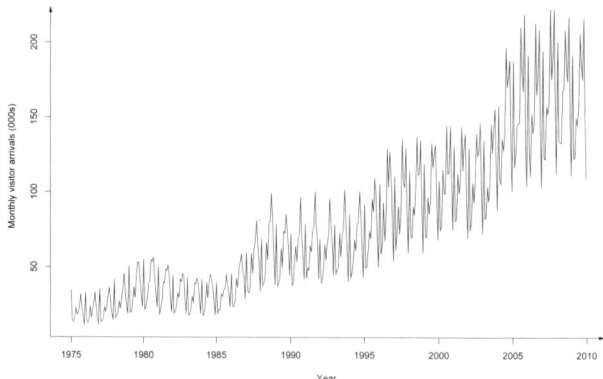

5.3 Summarising time series data

The minimum and maximum values (and the times at which they occurred) tell us something about the data. The sample mean and standard deviation are useful in some instances, and not so useful in others as illustrated by the next examples.

Example 5.5

The rainfall data in Table 5.1 have $\sum y_t = 7494.8$ and $\sum y_t^2 = 1926890$. These give $\bar{y} = 208.2$ mm and $s = 102.3$ mm (both to 1dp). The minimum rainfall was 44.0 mm in February 2007, and the maximum was 478.6 mm in July 2008. Summary statistics by year are given in the table below. These year-by-year means and standard deviations are roughly equal through time.

Year	$\bar{y}$	s
2007	220.17	101.74
2008	217.37	136.15
2009	187.03	60.82

Example 5.6

The total numbers of short-term visitors to New Zealand per year from 1960 to 2009 are given in the following table and time series plot (in millions, to the nearest thousand). The smallest annual arrivals were in 1960, while the largest were in 2008.

	1960s	1970s	1980s	1990s	2000s
0	0.040	0.121	0.472	0.727	1.270
1	0.047	0.130	0.435	0.783	1.303
2	0.051	0.155	0.387	0.739	1.303
3	0.058	0.195	0.364	0.784	1.382
4	0.070	0.247	0.382	0.808	1.739
5	0.084	0.247	0.374	0.905	1.880
6	0.094	0.248	0.481	1.082	1.877
7	0.106	0.275	0.633	1.115	1.984
8	0.101	0.332	0.746	1.177	1.985
9	0.107	0.429	0.721	1.210	1.936
$\bar{y}$	0.076	0.238	0.499	0.933	1.666
s	0.026	0.095	0.147	0.192	0.311

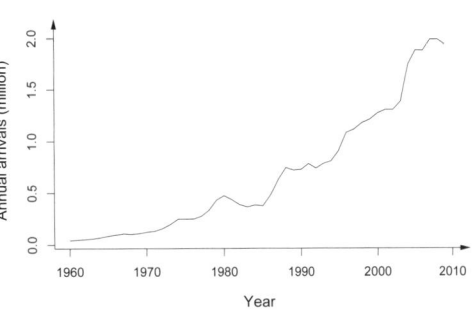

The overall sample mean is $\bar{y} = 0.6828 = 682{,}800$ visitors per annum. This is clearly very different from the decade means, and is a poor summary of the data. The overall standard deviation is $s = 0.6023 = 602{,}300$ and is much larger than the decade standard deviations given in the table.

The sample mean and standard deviation don't tell us much about trending data, but they are very useful for *stationary* series.

> A *stationary* time series has a constant mean and a constant variance through time. A *non-stationary* series has a changing mean, changing variance, or both.

When we plot a stationary time series, the data should oscillate about a constant level. If we imagine a trend component in the data, this should be the same at any point in the series. Similarly the variation in the data should be the same at any point in the series. The time series in Example 5.5 is approximately stationary, while the series in Example 5.6 is not. The standard statistics were much more useful for the stationary series than they were for the other, non-stationary, series.

In the remainder of this chapter, we present three simple ways of summarising a time series: a *linear time trend*, a *simple moving average*, and a *sample autocorrelation coefficient*.

5.3.1 Linear time trend

Some time series tend to grow by roughly the same amount each period, e.g. if someone spends roughly the same amount on groceries each month, their total grocery expenditure will grow by roughly the same amount each month.

> **Example 5.7**
>
> John spends approximately $200 at the supermarket each month. It varies according to discounts, changing needs, and other factors. After the first month, his total supermarket expenditure is approximately $200. After two months, it is approximately $400, after three approximately $600, etc.

Continuing this example, John's total expenditure starts at zero. Assume the amount spent in the first month is approximately β dollars. Then the total after the first month is approximately β dollars. In the second month, he again spends about β dollars, and so the total at the end of this month is approximately $\beta + \beta = 2\beta$. By the end of the third month the total is approximately 3β, by the 4th month it is approximately 4β and by the end of month t it is approximately βt. If the total started at α instead of zero (perhaps the carry-over from the previous year), the total spent by the end of month t would be $\alpha + \beta t$. If we acknowledge "approximately" by adding a small error to each total, we get the model

$$Y_t = \alpha + \beta t + \epsilon_t.$$

This is just the linear regression model that we saw in Chapter 4, using time t as the x-variable, and the observations of the time series as the y-variable. Referring back to the equations on page 68 we can estimate the parameters α and β using a and b, giving us an estimate of the *linear time trend*. We saw earlier that the sample average was not always a very good summary statistic of a time series. Similarly, a linear time trend will not always be a good summary of a time series. It will be appropriate only when the time series seems to grow (or fall) by roughly the same amount each period. Whether or not it is an appropriate model for a series is best established using a time series plot.

Example 5.8

Q: Following Example 5.7, supermarket expenditure data are below.

	Feb	Mar	Apr	May	Jun	Jul
Month spend	185	240	140	250	190	150
Total spend	185	425	565	815	1005	1155

Estimate a linear time trend for the total expenditure.

A: A graph of total expenditure against time indicates the linear time trend model is appropriate. Let the time variable be given by $t = 1, 2, 3, 4, 5, 6$. Entering the data into our calculator using the technique in Appendix B yields

$$\sum t = 21 \quad \sum y_t = 4150 \quad \sum t^2 = 91 \quad \sum y_t^2 = 3542350 \quad \sum t y_t = 17945$$

with $T = 6$ observations. Also

$$S_{XX} = \sum t^2 - \tfrac{1}{T}\left(\sum t\right)^2 = 91 - \tfrac{1}{6}(21)^2 = 17.5$$
$$S_{XY} = \sum t y_t - \tfrac{1}{T}\sum t \sum y_t = 17945 - \tfrac{1}{6}(21)(4150) = 3420.$$

These, and the sample means $\bar{x} = \tfrac{1}{T}\sum t = 3.5$ and $\bar{y} = \tfrac{1}{T}\sum y_t = 691.67$ (2dp), are used to find the regression line slope and intercept terms

$$b = \frac{S_{XY}}{S_{XX}} = \frac{3420}{17.5} = 195.4 \quad \text{and} \quad a = \bar{y} - b\bar{x} = 691.67 - 195.4 \times 3.5 = 7.77$$

giving the regression line $\hat{y}_t = 7.77 + 195.4\, t$.

Note that b is close to $200 and a is close to $0, as we would expect from the information provided in Example 5.7.

In the previous example, we simply entered the pairs $(1, y_1), (2, y_2), \ldots, (T, y_T)$ into our calculator, and let it do the job of calculating the coefficients. If we need to do the calculation by hand, the regular nature of the time index helps us simplify the equations. We label the time index X and the data Y. Some algebra, and the facts that

$$1 + 2 + \cdots + T = \frac{T(T+1)}{2} \quad \text{and} \quad 1^2 + 2^2 + \cdots + T^2 = \frac{T(T+1)(2T+1)}{6}$$

give

$$\bar{X} = \frac{T+1}{2} \quad \text{and} \quad S_{XX} = \frac{T(T^2-1)}{12}.$$

The slope and intercept estimates of the regression line are

$$b = \frac{12}{T(T^2-1)}\left(\sum t y_t - \tfrac{T+1}{2}\sum y_t\right) \quad \text{and} \quad a = \tfrac{1}{T}\sum y_t - b\left(\tfrac{T+1}{2}\right).$$

In order to evaluate these, we need $\sum y_t$, the sum of the time series, and $\sum t y_t = y_1 + 2y_2 + 3y_3 + \cdots + T y_T$.

5.3.2 Moving averages

We saw earlier that the sample average was not always a very good summary statistic of a time series. In particular, if the time series is not stationary, then the sample mean doesn't tell us a lot. Despite this, averaging time series data is something that *is* useful, it is just that we do it for short sections of the time series at a time. We call this process a *moving average*.

The idea behind a moving average is that although the entire time series is not stationary, over a short period (e.g. a few years for annual data) the time series is approximately stationary.

We illustrate the basic process through the following example.

> **Example 5.9**
>
> The monthly expenditure figures from Example 5.7 were: 185, 240, 140, 250, 190 and 150. The average of the first three observations is
>
> $$\tfrac{1}{3}(Y_1 + Y_2 + Y_3) = \tfrac{1}{3}(185 + 240 + 140) = 188.33 \text{ (2dp)}.$$
>
> Dropping the first observation, and adding the fourth, we now average the 2nd through 4th observations:
>
> $$\tfrac{1}{3}(Y_2 + Y_3 + Y_4) = \tfrac{1}{3}(240 + 140 + 250) = 210.$$
>
> Continuing,
>
> $$\tfrac{1}{3}(Y_3 + Y_4 + Y_5) = \tfrac{1}{3}(140 + 250 + 190) = 193.33 \text{ (2dp)}$$
>
> and
>
> $$\tfrac{1}{3}(Y_4 + Y_5 + Y_6) = \tfrac{1}{3}(250 + 190 + 150) = 196.67 \text{ (2dp)}.$$
>
> The numbers 188.33, 210, 193.33 and 196.67 are a three-point moving average of the monthly supermarket expenditure.

We saw in Example 5.9 that while we had six observations in the time series, we could calculate only four values in the moving average. This is called an *end-effect*, and in the example it affected two observations coinciding with three (the width of the moving average) minus one.

It is usual to "centre" a moving average, so that it coincides with an actual observation. In the example above, the average $\tfrac{1}{3}(Y_1+Y_2+Y_3)$ has Y_2 literally at its centre, and we say the average is centred on the second observation. In order for a moving average to be centred, we must have the same number of observations before and after the "middle" value and so we must be averaging an odd number of data points. If we use p observations on each side, then in total we average $2p + 1 = p + 1 + p$ observations: p in the past, 1 in the "present", and p in the future. Using this scheme, we cannot calculate the moving average for the first p observations nor the last p. These are the $2p$ end-effects, e.g. when $2p + 1 = 3$, $p = 1$ and $2p = 2$ averages can't be calculated.

> The centred moving average M_t calculated using $2p+1$ observations (data points) is given by
>
> $$M_t = \frac{1}{2p+1} \sum_{j=-p}^{p} Y_{t+j} \quad (t = p+1, \ldots, T-p).$$
>
> This is sometimes referred to as a $(2p+1)$-point centred moving average.

The formula for the centred moving average (M_t above) is the same as the usual sample mean formula. We are averaging $2p+1$ observations, and this takes the place of the sample size n. The summation index runs from $-p$ to p, which means the time index on the observations we add runs through, $t-p, t-p+1, \ldots, t-1, t, t+1, \ldots, t+p-1, t+p$. The first observation we can do this calculation for is the $(p+1)$st, where we average $Y_1, \ldots, Y_{2p+1}$. The last observation we can do this for is the $(T-p)$th, where we average the $Y_{T-2p}, \ldots, Y_T$. This is illustrated in the next example.

Example 5.10

Q: Calculate a 3-point moving average for the monthly supermarket expenditure.
A: Actually, this has already been done in Example 5.9, and we confirm it below.

Month	t	Y_t	$M_t = \frac{1}{3}(Y_{t-1} + Y_t + Y_{t+1})$
Feb	1	185	
Mar	2	240	$188.33 = \frac{1}{3}(185 + 240 + 140)$
Apr	3	140	$210.00 = \frac{1}{3}(240 + 140 + 250)$
May	4	250	$193.33 = \frac{1}{3}(140 + 250 + 190)$
Jun	5	190	$196.67 = \frac{1}{3}(250 + 190 + 150)$
Jul	6	150	

Note that we cannot calculate M_1 (since Y_0 is not available) nor M_6 (since Y_7 is not available). These two missing values are the end-effects for this 3-point moving average.

When we calculate a moving average, we trade off two things when we select p: smoothness and end-effects. The larger p is, the smoother the moving average M_t will be. On the other hand, the larger p is, the more observations of the moving average we miss out on at the ends of the series. There is no set rule for choosing p for non-seasonal time series data.

For *non-seasonal time series*, we assume the irregular component averages to zero, and so the moving average should give us an idea of the trend of the time series. If the irregular component has outliers, then the sample mean may not be a very reliable estimate, and so we may choose to calculate a moving median instead. This is done in the same way as above, but by calculating the median for each set of $2p+1$ points instead of the mean.

When the time series has a seasonal component, and we want to approximate the trend, we need to average exactly the number of "seasons" in the seasonal component (which should average to zero). If we don't do this, the moving average itself exhibits seasonal variation. The problem with averaging using the number of seasons is that this number is often even, i.e. 4 for quarterly data and 12 for monthly data. The solution is to calculate *two* moving averages, the second a moving average of the first, which is itself a moving average of the data.

Suppose we have quarterly data, then $\frac{1}{4}(Y_1 + Y_2 + Y_3 + Y_4)$ is centred on observation $\frac{1}{2}(1+4) = 2.5$. Similarly, $\frac{1}{4}(Y_2 + Y_3 + Y_4 + Y_5)$ is centred on observation $\frac{1}{2}(2+5) = 3.5$. So, to estimate the trend at observation 3, we average the averages at observations 2.5 and 3.5; i.e. we calculate a 2-point moving average of the 4-point moving average. This gives us

$$\frac{\frac{1}{4}(Y_1 + Y_2 + Y_3 + Y_4) + \frac{1}{4}(Y_2 + Y_3 + Y_4 + Y_5)}{2} = \tfrac{1}{8}Y_1 + \tfrac{1}{4}(Y_2 + Y_3 + Y_4) + \tfrac{1}{8}Y_5.$$

This equation shows we have two ways of calculating this moving average. On the left is the 2-point moving average of the 4-point moving average. On the right is a *weighted moving average* with the end observations getting half the weight of the middle observations. Note that the observations on the ends are both from the same quarter, and so all quarters have weight $\frac{1}{4}$ in the moving average.

> For monthly data, with $s = 12$, or for quarterly data, with $s = 4$, we estimate the trend using a $(2 \times s)$-point moving average. This is best calculated by first doing an s-point moving average of the data, and then a 2-point moving average of the first moving average.

Example 5.11

Q: Form quarterly data based on the monthly rainfall data in Table 5.1, and calculate a (2×4)-point moving average of it.

A: First, we combine observations to make the quarterly data:

$$Y_1 = 480.4 = 154.6 + 44.0 + 281.8 \qquad Y_2 = 610.8 = 127.6 + 148.2 + 335.0 \qquad \text{etc.}$$

Note that these are not moving sums, but sums of non-overlapping groups of observations. These numbers are shown in the table below, in the column marked Y_t.

Next, we calculate the 4-point moving average, labelled $M_{t+\frac{1}{2}}$ in the table. The average of the first four observations is

$$\tfrac{1}{4}(Y_1 + Y_2 + Y_3 + Y_4) = \tfrac{1}{4}(480.4 + 610.8 + 758.4 + 792.4) = 660.50$$

which has "centre" at $\frac{1}{2}(2+3) = 2.5$. This leads to the label $M_{t+\frac{1}{2}}$ and 660.50 sits on the row with $t = 2$. The next value in that column is

$$\tfrac{1}{4}(Y_2 + Y_3 + Y_4 + Y_5) = \tfrac{1}{4}(610.8 + 758.4 + 792.4 + 223.0) = 596.15$$

corresponding to $M_{3.5}$.

Finally, we calculate the 2-point moving average of the Y series, i.e. $W_t = \frac{1}{2}(M_{t-\frac{1}{2}} + M_{t+\frac{1}{2}})$ which is centred at t. The first average is

$$W_3 = \tfrac{1}{2}(M_{2.5} + M_{3.5}) = \tfrac{1}{2}(660.50 + 596.15)$$

and this is the first entry in the W column. The remaining values are calculated in similar fashion.

Quarter	t	Y_t	$M_{t+\frac{1}{2}}$	W_t
2007 Q1	1	480.4		
2007 Q2	2	610.8	660.50	
2007 Q3	3	758.4	596.15	628.33
2007 Q4	4	792.4	584.70	590.43
2008 Q1	5	223.0	669.10	626.90
2008 Q2	6	565.0	652.10	660.60
2008 Q3	7	1096.0	717.55	684.83
2008 Q4	8	724.4	697.65	707.60
2009 Q1	9	484.8	590.25	643.95
2009 Q2	10	485.4	561.10	575.67
2009 Q3	11	666.4		
2009 Q4	12	607.8		

We can confirm $W_3 = \frac{1}{8}Y_1 + \frac{1}{4}(Y_2 + Y_3 + Y_4) + \frac{1}{8}Y_5$ with

$$628.33 = \tfrac{1}{8}(480.4) + \tfrac{1}{4}(610.8 + 758.4 + 792.4) + \tfrac{1}{8}(223.0).$$

The following plot has both the quarterly data and the moving average (using the dashed line) plotted. As we can see, the moving average is much less erratic than the raw data. The end-effects are also apparent.

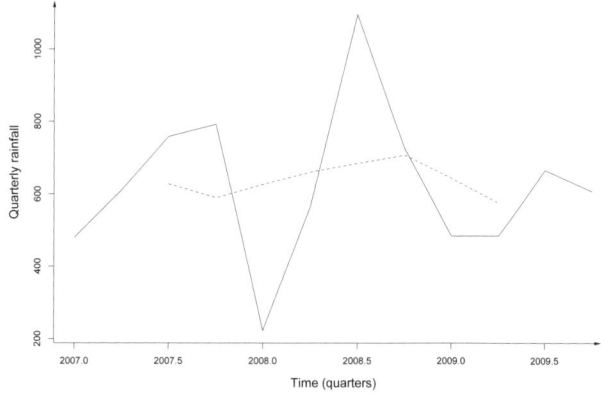

Moving averages are most easily calculated using a spreadsheet, because of the large amount of calculation involved. Instructions for a simple moving average are given in Appendix C.1.3.

5.3.3 Sample autocorrelation

In addition to the time series plot, a very important graphical tool in the analysis of time series data is the *sample autocorrelation function*, also known as the correlogram. This is built up from sample correlation coefficients of the time series with itself. The way this is done is to *shift* the time series to give us paired data, e.g. the pairs: y_2 with y_1, y_3 with y_2, y_4 with y_3 etc. Because the correlation is with itself, it is called an *autocorrelation*.

We first look at what this summary measure tells us, and then comment on how it is calculated.

Example 5.12

On the top row of the following collection of graphs are three time series plots. The first is the running data of Example 5.3 and the second is of the quarterly rainfall data used in Example 5.11, which was derived from the monthly figures in Table 5.1. The third series is new. The second row is of scatterplots with Y_{t-1} on the x-axis and Y_t on the y-axis.

For the running data, we have

t	1	2	3	4	5	6	7	8	9	10	11	12	13
Y_{t-1}		14.9	14.1	21.8	27.3	28.5	32.7	37.1	37.5	33.5	31.7	27.1	16.1
Y_t	14.9	14.1	21.8	27.3	28.5	32.7	37.1	37.5	33.5	31.7	27.1	16.1	

where the first observation of the first row ($Y_{1-1} = Y_0$) and the last of the second row (Y_{13}) are not available. This leaves $11 = T - 1$ pairs, and these are plotted in the first scatterplot.

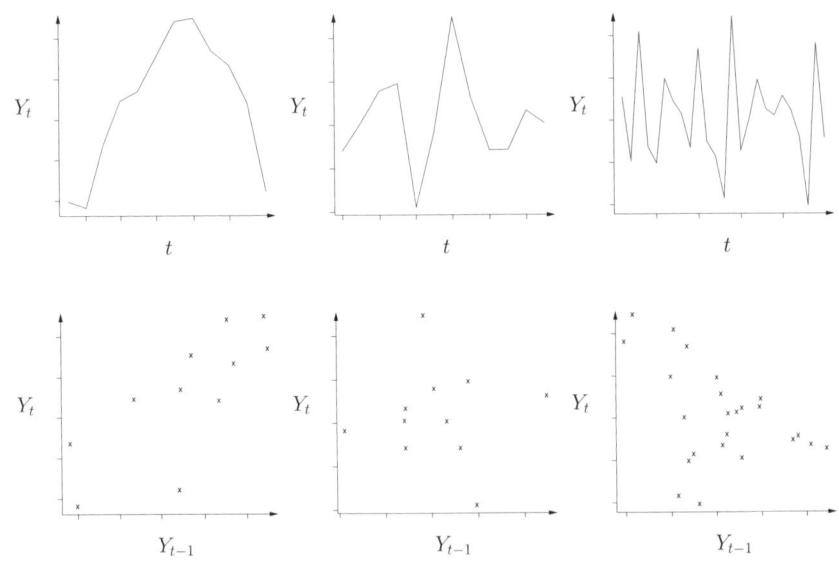

We see that the trending running data have a scatterplot indicating a large positive correlation. The approximately stationary rainfall data have a scatterplot indicating an approximately zero correlation. The third series tends to oscillate: a large value is followed by a small value and vice versa. The points in the scatterplot indicate a large negative correlation.

Pearson's correlation coefficients for the points in the three scatterplots are: 0.770 for the running data, -0.006 for the quarterly rainfall data, and -0.434 for the third series.

As the graphs in the previous example show, a series with a trend tends to have a large positive correlation between successive values in the time series. In contrast, a correlation close to zero corresponded to a "random looking" time series, and a negative correlation corresponded to a series which oscillates.

Recall that Pearson's correlation coefficient is calculated using

$$r = \frac{S_{XY}}{\sqrt{S_{XX}S_{YY}}}$$

where Y is one variable, and X is the other. If we consider a time series and its shifted values as per the following table:

t	1	2	3	$\cdots$	$T-1$	T
Y_t	Y_1	Y_2	Y_3	$\cdots$	Y_{T-1}	Y_T
Y_{t-1}		Y_1	Y_2	$\cdots$	Y_{T-2}	Y_{T-1}
Y_{t-2}			Y_1	$\cdots$	Y_{T-3}	Y_{T-2}

we see that we can collect pairs of observations, e.g. (Y_{t-1}, Y_t). This would give the collection of pairs: $(Y_1, Y_2), (Y_2, Y_3), \ldots, (Y_{T-1}, Y_T)$. As we saw in Example 5.12, while we started with T observations, we have only $T-1$ pairs when we shift the series by one observation. If we were to shift the series again and consider the pairs (Y_{t-2}, Y_t) we would have only $T-2$ pairs. This is one aspect of *autocorrelations* that makes them different from the correlations we saw in Chapter 4.

Another thing to note about correlation in the context of a time series, is that $\bar{X}$, $\bar{Y}$, S_{XX} and S_{YY} are all calculated for the time series itself. Since in this context, X and Y are the same variable, then $\bar{X}$ and $\bar{Y}$ must be the same, and S_{XX} and S_{YY} must be the same. Noting the definitions of S_{XY} and S_{XX} on page 58, Pearson's correlation formula simplifies to give the first-order autocorrelation – of Y_t with itself lagged *one* period, Y_{t-1}

$$r_1 = \frac{\sum_{t=2}^{T}(Y_{t-1} - \bar{Y})(Y_t - \bar{Y})}{\sum_{t=1}^{T}(Y_t - \bar{Y})^2}$$

where $\bar{Y} = \frac{1}{T}\sum Y_t$. Note that there are $T-1$ pairs in the calculation of the numerator, even though the denominator can be calculated from all T observations. Because we use the overall sample mean $\bar{Y}$, the numerator does not simplify nicely but it does take best advantage of the available data.

In general, we can calculate the kth-order autocorrelation which shifts the time series k observations, and then calculates the (amended) correlation between the $T - k$ pairs. The sample autocorrelation function calculates autocorrelations for $k = 1, 2, 3, \ldots, K$ for some upper value K, and then plots these values. With experience, the collection of autocorrelations provides useful insight into modelling the time series. This topic is beyond the scope of this book.

Exercises

Time series plots

5.1 The figures below are average monthly sunshine hours for Kaitaia (KAI), Gisborne (GIS), Nelson (NSN), Te Anau (TEU) and Scott Base (Antarctica, SBA), calculated from 1971-2000 data.

	KAI	GIS	NSN	TEU	SBA
Jan	212	241	266	196	380
Feb	201	201	229	179	110
Mar	197	185	212	156	46
Apr	155	156	188	118	5
May	140	148	173	85	0
Jun	123	124	143	64	0
Jul	137	130	157	68	0
Aug	159	154	171	112	0
Sep	162	173	186	137	37
Oct	182	210	212	170	160
Nov	187	215	225	192	339
Dec	219	234	245	184	337

(a) Draw a single time series plot with all five series shown.
(b) Summarise patterns in the data.
(*Data from* cliflo.niwa.co.nz)

5.2 A high level of per capita milk consumption is an indicator of a country's development. The following table gives fluid milk consumption (in kilograms per person per year) for various countries: developed (EU, Australia, NZ and US), developed Asian (Japan and South Korea), developing Asian (China, Indonesia, Viet Nam) over a ten-year period.

	1999	2000	2001	2002	2003
China (tot)	2.8	3.7	4.6	6.2	7.6
China (urb)	7.9	9.9	11.9	15.7	18.6
China (rur)	1.0	1.1	1.2	1.2	1.7
EU	76.0	76.3	76.7	72.5	72.8
US	95.6	94.3	93.3	93.0	92.8
Australia	104.5	101.4	99.5	99.9	100.1
NZ	96.9	91.5	91.1	90.4	89.8
Japan	39.1	39.2	38.8	39.3	39.5
Sth Korea	34.0	32.0	32.9	34.9	38.2
Indonesia	1.3	1.3	1.6	1.5	1.4
Viet Nam	1.2	1.0	1.1	1.3	1.8

	2004	2005	2006	2007	2008
China (tot)	8.0	7.7	8.2	9.2	9.6
China (urb)	18.8	17.9	18.3	17.8	15.2
China (rur)	2.0	2.9	3.2	3.5	3.4
EU	71.9	71.3	69.5	67.8	68.6
US	92.3	91.9	92.4	95.3	90.1
Australia	102.9	104.2	103.4	103.9	106.2
NZ	88.9	88.0	87.1	84.2	79.9
Japan	38.9	37.4	36.5	35.5	35.0
Sth Korea	33.4	32.1	34.9	34.2	35.1
Indonesia	1.4	1.4	1.4	1.4	1.4
Viet Nam	1.8	1.8	1.8	1.5	1.8

(a) Draw a time series plot of the Western (developed) milk consumption figures.
(b) Draw a time series plot with the Asian milk consumption figures.
(c) Comment on the features of your graphs.
(*Thanks to Carly Harker, RBNZ*)

5.3 The overnight cash rate (OCR, in % per annum) is the Reserve Bank of New Zealand's tool by which it is charged with maintaining inflation between 1 and 3% per annum. The OCR is reviewed approximately every 8 weeks. On 28 October 2004, the OCR was set to 6.50%. The following table lists the subsequent changes to this rate to the end of 2009. *Note that the changes are not equally spaced.*

Date	OCR	Date	OCR
10/3/05	6.75	24/7/08	8.00
27/10/05	7.00	11/9/08	7.50
8/12/05	7.25	23/10/08	6.50
8/3/07	7.50	4/12/08	5.00
26/4/07	7.75	29/1/09	3.50
7/6/07	8.00	12/3/09	3.00
26/7/07	8.25	30/4/09	2.50

Draw a time series plot of the OCR and comment on its features.
(*Data from* rbnz.govt.nz)

5.4 The full *Household Economic Survey* is conducted in NZ every three years. The following prices are median rents (in $ per week) for the years ended 30 June from a sample of 4500 households (N.I. and S.I. are North Island and South Island respectively).

	06/07	07/08	08/09
Auckland	280	280	300
Wellington	214	220	250
Rest of N.I.	169	181	200
Canterbury	192	212	245
Rest of S.I.	155	158	178
NZ	200	220	241

Draw these six time series on a single plot. Comment on any features.
(*Data from* stats.govt.nz)

5.5 Using the data from Exercise 4.24:
(a) Plot the agricultural output index and the agricultural patent density in separate graphs.
(b) Does either series appear stationary? Justify your comments.

5.6 The proportion of New Zealand's state highway network deemed "too rough" is given in the following table, corresponding to the years 1999 to 2008 inclusive.

| 1.78 | 1.87 | 1.60 | 1.44 | 1.42 |
| 1.38 | 1.41 | 1.40 | 1.63 | 1.51 |

(a) Plot the data in an appropriate graph.
(b) Calculate the sample mean and standard deviation of the data. Add horizontal lines to your plot at $\bar{y}$, $\bar{y} - s$ and $\bar{y} + s$.
(c) Do the data appear stationary?
(*Data from* nzta.govt.nz)

5.7 The following table summarises the medals won (Gold, Silver, Bronze, and overall) by New Zealand teams at the summer Olympic Games.

Year	Country	G	S	B	M
1920	Antwerp	0	0	1	1
1924	Paris	0	0	1	1
1928	Amsterdam	1	0	0	1
1932	Los Angeles	0	1	0	1
1936	Berlin	1	0	0	1
1948	London	0	0	0	0
1952	Helsinki	1	0	2	3
1956	Melbourne	2	0	0	2
1960	Rome	2	0	1	3
1964	Tokyo	3	0	2	5
1968	Mexico City	1	0	2	3
1972	Munich	1	1	1	3
1976	Montreal	2	1	1	4
1980	Moscow	0	0	0	0
1984	Los Angeles	8	1	2	11
1988	Seoul	3	2	8	13
1992	Barcelona	1	4	5	10
1996	Atlanta	3	2	1	6
2000	Sydney	1	0	3	4
2004	Athens	3	2	0	5
2008	Beijing	3	2	4	9

Noting the cancellation of the 1940 and 1944 games, draw a time series plot of the total number of medals won. Can you add the other time series in a way that displays their share of the total medals won?
(*Data from* wikipedia.org)

5.8 When pathogenic bacteria attack a human, the human immune system responds by sending specialised white blood cells to the site of infection. When the human cells first encounter the bacteria, the cells release concentrated bursts of hydrogen peroxide (H_2O_2) to kill the bacteria. In this study, a wild-type bacterium is compared with two mutant strains of the same bacterium in the presence of 25 mM H_2O_2. These bacterial strains are grown over time and measured using optical density. The following table summarises the growth reduction (in %).

Hour	Strain 1	Strain 2	Wild Type
0	116.4	103.2	105.5
1	81.8	83.3	72.8
2	76.3	63.7	58.8
3	80.4	52.4	63.7
4	88.1	37.3	54.3
5	86.0	32.9	55.3
6	87.5	27.7	62.6
7	84.0	25.0	60.1
8	85.2	22.2	61.4
9	87.6	21.5	66.1

(a) Draw a time series plot with all three series in it.
(b) Comment on any features evident from the plot.

(*Thanks to Laura Green, VUW, for the data*)

Moving averages

5.9 Real returns (in % per five years) on New Zealand rural land (adjusted for inflation) in the latter half of the 20th century were:

Period	%	Period	%
1955-59	8.5	1975-79	−9.3
1960-64	14.9	1980-84	34.8
1965-69	13.6	1985-89	−38.4
1970-74	34.3	1990-94	85.2

(a) Plot these returns, and comment on their features.
(b) Calculate a 3-point moving average and add it to your plot.

(*Data from Lally & Randal, Ground rental rates and ratchet clauses, Accounting and Finance, 2004*)

5.10 Repetitive abnormal behaviour is a problem in many captive animals. In the wild, servals (a medium-sized African cat) catch a variety of live prey, and are capable of catching birds in flight. Three servals at an urban zoo were monitored for one month, and the average amount of time spent engaged in abnormal behaviour (e.g. pacing and compulsive grooming) between 10 and 11 am each day was recorded. One week into the monitoring period, zoo staff initiated a behavioural enrichment programme which included automatic feeders firing meatballs into the enclosure at random times throughout the day. After two weeks the programme was halted due to budget constraints. The average times (in minutes per hour) were as follows:

M	T	W	T	F	S	S
		41	38	36	42	34
38	39	30	28	24	20	15
13	12	11	12	13	8	11
12	11	13	15	16	20	22
21	25	22	24			

(a) Draw a time series plot of these data.
(b) Calculate a 5-point moving average of the data and add it to your plot.
(c) Comment on patterns in the data.

5.11 The following is the partial record from a weather station at Wellington International Airport from 6 pm on 11 March 2010 to 10 am on 12 March. The first column gives the (date and) time, ddd denotes the wind direction (360 is from the north, 90 from the east, 180 from the south and 270 from the west), ff is the mean wind speed (in knots – multiply by 1.6 to get km/h), fm is significant gust speeds (in knots), ww is type of significant weather (+SHRA means heavy shower, SHRA means moderate shower, -SHRA means light shower and VCSH means showers within the vicinity), and TT is air temperature.

Time	ddd	ff	fm	ww	TT
18:00	140	6	16		13
19:00	40	4			11
20:00	40	8			13
21:00	10	7			14
22:00	40	9			15
23:00	10	12			16
00:00	360	15			17
01:00	10	17			18
02:00	20	13			17
03:00	10	15			20
03:30	320	17	56	+SHRA	12
03:35	210	50	66	+SHRA	9
04:00	190	42	58	SHRA	9
04:20	190	34	50	SHRA	10
04:45	180	32		SHRA	10
05:00	180	31		SHRA	11
05:15	190	29		SHRA	11
05:35	180	22		-SHRA	12
06:00	180	18		VCSH	12
07:00	200	23		VCSH	12
08:00	190	22		VCSH	13
09:00	190	22	32		13
10:00	200	22		-SHRA	13

(a) Draw time series plots of wind speed and temperature.
(b) Calculate a 3-point centred moving average of wind speed, and add this to your plot of wind speed.
(c) Draw a time series plot of wind direction. What special property of direction should be reflected in the y-axis for this particular variable (i.e. what can the wind direction of 360 be recoded as)?
(d) Can you explain what meteorological phenomenon occurred during the sample period?

(*Thanks to Paula Acethorp and the Meteorological Service of New Zealand Ltd* metservice.com *for the data*)

5.12 In the year ending March 2007, there was a total of 152.2 petajoules (PJ) of electricity generated in New Zealand, with hydro accouting for the largest share (57.4%). Non-renewable sources, gas and coal, were 20.6% and 11.2% respectively, and the remainder was supplied by renewable geothermal, wind, wood and biogas sources. The following table gives the total observed renewable energy (in PJ), its share of total electricity ($\%^1$), and its share corrected for variation in rainfall ($\%^2$).

Year	Total	$\%^1$	$\%^2$
1998	94.0	72.8	70.9
1999	99.5	75.9	69.3
2000	98.7	73.2	73.2
2001	103.9	74.5	72.0
2002	92.9	67.8	75.8
2003	102.6	71.4	73.3
2004	101.6	70.8	70.8
2005	109.3	74.4	71.9
2006	95.9	64.0	72.7
2007	103.7	68.2	70.6

(a) Draw a time series plot of the total observed renewable energy. Does this series seem stationary?
(b) Draw a time series plot of the (uncorrected) share ($\%^1$). Does this series seem stationary? What does this imply about the total observed energy, renewable plus unrenewable?
(c) Add the corrected share ($\%^2$) time series to your plot in (b). Does this series seem stationary?

(d) Calculate a 3-point moving average of the corrected share time series, and add this to your plot.
(*Data from* eeca.govt.nz)

5.13 The following table gives the numbers of oil spills per year for three decades 1970-2009. Medium spills are between 7 and 700 tonnes (with number N_m) and large spills are more than 700 tonnes (with number N_ℓ).

Year	N_m	N_ℓ	Year	N_m	N_ℓ
1970	7	29	1985	31	8
1971	18	14	1986	28	7
1972	48	27	1987	27	10
1973	28	32	1988	11	10
1974	89	28	1989	33	13
1975	96	23	1990	50	14
1976	67	27	1991	30	7
1977	68	17	1992	31	10
1978	59	21	1993	31	11
1979	60	35	1994	26	9
1980	52	13	1995	20	3
1981	54	7	1996	20	3
1982	45	4	1997	28	10
1983	52	13	1998	26	6
1984	26	8	1999	20	6

(a) Plot the two time series in a single graph.
(b) Calculate a 3-point moving average of the number of large spills and add this to the time series plot.
(c) Summarise the features of the data. (*Data from* iotpf.com)

5.14 The following data are for rainfall (in inches per year) on Niulakita, the smallest island of Tuvalu, for the period 1963-1974.

Year	Rainfall	Year	Rainfall
1963	147.9	1969	128.8
1964	111.2	1970	147.9
1965	135.3	1971	184.2
1966	139.8	1972	96.8
1967	120.7	1973	139.4
1968	186.5	1974	111.7

(a) Plot these data in a time series plot.
(b) Calculate the sample mean of the data, and draw this on the plot.
(c) Calculate a 3-point moving average of the data, and add this to your plot.
(d) Does the series appear to be stationary?
(*Data from* Morrison, *Teaching exploratory data analysis*, New Zealand Journal of Geography, 1983)

5.15 The table below gives the number of sentences issued in the New Zealand justice system annually from 1980 to 2009 (in thousands; the top row is 1980 to 1985 inclusive, the next 1986 to 1991, etc.).

132.0	143.4	94.3	85.7	84.7	84.6
86.0	90.8	103.8	110.9	83.4	84.9
80.8	81.8	82.7	82.9	82.8	75.3
77.6	76.4	74.7	75.1	74.7	77.8
78.9	79.1	82.3	88.5	95.1	102.3

(a) Plot the time series in a suitable graph.
(b) Calculate a 3-point moving average of the numbers and add this to the plot.
(c) Summarise the features of the data. (*Data from* stats.govt.nz)

5.16 Economists have begun to investigate using google.com's search data to predict economic phenomena. The data reflects how often individual search terms have been used on Google over time. As at July 2010, New Zealand has the highest number of searches for "depression diagnosis", ahead of Australia and the United States. The following table contains a weekly index of the frequency of this particular search from 3 January to 27 June, 2010.

1.10	1.70	1.64	1.88	1.70	2.02	2.04
2.44	2.32	2.02	2.10	2.20	2.22	1.88
2.06	2.12	2.22	1.82	1.64	1.70	1.74
1.60	1.54	1.54	1.44	1.36		

The second observation in the series is 1.70, and it continues along the rows.
(a) Plot a time series plot of the data.
(b) Calculate a 5-point moving average of the time series and add it to your plot.
(c) Based on your evidence so far, does the series seem stationary?
(d) Below is a time series plot of this index from 5 August 2007 (the last 26 observations are the data above). Based on *this* graph, does the series seem stationary?

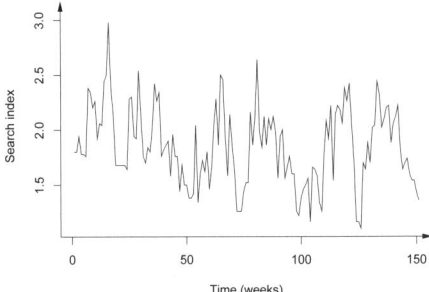

(*See Choi & Varian, Predicting the present with google trends, Google Inc., 2009*)

Linear time trend

5.17 The New Zealand Hector's dolphin is one of the rarest species of marine dolphin. Genetic diversity of this population is under severe threat from fisheries and other human activities. Over the last 25 years, the North Island population has been reduced to only one matrimonial lineage. Historical genetic samples were derived from museum specimens and analysied in conjunction with modern dolphins caught as bycatch or sampled using non-lethal means. Estimated haplotype diversity (HD) for this population is tabulated below using an irregularly spaced time series.

Year	HD	Year	HD
1960	0.75	1990	0.58
1980	0.68	1992	0.58
1984	0.65	1994	0.57
1985	0.68	1995	0.52
1987	0.70	1996	0.51
1988	0.67	2000	0.50
1989	0.59		

(a) Plot a time series graph of these data, taking care to account for the irregularly spaced observations.
(b) Estimate a linear time trend and add it to your graph.
(c) Do you think the linear time trend is an appropriate summary of the data?

(*Based on Pichler & Baker, Loss of genetic diversity in the endemic Hector's dolphin due to fisheries-related mortality, Proceedings: Biological Sciences, 2000*)

5.18 Part of a report investigating the effects of lowering the drinking age provided the following annual figures on the volume of spirit-based drinks (in million litres) available for purchase for each calendar year.

Year	Volume	Year	Volume
1995	2.0	2000	24.5
1996	3.9	2001	28.3
1997	9.2	2002	31.4
1998	18.3	2003	34.5
1999	20.1	2004	37.0

(a) Draw a time series plot of the volume data.
(b) Estimate a linear time trend for these data, and add this trend to your plot.
(c) Briefly summarise the patterns in spirit-based drink availability.

(*Data from* courts.govt.nz)

5.19 Reforestation at Wellington's Otari/ Wilton's Bush since 2002 has been supported by an active group of volunteers. The efforts included data collection and analysis to see which native species best suited planting sites within the area. The following table presents statistics on average heights of plants of the following species, observed at one site:

Date	Kanuka	Miro	Totara
Jun 2004	0.60	0.64	0.52
Aug 2004	0.57	0.45	0.54
Oct 2004	0.58	0.62	0.58
Dec 2004	0.56	0.62	0.61
Feb 2005	0.68	0.60	0.69
Apr 2005	0.91	0.68	0.74
Jun 2005	0.95	0.60	0.80
Aug 2005	1.00	0.70	0.75
Oct 2005	1.00	0.74	0.80
Dec 2005	1.00	0.70	0.89
Feb 2006	1.26	0.76	1.05
Apr 2006	1.45	0.85	1.10
Jun 2006	1.50	0.84	1.12
Aug 2006	1.51	0.90	1.15
Oct 2006	1.51	0.81	1.07

(a) Plot the average heights of these three species in a single time series plot.
(b) Estimate a linear time trend for the Miro heights. Is this a suitable summary of the heights?
See also Exercise 10.12.
(*Many thanks to Jonathan Kennett and the Otari volunteers for the data*)

5.20 The following table gives retail diesel prices at month end in New Zealand for the period 2008-2009 (in $).

	2008	2009		2008	2009
Jan	1.25	1.04	Jul	1.84	1.02
Feb	1.26	1.01	Aug	1.60	1.07
Mar	1.37	0.99	Sep	1.54	1.00
Apr	1.57	0.99	Oct	1.32	1.08
May	1.74	1.02	Nov	1.21	1.08
Jun	1.84	1.11	Dec	1.04	1.08

(a) Draw a time series plot of the prices.
(b) Is a linear time trend an appropriate summary of the data?
(c) Calculate a 5-point moving average of the prices, and add it to your plot. Do you think this is a reasonable estimate of the trend of the data? (*Data from* aa.co.nz)

5.21 The Karapoti Classic is one of the longest-running mountainbike events in the world. It has been held in the Akatarawa Forest every year since 1986. The table below gives the number of finishers by year (N_t):

Year	N_t	Year	N_t	Year	N_t
1986	45	1995	652	2004	730
1987	42	1996	729	2005	968
1988	50	1997	692	2006	788
1989	114	1998	800	2007	872
1990	151	1999	767	2008	930
1991	223	2000	643	2009	871
1992	318	2001	604	2010	835
1993	294	2002	672		
1994	582	2003	675		

(a) Draw a time series plot of the number of finishers.
(b) Estimate a linear time trend to the number of finishers, and add this trend to your plot.
(c) Do you think the linear time trend is a good way of describing the number of finishers? What is the predicted number of finishers in 2020? (*Data from* karapoti.co.nz)

5.22 A person suffering from a major depression fills in a mood diary, recording their mood at midday on an 11-point Likert scale (0 to 10, with 0 corresponding to the most severe symptoms). Her record is below, by day of the week:

	M	T	W	T	F	S	S
Week 1	2	3	4	5	5	6	6
Week 2	6	6	5	5	2	5	4
Week 3	4	4	4	5	6	7	7

(a) Plot these scores as a time series.
(b) Calculate a 3-point moving average for the scores, and add it to your plot.
(c) Now calculate a 3-point moving *median* by taking the median value of each group of three numbers. Add this to your plot.
(d) Is the moving mean or the moving median more appropriate in this instance? Why is this?

5.23 Following Exercise 5.2, the level of per capita milk consumption for China in the years 1999 to 2008 is as follows:

$$\begin{array}{ccccc} 2.8 & 3.7 & 4.6 & 6.2 & 7.6 \\ 8.0 & 7.7 & 8.2 & 9.2 & 9.6 \end{array}$$

(a) Plot these data.
(b) Estimate a linear time trend and add this to your plot.
(c) Comment on the suitability of this linear time trend to forecast future per capita milk consumption both in the short and in the long term.

(*Thanks to Carly Harker, RBNZ, for the data*)

5.24 The following table gives the number of large-magnitude earthquakes in the world each year since 1980 (being over 6.0 on the Richter scale).

Year	N_t	Year	N_t	Year	N_t
1991	112	1997	136	2003	155
1992	179	1998	129	2004	157
1993	149	1999	134	2005	151
1994	159	2000	161	2006	153
1995	203	2001	137	2007	196
1996	164	2002	140	2008	180

(a) Plot these counts in a time series plot.
(b) Estimate a linear time trend for these counts. Is this a reasonable description of the data?
(c) What is risky about using this linear time trend to predict the number of large earthquakes in the future?
(d) What explanations could there be for an increase in the number of recorded earthquakes (other than earthquakes actually being more frequent)?

(*Data from* earthquake.usgs.gov)

Autocorrelation

For the following exercises, use Pearson's correlation coefficient to approximate the first-order autocorrelation of the specified time series. Use the size and sign of your estimate to comment on the likely features of the time series.

5.25 Real rural land returns (5.9)

5.26 Total renewable energy (5.12)

5.27 Genetic diversity of Hector's dolphins (5.17)

5.28 Miro average heights (5.19)

5.29 Mood scores (5.22)

5.30 Number of large earthquakes (5.24)

5.31 The digits in Table D.1 (on page 324) should be completely random. Not only does this imply equal frequencies of occurrence of each digit, but also lack of autocorrelation within any sequence.
(a) Select any sequence of 30 digits from the table.
(b) Use Pearson's correlation coefficient to approximate the first-order autocorrelation of the sequence.
(c) Comment on your finding.

Part II

Introduction to probability

Chapter 6

Working with probabilities

Probability is the way we describe the chance of an outcome occurring. Most people have had exposure to some of the ideas that follow, and are familiar with the concepts, even those with no formal training in probabilities. Consider the following everyday situations:

- What is the chance of getting a head when I toss a coin? What about the chance of getting a tail?
- What is the chance that I get a six when I roll a fair die? What about a four or a six?
- What is the chance that I win Lotto?
- If I smoke cigarettes, what is the chance I will get lung cancer?

We probably have a sense that the chance of getting a head is 0.5, and the same for getting a tail. We'll get a six one-sixth of the time, and a four or a six one-third of the time. The chance of winning the Lottery is not very large at all (one chance in 3,838,380 in New Zealand's six-balls-from-40 version of the game) but it is still fun to try! If I smoke, I am more likely to get lung cancer than if I don't, and the actual probability is probably sizable. These everyday events all have probabilities associated with them, and our everyday experience equips us reasonably well to answer probability questions. We will formalise some of our knowledge in the following material.

6.1 What is a probability?

Whenever outcomes are uncertain we have what is called a random experiment. Probability helps us describe the results of the random experiment, e.g. we can describe the reaction time to some stimulus, the amount of leaf browse in mistletoe in a particular location, the price of a stock, etc.

There are several different ways to look at probability – as:

- a number that gives a measure of the chance of some event
- a number that sums up your degree of belief in the occurrence of an event
- the relative frequency of an event in a very large number of trials.

We will use the following definition of probability:

> The probability of a particular outcome in a random experiment is the *proportion* of times that the outcome would occur in a large number of observations (or in a large number of repetitions of the experiment).

So a probability is *just a number*, and it applies to an *outcome*. An outcome is just something which happens (and which has some associated uncertainty), and we commonly give these outcomes abbreviated names, e.g. outcome A or B. A statement like, "the probability of A happening is equal to 0.25" just means that the outcome A would happen 25% of the time if we made many observations. A could be the outcome that we get two heads in two tosses of a coin.

It is useful to look at a few technical terms that are often used interchangeably: *incidence*, *proportion*, and *probability*. Incidence is used a lot in biomedical science, especially in genetics, in statements like: "the incidence of Huntingdon's chorea (HC) in the general population is 5 in 100,000", meaning 5 cases in 100,000 people. In this situation we could say the proportion of the population with HC = 5/100,000, i.e. the *fraction* of the population with HC. We could report the incidence as 0.00005 or, in per cent, as 0.005%. In this same situation the probability of a person chosen at random from the general population having HC is 0.00005.

We will start by looking at some very simple situations that we will use to develop a common language relating to probability.

Example 6.1

Imagine tossing a coin three times. If we use H for getting a head and T for getting a tail and we write down what happens by noting the outcome of the first toss followed by the second followed by the third, then there are eight possible outcomes:

$$HHH, \; HHT, \; HTH, \; THH, \; TTH, \; THT, \; HTT, \; TTT.$$

The first of these corresponds to getting three heads, the second two heads followed by a tail, etc.

The eight outcomes listed in the coin-tossing example can be grouped to form *events*. An example is the event that we get one head, which consists of the outcomes TTH, THT and HTT.

> An *event* is one single possible outcome or a group of outcomes.

It is conventional to label events with capital letters: e.g. $A = \{TTT\}$ is the event that all three tosses come up tails (and consists of only one outcome), whereas the event $B = \{HHT, HTH, THH\}$ is the event that we get two heads in three tosses, and there are three possible outcomes in this event. If any one of these three outcomes occurs, we say that B occurs.

> We say that A or B occurs if any one of the outcomes from A or B occurs.

Another event is $C = \{HHH, HHT, HTH, THH\}$, which is the event that at least two of the tosses come up heads. A and C cannot both occur, since no outcomes are common to both events. In contrast, B and C can occur together, since all the outcomes in B are in C.

> An *impossible* event is an event that cannot occur!

Such an event occurs a proportion (or relative frequency), 0, of the time. This means that the probability of an impossible event is zero. For most people this is intuitively obvious: if the horse Phar Lap is not in the race, then the probability of its winning is zero. We distinguish between impossible events and those which are very unlikely, e.g. winning Lotto is unlikely (it has a very small probability), but it is not impossible.

> Two events are *mutually exclusive* if they can't both occur.

We saw that A and B above were mutually exclusive, as we cannot get both three tails and two heads in three tosses of a coin. If two events are mutually exclusive, then, like A and B, they share no common outcomes.

Probabilities have to conform to certain mathematical rules, again which many people know intuitively.

6.2 Probability rules

A probability is a number attached to an event. We will write $P(A)$ to mean "the probability of A occurring". What can we say about $P(A)$ in a general sense?

Rule 1. $0 \leqslant P(A) \leqslant 1$.

Proportions must lie between zero and 1 (inclusive), and since all probabilities are proportions, the same follows for probabilities. Alternatively, we cannot have a relative frequency less than zero or greater than 1.

Rule 2. The sum of the probabilities of all possible outcomes in a situation must be 1.

If we list all the individual outcomes of an experiment along with their probabilities then these probabilities must add to one (just as a table of relative frequencies must add to one). For example, if we toss a die, there are six possible outcomes: 1, 2, 3, 4, 5, and 6. These are all equally likely if we have a fair and balanced die so we have a chance of 1 in 6 for each of these outcomes and of course, six of these add to 1.

Rule 3. $P(A) = 1 - P(A \text{ doesn't occur})$.

For example, if there is a probability of 20% that your suspicious freckle is cancerous, then there is a probability of 80% that it's not.

Rule 4. If A and B are mutually exclusive events, then $P(A \text{ or } B) = P(A) + P(B)$.

This is also reasonably obvious. If in a particular alpine environment 40% of all poa grasses are *Poa kirkii* and 15% are *Poa exigua*, then 55% of the poa grasses in this area are either *kirkii* or *exigua* (a single plant cannot be both).

Rule 5. The probability of an event is the sum of the probabilities of the outcomes making up that event.

Consider the experiment of tossing a coin three times. There were eight outcomes. The probabilities must all add up to 1 (Rule 2). If the coin is fair, then no outcome is any more likely than any other so each individual outcome has probability $1/8 = 0.125$ and therefore $P(A) = 0.125$, $P(B) = 0.375$ and $P(C) = 0.5$.

In the coin tossing experiment A and B are mutually exclusive as it is not possible to get both three tails and two heads in three tosses of a coin. This means that

$$P(A \text{ or } B) = P(A) + P(B) = 0.125 + 0.375 = 0.5$$

by Rule 4. By the same reasoning, A and B cannot both happen, so $P(A \text{ and } B) = 0$.

There is another property of probabilities that you will find useful and it needs one more idea, the idea of *independence*. We have seen this term before, when we required the regression errors to be independent of one another. We basically required that the errors be *random*. Most people have a fairly good idea of what this means.

> **Example 6.2**
>
> Imagine you are sitting in a locked, soundproof, and windowless room. You will be let out in half an hour if you can answer the following question correctly: "*will it be raining outside in five minutes' time?*" You are given one of the following pieces of information:
> A: It was raining outside an hour ago.
> B: It was sunny outside an hour ago.
> C: It is raining in London.
> If you were being smart, A and B would affect your choice, but not C. We say that your

choice is *independent* of event C, but not of A or B. This leads us to the final rule.

Rule 6. If A and B are independent events, then $P(A$ and B occur$) = P(A) \times P(B)$.

We will look at independent events in more detail in Section 6.3.1.

6.3 Conditional probability

As we saw in Example 6.2, rain outside your building and rain in London are independent events (unless your building is in London). On the other hand, the weather outside an hour ago would have an effect on the probability of rain. These events (weather an hour ago, and rain in five minutes) are not independent, and we say they are dependent. The way we typically describe dependent events is using *conditional probabilities*.

A conditional probability is similar to a regular probability in that it obeys the probability rules. We'll most often use Rules 1 to 3 for conditional probabilities.

On the other hand, a conditional probability is unusual in that the probability we give *is conditional on* another event happening.

Example 6.3

Continue the set up of Example 6.2. Suppose you are told that A is true, i.e. it was raining one hour ago. You realise if this was the case, it probably is still raining, and so you guess the probability of rain in five minutes' time is 0.8. We write $P(\text{rain, given } A) = 0.80$.

With a sense of *déjà vu* you find yourself in the same predicament, but this time B is true, i.e. it was sunny outside an hour ago. You realise if this was the case, it is very unlikely to be raining in five minutes, so guess the probability of rain is 0.05. We write $P(\text{rain, given } B) = 0.05$.

The two probabilities of rain are *conditional* on whether A or B has occurred.

In this example, we have two probabilities of rain, and both of them are conditional. We gave the chance of rain if it was raining, and the chance of rain if it was fine. Neither of them is *the* chance of rain, because they both depend on some other information, whereas the (unconditional) probability of rain would be the chance of rain in five minutes' time regardless of the conditions an hour ago.

A conditional probability always gives the probability of something happening, conditional on something else happening. If we're interested in the probability of B happening and we know A has happened, the conditional probability is written $P(B$ given $A)$ as in the above example. Sometimes the word "given" is replaced by a vertical bar, as follows: $P(B|A)$.

There is much more variation in how conditional probabilities are described in words. Consider the following:

> **Example 6.4**
>
> Q: The following statements all involve conditional probabilities. What event is being conditioned on? What event is of interest? Write in the form $P(B \text{ given } A)$.
>
> (i) If a person is male, what is the probability he is a smoker?
> (ii) You select a large business. What is the probability it has more than 100 employees?
> (iii) Of all Japanese import cars to New Zealand, what proportion have had their odometers tampered with?
> (iv) What is the probability of rain in Wellington today?
> (v) What is the probability of improved symptoms for a patient taking antibiotics?
>
> A:
>
> (i) We condition on {male} and are interested in {smoker}, i.e. we want P(smoker given male).
> (ii) We condition on {large business} and are interested in {more than 100 employees}, i.e. we want P(more than 100 employees given large business).
> (iii) We condition on {Japanese import} and are interested in {tampered odometer}, i.e. we want P(tampered odometer given Japanese import).
> (iv) We condition on {Wellington today} and are interested in {rain}, i.e. we want P(rain given Wellington today).
> (v) We condition on {antibiotics} and are interested in {improved symptoms}, i.e. we want P(improved symptoms given antibiotics).

As is clear in the example above, there are many ways of describing conditional probabilities. There are no set rules, although use of "if" and "given" are common.

6.3.1 Probability trees

Probability trees, or tree diagrams, are useful graphs to help understand what is going on in some situations involving probability. They are best used when there is a sequence of events, whose probabilities may or may not depend on what has already happened, i.e. when the probabilities are *conditional probabilities*. All tree diagrams have the same basic structure, and this is shown in Figure 6.1. From a particular starting point

- a pathway of *branches* leads to a possible *outcome*

- each branch is labelled with its probability

- the probability of any particular outcome is found by multiplying the probabilities along the pathway of branches leading to that particular outcome.

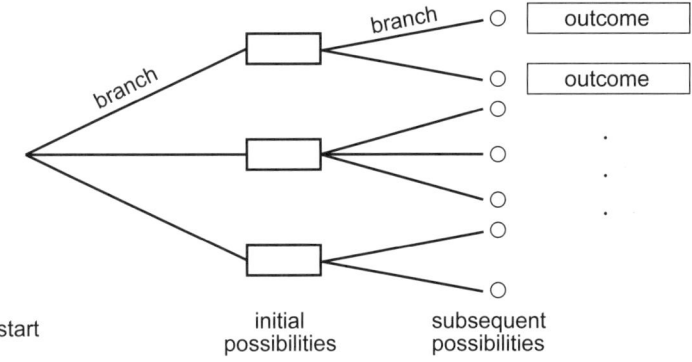

Figure 6.1 Basic tree diagram structure

Generally the probabilities on the first set of branches do not depend on any prior information, and are unconditional probabilities, of the type $P(A_1)$, $P(A_2)$, etc. Probabilities on the subsequent sets of branches are generally conditional probabilities, and are of the type $P(B$ given $A_1)$, $P(B$ given $A_2)$, etc. The outcomes are of the type $\{B$ and $A_1\}$, $\{B$ and $A_2\}$ and their probabilities are again unconditional. These probabilities are found by multiplying along the branches.

The function of probability trees is illustrated in the following example.

Example 6.5

Q: One in 120 men over 50 will develop prostate cancer. A blood test exists that identifies prostate-specific antigens. It will come up positive (+) or negative (−); however, like many medical tests, it is not foolproof. The test comes up positive in 90% of cases where the individual *does* have the disease and it registers a negative in 80% of healthy men. Many medical tests behave like this, throwing up false negatives and positives.

(i) What is the probability that a healthy male gives a positive test?
(ii) What is the probability a randomly selected male over 50 will be healthy and yet give a positive blood test?
(iii) What is the probability that a randomly selected male over 50 will give a positive test?
(iv) What is the probability that a randomly selected male over 50 will give a negative test?

A: Let D be the event that the man has the disease, and H be the event that he is healthy. The events occur first (i.e. before the test can be conducted), and we allocate them to the first set of branches in the following diagram. $P(D) = \frac{1}{120}$ and as a consequence $P(H) = 1 - \frac{1}{120} = \frac{119}{120}$. Subsequently, the test is conducted, and so we allocate the outcomes of the test to the next set of branches. We are given $P(+$ given $D) = 0.9$ and $P(-$ given $H) = 0.8$. The remaining probabilities are found by subtracting these from one, and these are shown in the boxes in the tree.

122 Chapter 6 Working with probabilities

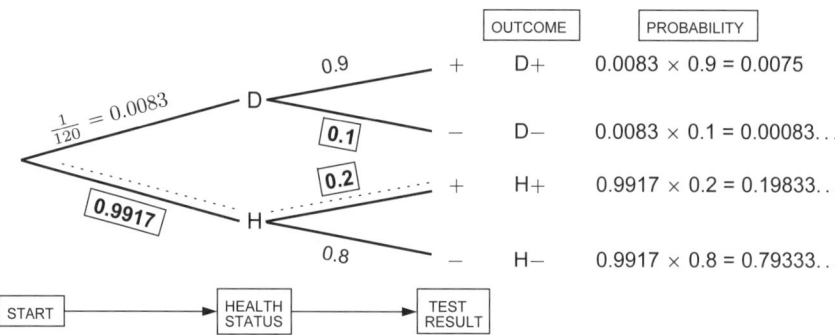

The final outcomes are found by following the path to that outcome. Their probabilities are obtained by multiplying the probabilities along the path to the outcome. Note that the sum of these four probabilities is 1, since the four outcomes are mutually exclusive, and they cover all possibilities.

(i) We start with a healthy male (node H in the tree) so we begin our path at H and go to +. So P(healthy male is +) = 0.2 from the tree.

(ii) We require P(the man is healthy and gives a positive test), i.e. $P(H$ and $+)$. This is the outcome for the dotted pathway, with probability $0.9917 \times 0.2 = 0.1983$ (4dp).

(iii) Two pathways lead to positive tests, so
$$P(+) = P(D+ \text{ or } H+) = P(D+) + P(H+)$$
$$\text{i.e. mutually exclusive outcomes (Rule 4)}$$
$$= 0.0075 + 0.19833\ldots$$
$$= 0.2058 \text{ (4dp)}.$$

(iv) $P(-) = 0.00083\ldots + 0.79333\ldots = 0.7942$ (4dp), or alternatively, from Rule 3, we have $P(-) = 1 - P(+) = 0.7942$ (4dp).

This next example demonstrates a situation where we do not necessarily face two choices at a node.

Example 6.6

Q: A rat in a maze has the following options: initially, it is faced with a choice of four pathways, which lead to doors A, B, C and D respectively. The rat has no preference for any of these paths over another. The doors are like cat doors but they are such that once through them the rat cannot go back. Door A leads to a path with a single door at the end which leads to a food source. Door B leads to a choice of two further doors, one leading to food and one to a water source. Door C also leads to a choice of two further doors, but one has water behind it and the other nothing. Door D leads into the same path as door A. Given a choice of doors leading to food or water the rat has a probability of 0.75 that it will choose to go toward food. Given a choice of doors leading to water or nothing the rat has a probability of 0.9 that it will choose to go toward the water. What is the probability that a rat started at the beginning of this system ends up with food?

A: We draw the tree representing this situation, and allocate to it the probabilities we are given. Remaining probabilities on the branches are found using Rule 3 and the probabilities of the final outcomes are found by multiplying along the branches. These final probabilities add to 1.

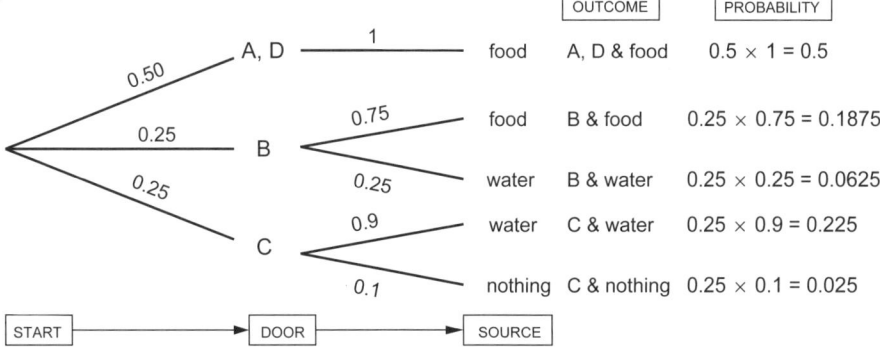

The probability that the rat ends up with food is found by adding the probabilities of all the final outcomes which have led to food. Note that this does not include getting food via C, which has zero probability. We find

$$P(F) = P([A, D \text{ \& food}] \text{ or } [B \text{ \& food}]) = 0.5 + 0.1875 = 0.6875.$$

6.3.2 Bayes' Rule

One common use of probability trees is to help us apply a further rule of probability, known as *Bayes' rule*.

Rule 7. If we assume an event B occurs, then we can calculate the probability of another event A occurring. This is given by

$$P(A|B) = P(A \text{ given } B) = \frac{P(A \text{ and } B \text{ occur})}{P(B \text{ occurs})}$$

and we can use the probability tree to help us find both $P(A \text{ and } B \text{ occur})$ and $P(B \text{ occurs})$.

The probability we are trying to find is a *conditional probability*. We have seen these before, e.g. in Example 6.5, the probability of a man testing positive for prostate cancer $+$, given that he has prostate cancer D, was 0.9. This is not the probability we are most interested in, though. We would probably like to know what the probability of having prostate cancer is, if we have just tested positive for it. We use Bayes' rule to help us.

Example 6.7

We continue Example 6.5, and find the probability of having prostate cancer, given a positive test. From the probability tree in Example 6.5, we see $P(D \text{ and } +) = 0.0075$. $P(+)$ was given by $P(D \text{ and } +) + P(H \text{ and } +) = 0.2058$. The probability we want is given by Bayes' rule

$$P(D \text{ given } +) = \frac{P(D \text{ and } +)}{P(+)} = \frac{0.0075}{0.2058} = 0.036 = 3.6\%.$$

This very low probability is due to the fact that the test is not very good for healthy males, where 20% of the time, it gives a false positive.

As demonstrated in the previous example, use of Bayes' rule is made much easier by first drawing a probability tree for the situation.

Example 6.8

Q: Continuing Example 6.6, if the rat actually did find food, what is the probability that it went straight to the food, i.e. through doors A or D?

A: Here, we need to calculate $P(A, D$ given food), so from Bayes' rule, we need to find $P(A, D$ and food) and also P(food). The first of these is read directly off the tree, and is $P(A, D$ and food) $= 0.5$. The second was found in the earlier example, and was P(food) $= 0.6875$. Using Bayes' rule, we find:

$$P(A, D \mid \text{food}) = \frac{P(A, D \text{ and food})}{P(\text{food})} = \frac{0.5}{0.6875} = 0.727 = 72.7\%.$$

So, if the rat found food, it is most likely that it found it by going through either of doors A and D.

6.4 Probability distributions and random variables

Random variables are variables whose value depends on the outcome of a random experiment, e.g. the number of people in a certain lecture room in seven minutes' time, the height of the next person to come into the room, the number of clinically depressed people in Sydney, etc. Random variables are usually labelled with capital letters (X, Y, etc.) so that we can talk about them and use them to evaluate probabilities easily. They can be discrete or continuous.

As we saw in Section 2.1, the first and third of the above examples are counts and are discrete random variables while the second (the height) is a measurement and is a continuous random variable.

We are often interested in the outcome of a random variable. We consider the possible outcomes, and their probabilities. This information combines to make the *probability distribution* of the random variable. The following example is about a simple discrete random variable.

Example 6.9

Q: Consider rolling two standard unbiased (six-sided) dice. What is the probability distribution of the sum of those two dice?

A: Each die has the outcomes $1, 2, 3, 4, 5$ or 6 with equal probability since the dice are unbiased. When we add these, we obtain numbers between 2 and 12. The 36 possible outcomes are shown in the table below. Since the outcomes are equally likely, each has probability $\frac{1}{36}$. Let X = the sum of the dice, i.e. when we throw $(1, 1)$, then $X = 2$;

when we throw $(6,3)$, then $X = 9$, etc. We can also find the probability distribution of X, given by $P(X = x)$ in the following table.

Outcomes	x (sum of dice)	$P(X = x)$
$(1,1)$	2	$\frac{1}{36}$
$(1,2),(2,1)$	3	$\frac{2}{36}$
$(1,3),(3,1),(2,2)$	4	$\frac{3}{36}$
$(1,4),(4,1),(2,3),(3,2)$	5	$\frac{4}{36}$
$(1,5),(5,1),(2,4),(4,2),(3,3)$	6	$\frac{5}{36}$
$(1,6),(6,1),(2,5),(5,2),(3,4),(4,3)$	7	$\frac{6}{36}$
$(2,6),(6,2),(3,5),(5,3),(4,4)$	8	$\frac{5}{36}$
$(3,6),(6,3),(4,5),(5,4)$	9	$\frac{4}{36}$
$(4,6),(6,4),(5,5)$	10	$\frac{3}{36}$
$(5,6),(6,5)$	11	$\frac{2}{36}$
$(6,6)$	12	$\frac{1}{36}$

The probability distribution is one summary of the properties of the sum, e.g. we notice that the most likely value of X is seven.

Tree diagrams can be useful for deriving probability distributions. Actually, we could have used one in the previous example; however, we would have had 36 outcomes and the whole thing would have looked pretty daunting.

Example 6.10

Q: Stephen's Island is a small uninhabited island in Cook Strait (between the North and South Islands of New Zealand), on which conservation groups are re-establishing tuatara. Three locations on the island are regularly checked for tuatara. It has been noticed for some years now that on each visit, tuatara have been found at each of the locations with probability 0.4. At each visit let X be the random variable representing the number of locations at which tuatara are observed. X can take the values 0, 1, 2, or 3.

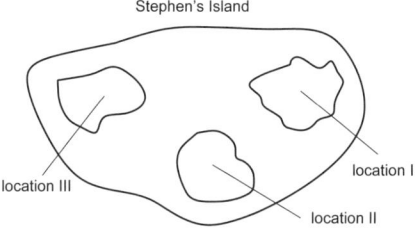

Find the probability distribution of X, and determine the probability that tuatara are found at at least one location.

A: Let T label a location where tuataras are found to be present, and N a location when they are not found. Drawing the tree diagram we get:

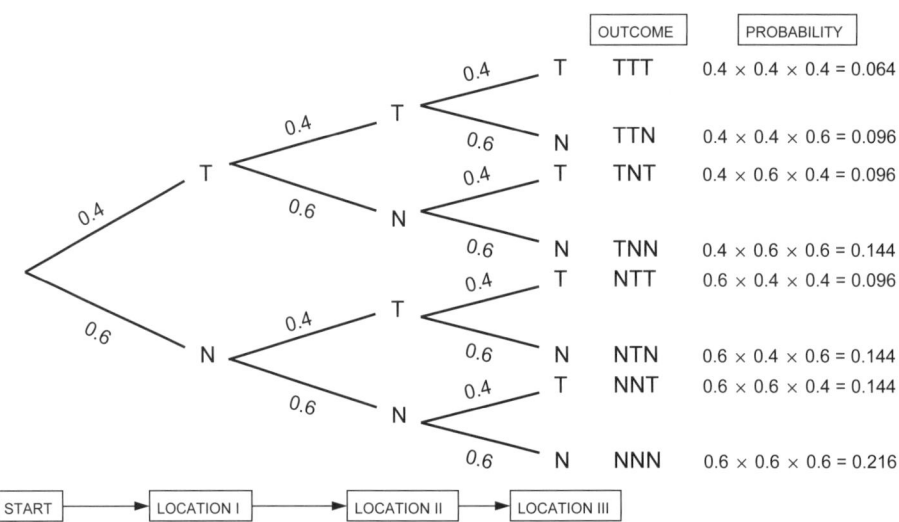

We note that whether tuatara are found at location I has no bearing on whether they are found at location II or location III. These events are independent, and result in the basic structure of the probability tree (i.e. the T and N branches and probabilities) being repeated over and over again.

At each visit we count X, the number of locations at which tuatara are found, then:

$P(X = 0) =$ none at location one and none at two and none at three $= P(NNN)$
$\qquad = 0.6 \times 0.6 \times 0.6$ (from tree – bottom pathway) $= 0.216$
$P(X = 1) = P(TNN \text{ or } NTN \text{ or } NNT) = 0.144 + 0.144 + 0.144 = 0.432$
$P(X = 2) = 0.096 + 0.096 + 0.096 = 0.288$
$P(X = 3) = P(TTT) = 0.4 \times 0.4 \times 0.4 = 0.064$

It is more convenient to make up a table of these results and we get:

x	0	1	2	3
$P(X = x)$	0.216	0.432	0.288	0.064

Since we have covered all possible outcomes for X, the probabilities add to 1. We can use this distribution to find the probability that tuatara are found at at least one location:

$$P(X \geq 1) = P(X = 1) + P(X = 2) + P(X = 3) = 0.784.$$

Alternatively, we could have used $P(X \geq 1) = 1 - P(X = 0) = 0.784$.

Different situations and experiments give rise to different distributions. We will meet some of the major ones in the rest of the book. A particularly common and important family of distributions is the binomial distribution. This distribution is so important we devote the next chapter to it.

Exercises

Events

6.1 Which of the following events are independent, or almost independent? Explain your choices.
  (a) Global warming and sea levels.
  (b) Immunisation rates and levels of cervical cancer in young women.
  (c) Levels of overcrowding and rates of meningitis in South Auckland.
  (d) The Black Ferns' results in the next Canada Cup and World Championship tournaments.
  (e) Size of student loan and degree.
  (f) A patient having a substance abuse disorder and another psychiatric disorder.
  (g) Simultaneous warm weather in Christchurch and Hobart.
  (h) Rata health in the Orongorongo Valley and level of possum control.

Probability trees

Questions using Bayes' rule are identified with an asterisk*.

6.2 The following statements are from an advertisement for Fairfax Media. For each of the following claims, identify the conditional probability stated, draw a probability tree with events shown on it, and add the conditional probability to the tree. The advert claims "Fairfax Magazines reach":
  (a) 62.2% of all Main Household shoppers;
  (b) 71% of all people with household incomes over $120,000 p.a.;
  (c) 2 out of 3 females 40+;
  (d) 62% of people in Socio's 1-3

6.3 Survey evidence indicates 15% of New Zealanders have been scammed or tricked out of money. Of these, 60% report losing less than $1000.
  (a) Draw a probability tree representing the various outcomes (not scammed, scammed with small loss, scammed with large loss).
  (b) Allocate probabilities to the branches of your tree, and calculate probabilities of the three final outcomes.
  (c) What is the probability that a randomly selected New Zealander has been scammed of a small (less than $1000) amount?
  (d) What is the probability that a randomly selected New Zealander has lost a large amount through a scam?
  (Data from consumeraffairs.govt.nz)

6.4 The Campaign for Action on Family Violence conducted a survey to estimate the effect of its "It's not OK" television advertisements. 58% of the respondents were female. Of these females, 75% reported discussing the advert with their family or friends. Of the males who had seen the advert, 60% reported discussing it.
  (a) Draw a probability tree for this situation.
  (b) Calculate the probability of a randomly selected respondent discussing the advert.
  (c)* If the advert was discussed, what is the probability that the discussant was male?
  (Data from msd.govt.nz)

6.5 Food-hoarding animals often use behavioural strategies to reduce theft of "cached" food. Researchers provided a sample of New Zealand robin pairs (male and female) with an unlimited supply of food, and observed the caching behaviour of the birds. The following probabilities were observed:

- 69% of the time, the food was obtained by the male. The male ate the food immediately in 30% of these instances, and otherwise stored it.
- When the food was obtained by the female, she stored it 45% of the time.
- Food stored by the male was retrieved 28% of the time by one or other of the birds, and eaten 50% of the time and otherwise stored at another location.
- Food stored by the female was retrieved 20% of the time by one or other of the birds, and eaten 42% of the time.

Consider a randomly selected item of food.
(a) Draw a probability tree representing all possible outcomes of the food item (which bird, eaten or stored, retrieved, eaten or stored).
(b) Allocate probabilities to the branches of your tree, and calculate the probabilities of all final outcomes.
(c) What is the probability that the food item is eaten without being stored at all?
(d) What is the probability that the food item is initially stored?
(e) What is the probability that the food item is ultimately stored, i.e. not eaten?
(f) What is the probability that the food item is eaten, i.e. not stored?
(g)* If the food item is eaten, what is the probability that it was never stored?

See also Exercise 12.17.
(Data from Burns & van Horik, Sexual differences in food re-caching by New Zealand robins, Journal of Avian Biology, 2007)

6.6 Researchers compared people's memories of an event after they discussed the event either with their romantic partner or a stranger. Pairs of subjects watched slightly different versions of a movie, and they discussed some details from the movie, but not others. Because the movie versions were different, conversations often involved misinformation about the other's movie. 51% of the time, the misinformation arose in conversations between stranger pairs. The remaining instances were between romantic partners. Among stranger pairs, 46% of the time, the misinformation was disputed, while it was disputed 51% of the time among romantic partners.
(a) Draw a probability tree for this situation.
(b) Find the probability that misinformation was disputed.
(c)* If the misinformation was disputed, what is the probability that it was disputed by a romantic partner?

(Data from French, Garry & Mori, You say tomato? Collaborative remembering leads to more false memories for intimate couples than for strangers, Memory, 2008)

6.7 Tuatara are known to exhibit various aggressive behaviours, including posing, chasing, and aggressive tail biting. Among a population of tuatara, 60% are male. When aggressive, the males pose with probability 21%, chase with probability 7%, and tail bite with probability 3%. In contrast, females pose with probability 22%, chase with probability 9%, and they never tail bite. Otherwise, the tuatara pushes, follows or attacks. Consider a randomly selected tuatara from this population. It is observed exhibiting aggressive behaviour.
(a) What is the probability that this tuatara is posing?
(b)* If this tuatara is observed posing, what is the probability it is male?

(c) What is the probability the tuatara is tail biting?

(d)* If the tuatara is observed tail biting, what is the probability it is male?

(*Data taken from Wörner, Aggression and competition for space and food in captive juvenile tuatara, VUW thesis, 2009*)

6.8 A random sample of students is asked to recall words from a list of "ambiguous" words – words with a mix of violent and kitchen themes, e.g. *cut, whip, mug, butcher*. 37% of subjects report at least one word that wasn't in the original list. Of these words, 64% are judged to be violent, e.g. *stab, hit, slap*, 27% are kitchen-related, e.g. *bread, eggs, spoon*, and the remainder are ambiguous, e.g. *chop, slice, sharp*. Consider a randomly selected word from the students' lists.

(a) Draw a probability tree for this situation.

(b) What is the probability it is ambiguous?

(c)* If the word is ambiguous, what is the probability it was on the original list?

See also Exercise 9.2.

(*Data from Takarangi, Polaschek, Hignett & Garry, Chronic and temporary agression causes hostile false memories for ambiguous information, Applied Cognitive Psychology, 2008*)

6.9 A random sample of New Zealand families was used to study holiday travel patterns. Of the sample, 35% of families had an international holiday in the current year, with 65% of these trips to Australia. Of those who did not travel overseas in the current year, 3% had already confirmed an international holiday for the following year, with 67% of these to Australia. Assume none of those who travelled internationally in the current year have a confirmed international trip for the following year.

(a) Draw a probability tree showing those with an international holiday in the current year or a confimed holiday in the following year.

(b) What is the probability that a randomly selected family neither travelled overseas, nor had a confirmed trip planned?

(c) What is the probability that a randomly selected family either travelled to Australia or has a confirmed trip to Australia planned?

(d)* If the selected family did have a trip to Australia (taken, or confirmed), what is the probability they had already made the trip?

(*Data adapted from Schänzel, Family time and own time on holiday: generation, gender, and group dynamic perspectives from New Zealand, VUW thesis, 2010*)

6.10 New Zealand Sign Language is one of many languages around the world that exhibits variation in its numbering system. Among the Deaf population in New Zealand, three major variants (A, B and C below) are observed for the number 9:

Variant A	Variant B	Variant C

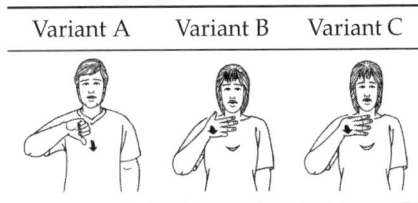

Use of these variants occurs with probabilities 58%, 20% and 22%. Among those who use variant A, 34% were schooled in the North region; among those who use variant B, 86% were schooled in the North region; while among those who use variant C, 26% were schooled in the North region. The others were schooled in the

Central or South regions. Consider a randomly selected person from the NZ Deaf community.
(a) Draw the probability tree showing combinations of variant and school region;
(b) Calculate the probabilities of each variant and school combination;
(c) Find the probability that the person was schooled in the North region.
(d)* If the person was schooled in the North region, what is the probability they use variant A?
(e)* If the person was schooled in the North region, what is the probability they use variant B?
(f)* If the person was schooled in the North region, what is the probability they use variant C?
(*Data from McKee, McKee and Major, Sociolinguistic Variation in NZSL Numerals, 2008. Line drawings thanks to A Dictionary of New Zealand Sign Language, 1997, Graeme Kennedy ed.*)

6.11 The report *Seen and heard: children's media use, exposure and response* describes the results of a survey of 604 children aged 6 to 13 years and their primary caregivers which was commissioned by the Broadcasting Standards Authority. 13% of the younger children (6-8 years) are watching TV after 8:30 pm. In total, 4% of younger children are also watching TV after 9:30 pm.
(a) Draw a probability tree representing this scenario and allocate probabilities to all branches.
(b) Calculate the probabilities of the final outcomes.
(c)* If a randomly selected child in this age group is watching TV after 8:30 pm, what is the probability they are also watching TV after 9:30 pm? (*Data from* bsa.govt.nz)

6.12 Littering patterns on the steps area of the Victoria University Quad were monitored for a period of time. Litter was classified as either cigarette or non-cigarette litter; littering was classified either as active (e.g. discarded while leaving the area, on the ground, or in the bin) or passive (e.g. put on the ground while seated, and left, or picked up). A total of 452 people was observed, and they either littered, or disposed of their litter responsibly (nonlittered). Consider the following facts:
• 181 people disposed of cigarettes;
• of those with cigarettes, 122 littered (44 actively), the remainder nonlittered (51 actively);
• of those with non-cigarette litter, 57 littered (10 actively), the remainder nonlittered (120 actively).
Now consider a randomly selected "litterer".
(a) Draw a probability tree summarising this scenario.
(b) Calculate the probabilities of the eight final outcomes.
(c) What is the probability that the selected person actively litters?
(d) What is the probability that the selected person actively nonlitters?
(e) What is the probability that the selected person litters?
(f)* If the person litters, what is the probability they discarded a cigarette?
(g)* If the person nonlitters actively, what is the probability they discarded a cigarette, i.e. if they put a butt on the ground, what is the probability they pick it up when leaving?
(*Data from Sibley & Liu, Differentiating active and passive littering: a two-stage process model of littering behavior in public spaces, Environment and Behavior, 2003*)

6.13 The New Zealand General Social Survey investigated a range of social and economic outcomes in 2008. 18.2% of the participants were "young adults" (15-24 years), 34.9% were "prime working age" (25-44), 31.6% were "middle aged" (45-64), and the rest were "older people". Among each of these groups, the proportion that had experienced discrimination in the last 12 months were: 15.1%, 11.6%, 9.3% and 2.7% respectively. Consider a randomly selected participant.
  (a) Draw a probability tree summarising this scenario.
  (b) Calculate the probabilities of the eight final outcomes.
  (c) What is the probability the individual selected was discriminated against in the last 12 months?
  (d)* If the participant had been discriminated against, what is the probability they were in each of the four age bands?
  (Data from stats.govt.nz)

6.14 Following Exercise 6.13, among each of the age groups, the proportions that did not have enough money to meet everyday needs were 0.202, 0.158, 0.142 and 0.078 respectively.
  (a) Draw a probability tree summarising this scenario, using the age-group proportions from Exercise 6.13.
  (b) Calculate the probabilities of the eight final outcomes.
  (c) What is the probability the individual selected did not have enough money to meet everyday needs?
  (d)* If the participant did have enough money to meet everyday needs, what is the probability they were in each of the four age bands? *(Data from* stats.govt.nz)

6.15 The New Zealand General Social Survey investigated a range of social and economic outcomes in 2008. 30% of the respondents were couples without children, 42.6% were couples with child(ren), 8.4% were one parent with child(ren), while the rest were not in a family. All were asked whether or not their household had stored emergency water. 46% of the couples without children had, 40.4% of the couple with child(ren) had, 33.7% of the one parent with child(ren) had, and 36.2% of those not in a family had.
  (a) Draw a probability tree representing this scenario.
  (b) What is the probability that a randomly selected respondent had stored emergency water?
  (c)* If the randomly selected respondent had *not* stored emergency water, what is the probability they were in a couple without children?
  (d)* If the randomly selected respondent had *not* stored emergency water, what is the probability they were either a couple or (solo) parent with child(ren)?
  (Data from stats.govt.nz)

6.16 A random sample of students is asked to recall words from a list of "ambiguous" words – words with both violent and kitchen usage, e.g. *cut, whip, mug, butcher*. 37% of subjects report at least one word that wasn't in the original list. Of these words, 64% are judged to be violent, e.g. *stab, hit, slap*, 27% are kitchen-related, e.g. *bread, eggs, spoon*, and the remainder are ambiguous, e.g. *chop, slice, sharp*. Consider a randomly selected word from the students' lists.
  (a) What is the probability it is ambiguous? (Recall that all the "correct" words are ambiguous.)

(b)* If the word is ambiguous, what is the probability it was on the original list?

See also Exercise 9.2.

(*Data from Takarangi, Polaschek, Hignett & Garry, Chronic and temporary agression causes hostile false memories for ambiguous information, Applied Cognitive Psychology, 2008*)

6.17 Following Exercise 6.13, among the different household units (i.e. 30% were couples without children, 42.6% were couples with child(ren), 8.4% were one parent with child(ren), while the rest were not in a family), the proportions which recycled all or most of their recyclable waste were: 0.746, 0.749, 0.702 and 0.722 respectively.
   (a) Draw a probability tree representing this scenario.
   (b) What is the probability that a randomly selected respondent recycled all or most of their recyclable waste?
   (c)* If the randomly selected respondent did recycle, what is the probability they were living without children? (*Data from* stats.govt.nz)

6.18 On 31 May 2010, the TAB listed New Zealand's All Whites' probability of winning the 2010 Football World Cup in South Africa as 1/1000. At the same time, New Zealand's All Blacks were being given a 30.7% probability of winning 2011's Rugby World Cup. Suppose that if the All Whites win the Football World Cup, the All Blacks' probability of winning the Rugby World Cup doubles, i.e. becomes 61.4%. Assuming these probabilities are correct, and you are at 31 May 2010:
   (a) Draw a probability tree representing this scenario.
   (b) Find the probability that the All Blacks win the Rugby World Cup.
   (c) Find the probability that both teams win their respective World Cups.
   (d)* Find the probability that the All Whites lost if the All Blacks won?
   (*Data from* tab.co.nz)

6.19 51.2% of New Zealanders are female. 1 in 3 New Zealand women has reported assault by their partner, while 14% of men have reported assault by their partner.
   (a) Draw a probability tree representing this situation, and label all final outcomes.
   (b) Label all branches with their probabilities and calculate the probability of the final outcomes.
   (c) What is the probability that a randomly selected New Zealander has been assaulted by their partner?
   (d)* If the selected person has been assaulted, what is the probability that they are female?
   (*Data from* areyouok.org.nz)

6.20 Patients in a mental health system were approximately two-thirds women. Of the women, 72% were admitted for manic episodes, 15% for depression and the remainder mixed (i.e. both). Of the males, 75% were manic, 6% were for depression and the remainder mixed.
   (a) Draw a probability tree representing this scenario.
   (b) Find the probability that a randomly selected patient was manic only.
   (c) Find the probability that a randomly selected patient was manic (possibly with depression also).
   (d)* If a selected patient is manic, find the probability that they have not also been admitted for depression.
   (*Data from Harris et al., The impact of mood stabilizers on bipolar disorder: the*

1890s and 1990s compared, History of Psychiatry, 2005)

6.21 *Advanced:* Based on survey evidence, 18% of New Zealanders can correctly answer five questions testing their consumer rights awareness. 26% of those surveyed who earned at least $60,000 per annum were able to answer correctly, while 17% of those who earned below $60,000 were able to answer correctly. Can you deduce the proportion of those surveyed who earned the higher income?

Hint: let this number be p, and draw the probability tree, with income on the first branches, and obtain the equation $0.26p + 0.17(1-p) = 0.18$. Solve this for p.

(*Data from* consumeraffairs.govt.nz)

6.22 A random sample of 2484 individuals was asked whether they think they are better than average, or worse than average, at their job. The responses are summarised in the frequency table below:

	Female	Male
Worse than average	691	286
Better than average	761	746

(a) Calculate the relative frequency of each of the four combinations in the table, i.e. calculate the proportion that are male *and* think they are worse than average at their job, etc.

(b) How many males were in the sample? How many females? What is the probability that a randomly selected person from the sample is male? Female?

(c) Of the males alone, what is the proportion that think they are better than average at their job? Worse?

(d) Of the females alone, what is the proportion that think they are better than average at their job? Worse?

(e) Of those who think they are worse than average at their job, what is the proportion that are male? Female?

(f) Of those who think they are better than average at their job, what is the proportion that are male? Female?

(g) Redo this analysis using a probability tree, with gender on the first branches.

(*Special thanks to Marc Wilson for the data*)

Chapter 7

Proportions and the binomial distribution

Occasionally, we are asked to solve a very specific sort of probability problem, for which a tree diagram is rarely practical. It concerns a collection of repeated *trials*, that are held under identical conditions, the outcomes of which, "success" or "failure", are unrelated. The classic example of this is tossing a coin, since each toss is the same, and the outcome of one toss is completely unrelated to the outcomes of previous or future tosses. For each toss of the coin, we would have a pair of branches in the tree, and this pair of branches would be repeated over and over again in the completed probability tree. The pair of branches is like that shown in Figure 7.1.

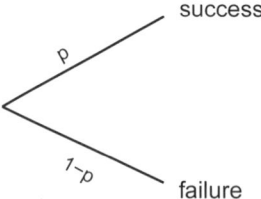

Figure 7.1   Basic unit of the binomial probability tree

If we are interested in the number of successes from our collection of trials (or alternatively, the number of failures), we can use the *binomial distribution* to help us. This distribution provides some shortcuts when it comes to analysing the possible outcomes of the binomial experiment. In particular, we take advantage of the repetitive nature of the tree, and using some mathematics (which we will not present here) we completely avoid the use of a probability tree. This is most helpful when we have a large number of trials, since a binomial experiment with a mere ten trials would have 1024 endpoints, and one with 15 trials would have 32,768! This would be very difficult to draw!

7.1 Properties of the binomial distribution

In order to use the binomial distribution, we need some very specific conditions to be met.

> A binomial random variable counts the number of successes in a fixed number of trials.

The random variable we are interested in will usually be labelled X. It is the number of successes in a fixed number of trials, and as X is counting the number of successes, it must be *discrete*. Note that the result called a "success" does not have to be successful in conventional terms, e.g. X could be counting the number of students in a group of 3000 who have an STD (sexually transmitted disease). The "success" is usually some characteristic that the experimenter is interested in, and often each trial represents a person.

In addition to requiring that we are counting the number of successes in a fixed number of trials, we insist that the trials satisfy certain conditions.

> X is a binomial random variable if X counts the number of successes in n trials where:
>
> - n is fixed
> - each trial has only two possible outcomes which are called "success" and "failure"
> - the probability of success on any one trial equals a value p which is constant throughout the trials
> - the trials are independent of one another and carried out under identical circumstances.

These conditions are so that the branches in Figure 7.1 repeat throughout the tree. Having n fixed is important so that we know exactly how many times to replicate Figure 7.1 in the tree. If we were tossing a coin, it is essential that we know how many times we are going to toss it when we are counting.

The name "binomial" comes from the fact that at each trial we must have only two outcomes. A coin obviously satisfies this: it has a "head" side, and a "tail" side. Animals are either male or female, and students either pass or fail an exam. Other interesting things have many outcomes, but we can construct a binomial situation by forming categories. An example is a person's height, which is a continuous random variable with infinitely many outcomes. We can construct two outcomes from this variable by combining, e.g. we could label a trial a success if the height is greater than 1.65 m and a failure if it is less than 1.65 m. Similarly, we can combine the six outcomes on a die as "odds" or "evens", "1" or "2, 3, 4, 5 or 6" and others. The important thing is that each trial is defined to have only two outcomes.

In addition to having only two outcomes, the trials must have a fixed probability of success, and this is labelled p. When we toss a coin, the physical properties of the coin do not change, and we toss it in such a way that the chance of getting a head is unchanged throughout the experiment. If we were interested in people's heights, a team of NBA basketball players mixed in with a few "normal" people would cause problems, as the basketball players are more likely to be tall. We can remember this condition by again thinking of drawing our large probability tree, with the repeated branches in Figure 7.1. This repeated structure must be unchanged throughout the tree, so we need the probability of success to be fixed.

The final condition is that the trials need to be *independent* of one another. When events are not independent, the probability of one event depends on what happened in the other trials. If this is the case, the probability tree will be much more complicated since the probabilities will change depending on where we are in the tree. Also, if the trials are not independent and the outcome of one trial affects the others, this means the experiment is not the same each time. For meaningful conclusions we require that the conditions under which an experiment is held be the same each time.

Some of these conditions are only approximate in many social science experiments involving human subjects. In particular, the independence condition can be hard to achieve.

Since saying "the number of successes in n trials is binomial with probability of success p" is a bit of a mouthful, and since the binomial is a very common distribution, we develop some shorthand notation.

> If X is a binomial random variable with n trials, and probability of success equal to p, we write
>
> $$X \sim Bin(n, p)$$
>
> and we say "X has a *binomial distribution* with parameters n and p."

The squiggle "$\sim$" in the shorthand $X \sim Bin(n, p)$ means "has the distribution", but we can usually think of it simply as "is". Bin is short for *binomial*, and n and p are important features of the experiment, known as the *parameters* of the binomial distribution. If we know that X satisfies the binomial conditions, and it has parameters n and p, we will see that we know everything about the possible outcomes of X.

Example 7.1

Q: Which of the following situations is binomial?
  (i) Select 10 men at random, count X, the number who are colour-blind.
  (ii) Select a person at random and record which of the age groups: <20, 20-30, >30 they belong in.

(iii) X = the number of attempts at goal before a goal occurs.
(iv) Interview 50 people consisting of 25 couples and ask them whether or not there should be paid paternity leave.
(v) Test the reaction time of a student to a stimulus tested 100 times in a psychology laboratory. A success consists of reacting within 0.1 second.

A: (i) Yes, $n = 10$ is fixed, there are two outcomes: colour-blind or not, men are independent (unless you have brothers or other relations in your group of ten), the test doesn't change from person to person, probability of a male being colour-blind is fixed at approximately 8%.
(ii) No, there are three outcomes in this experiment.
(iii) No, n is not fixed in advance. Some of us would take one or two attempts, others 20 or 30.
(iv) No, the views of two members of a couple are unlikely to be independent.
(v) No, there is a possible learning effect or tiredness effect here, which would make the probability of success unlikely to stay the same throughout the experiment.

Example 7.2

The situation counting tuatara in Example 6.10 is likely to be a binomial one:
- $n = 3$ sites are examined on each visit to the island
- at each site tuatara are either observed (success) or not
- P(observe tuatara at a site) = 0.4 and stays constant
- experiments are conducted in the same way each time and sites are probably independent.

So, if X = number of sites where tuatara are observed on a visit, then $X \sim Bin(3, 0.4)$.

7.2 Finding binomial probabilities

Since X is counting the number of successes in n trials, X will be one of the following outcomes:

$$0 \quad 1 \quad 2 \quad 3 \quad \cdots \quad n-1 \quad n.$$

Note that the smallest number of successes you can have in n trials is zero, and the largest number is n. Often we need to find the probabilities $P(X = x)$ where x represents one of the above values. The list of probabilities $P(X = x)$ for $x = 0, 1, 2, \ldots, n$ is called the *binomial distribution*.

Probabilities for specific outcomes of a binomial random variable are given in Table D.2 on pages 325 and 326 for selected n and p. To use the table, we firstly find the n we are interested in, and then identify the appropriate column by choosing p. We can then read off $P(X = x)$ by specifying a value for x, and reading along that row.

Example 7.3

Q: Suppose $X \sim Bin(4, 0.3)$, what is the probability distribution for X?

A: Looking at the binomial table on page 325, and finding the block corresponding to $n = 4$ and $p = 0.3$, we find the probabilities:

$$P(X = 0) = 0.2401, P(X = 1) = 0.4116, P(X = 2) = 0.2646, P(X = 3) = 0.0756$$

and $P(X = 4) = 0.0081$.

Example 7.4

Q: Suppose we toss a fair coin 12 times. Find:
   (i) the probability of getting exactly six heads;
   (ii) the probability of getting exactly six tails;
   (iii) the probability of getting no heads.

A: Since the coin is fair, we have $p = 0.5$, and the number of heads in the 12 tosses is $X \sim Bin(12, 0.5)$.
   (i) $P(X = 6) = 0.2256$.
   (ii) If six tails were observed, then the other tosses must have given heads, i.e. getting six tails is the same as getting six heads. The answer is also 0.2256.
   (iii) $P(X = 0) = 0.0002$.

We can also calculate probabilities for events which comprise more than one outcome. Examples include $X \geqslant 5$, $X < 2$, or $3 < X < 12$. The easiest way of solving these problems is to look at the individual outcomes which are indicated, and add the probabilities for these outcomes. In the first of the above examples we would have

$$P(X \geqslant 5) = P(X = 5, 6, 7, \ldots, \text{ or } n) = P(X = 5) + P(X = 6) + \cdots + P(X = n).$$

Note that the individual outcomes $5, 6, 7$, etc. are mutually exclusive: if we get a total of five successes, we cannot also get six successes, and so we add their individual probabilities (see Rule 4). The remaining probabilities are:

$$P(X < 2) = P(X = 0, 1) = P(X = 0) + P(X = 1)$$
$$P(3 < X < 12) = P(X = 4, 5, \ldots, 10, 11) = P(X = 4) + \cdots + P(X = 11).$$

An alternative to this is to use the cumulative binomial probability table on pages 327 and 328 by rewriting the probabilities into the form $P(X \geqslant x)$. This can be much more complicated than using the individual terms table. $P(X \geqslant 5)$ is already in the right form, but the others need more work:

$$P(X < 2) = P(X = 0, 1) = 1 - P(X = 2, 3, \ldots, n) = 1 - P(X \geqslant 2)$$
$$P(3 < X < 12) = P(X = 4, \ldots, 11) = P(X \geqslant 4) - P(X \geqslant 12).$$

Example 7.5

Fifteen elderly individuals take a test of cognitive function. With large numbers of subjects it has been observed that an individual has a chance of 1 in 20 of completing this test successfully within half an hour.

Let X = the number who successfully complete in half an hour, $n = 15$, $P(\text{success}) = p = 1/20 = 0.05$.

This seems to be binomial: either they do complete in the half hour or not (two outcomes), fixed number (15) of subjects, $P(\text{success})$ roughly constant, individuals likely to be independent unless they are siblings. So $X \sim Bin(15, 0.05)$ and using the table for $Bin(15, 0.05)$ we see:

$P(X = 0) = 0.4633 \quad P(X = 1) = 0.3658 \quad P(X = 2) = 0.1348 \quad P(X = 3) = 0.0307$
$P(X = 4) = 0.0049 \quad P(X = 5) = 0.0006 \quad P(X = 6) = ?$

This last entry does not mean strictly that $P(X = 6) = 0$ but that the probability of this happening is 0 to four decimal places, i.e. its probability is less than 0.00005. Likewise for $X = 7, 8, \ldots, 15$.

We can draw the probability distribution with a bar graph (remember this is discrete data) with the thin bars shown below, although it is often convenient to use a histogram (this would correspond to the dotted lines below). Just don't forget that you can't really have three-and-a-half women completing within 30 minutes!

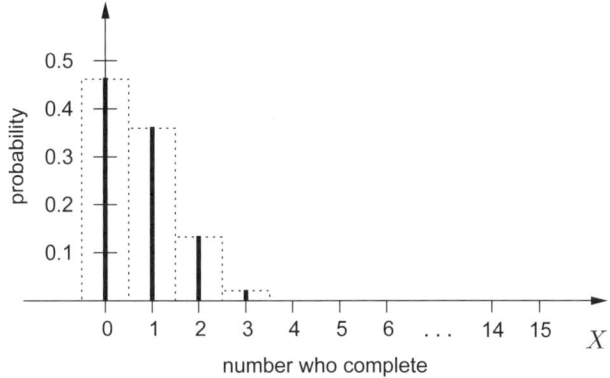

For example suppose you wished to know $P(X \geqslant 4)$. You can work it out from the individual tables by observing

$$\begin{aligned} P(X \geqslant 4) &= P(X = 4, 5, 6, \ldots, 15) \\ &= P(X = 4) + P(X = 5) + \cdots + P(X = 15) \\ &= 0.0049 + 0.0006 + 0 + \cdots + 0 \\ &= 0.0055. \end{aligned}$$

Alternatively, using cumulative tables with $n = 15$ and $x = 4$ you can read off $P(X \geqslant 4)$ directly.

In some instances it can be useful to relabel "success" and "failure." The most common reason for doing this is that $p > 0.5$, and the tables do not have listings for $p > 0.5$. If the probability of success is p, then because there are only two outcomes at each trial, the probability of failure is $1 - p$. It follows that if $p > 0.5$, then $1 - p < 0.5$.

> If $X \sim Bin(n,p)$, then $Y =$ the number of failures in n trials, has
>
> $$Y \sim Bin(n, 1-p)$$
>
> with $X + Y = n$.

This really means we can interchange the labels "success" and "failure". Counts of both will have binomial distributions, and if we add their counts, we must get the number of trials n. If we toss a coin n times, we do not know in advance how many heads we will get, but we do know that the number of heads, plus the number of tails, will be exactly n. Finding probabilities using this technique is illustrated in the following example.

Example 7.6

Q: In the 14-year-and-under age group, 70% of all first admissions to mental health facilities have mental retardation as the leading cause of admission. Of 20 randomly selected first admissions to a mental health facility in this age group, what is the probability that fewer than 10 will be as a result of mental retardation?

A: Here we have $n = 20$, $X =$ number in sample admitted due to mental retardation, $p = P(\text{admission due to mental retardation}) = 0.7$, and the problem is to find $P(X < 10)$.

X is binomial (fixed number of trials, constant p, two outcomes – due to mental retardation or not, and independent subjects), $X \sim Bin(20, 0.7)$. The tables do not list $p = 0.7$ so instead we count the number admitted for other reasons, call it Y, with $Y \sim Bin(20, 0.3)$. Here $p' = P(\text{admit for other reasons}) = 0.3$, and we note that $X + Y = 20$.

We now turn our problem about X into an equivalent problem about Y. If $X < 10$, then $X = 0, 1, 2, \ldots, 9$ and so $Y = 20, 19, 18, \ldots, 11$. Why? When $X = 1$ that means that 1 of the 20 was admitted due to mental retardation and therefore the other 19 were admitted for other reasons, etc.

So $X < 10$ is equivalent to $Y \geq 11$. Using the $Bin(20, 0.3)$ tables $P(Y \geq 11) = 0.0171$.

In some instances, the combination of n and p we require is not available in the tables. In situations when n is large (usually considered to be greater than 30), we can use an approximation for the binomial probabilities. This is covered later, in Section 8.4. If n is not large, or we want to be really exact, we may wish to use a formula for the binomial probabilities.

> If $X \sim Bin(n, p)$, then
>
> $$P(X = x) = \frac{n!}{x!(n-x)!} p^x (1-p)^{n-x}$$
>
> where $n! = n \times (n-1) \times \cdots \times 2 \times 1$ and is called n factorial.

This is the formula that is used to determine the probabilities given in the binomial individual terms table on pages 325 and 326. An alternative to using this formula is

7.3 Mean and variance of a binomial random variable

to use a spreadsheet (which evaluates the formula for you). Instructions for how to compute binomial probabilities in a spreadsheet are given in Appendix C.

7.3 Mean and variance of a binomial random variable

One property of a binomial random variable that we often wish to know is its mean, i.e. what will happen in our binomial experiment *on average*. Rather than repeating the experiment many times (which may be possible in the case of coin tossing, or impossible in other examples) and then calculating the average, we can write down a formula for the mean. Luckily, this formula agrees with our intuition.

Example 7.7

If a room full of people all toss a fair coin 100 times (i.e. probability of tossing a head = 0.5), and we list the number of heads each person gets, most people are comfortable with the statement that *on average we will get about 50 heads in 100 tosses*. Some people will get a few more than 50 and some a little less. Notice that $50 = 100 \times 0.5$.

Since the probability of success is equal to the proportion of trials in which we would expect to get a success, the average number of successes is just this proportion of the number of trials n.

> The mean of $X \sim Bin(n,p)$ is $\mu = np$.

To find the variance of a binomial random variable, we multiply the mean np by $(1-p)$. If we wish to find the standard deviation of X, this is just the square root of the variance, i.e. $\sigma = \sqrt{np(1-p)}$.

> The variance of $X \sim Bin(n,p)$ is $\sigma^2 = np(1-p)$.

Example 7.8

Q: What are the mean and standard deviation of the number of heads in 30 tosses of a fair coin?

A: Since the conditions for a binomial experiment are met in this case, X = the number of heads is binomial with $n = 30$ and $p = 0.5$, i.e. $X \sim Bin(30, 0.5)$. The mean of X is $\mu = np = 30 \times 0.5 = 15$. The variance is $\sigma^2 = np(1-p) = 15 \times 0.5 = 7.5$, giving the standard deviation $\sigma = \sqrt{7.5} = 2.74$ (2dp).

Example 7.9

Q: Going back to Example 6.10, what are the mean and standard deviation of the number of locations we would find tuatara if we visit Stephen's Island on many occasions?

A: The number of locations with tuatara present on any one visit is $Bin(3, 0.4)$. Over many visits the mean number of locations we will find tuatara at is $\mu = 3 \times 0.4 = 1.2$. Looking at all our visits over time, on average just over one location per visit will have tuatara present.

The variance of the number of locations with tuatara present on any one visit is $\sigma^2 = 3 \times 0.4 \times (1 - 0.4) = 0.72$ and the standard deviation is $s = \sqrt{0.72} = 0.85$.

7.4 Proportions

One of the difficulties with the binomial random variable is that we must keep track of n in order to make sense of the outcome. For example, if we are tossing a coin, and get 63 heads, we are not sure what to make of this unless we are also told the number of times that coin was tossed. If it was 80 times, we would be surprised, but if the coin was tossed 120 times, the outcome seems about right (it is close to the mean of $np = 120 \times 0.5 = 60$). We can simplify things a little by reporting a *sample proportion*.

To calculate the sample proportion, we divide the number of successes in the experiment (the binomial random variable) by the number of trials. We denote the sample proportion by $\hat{p}$, said "p hat."

> The sample proportion is given by
> $$\hat{p} = \frac{X}{n} = \frac{\text{number of successes}}{\text{number of trials}}$$
> where $X \sim Bin(n, p)$.

Example 7.10

Ten per cent of the general population is left-handed. Suppose that in a random sample of 50 marketing majors we count X, the number that is left-handed. This is a binomial situation: $n = 50$ (fixed), 2 outcomes (left-handed or not), $p =$ probability of being left-handed $= 0.1$ is fixed and the students are independent of one another). So $X \sim Bin(50, 0.1)$. X has mean $50 \times 0.1 = 5$, so in many groups of 50 people, we would expect on average five to be left-handed. The variance is $50 \times 0.1 \times 0.9 = 4.5$. It turns out that 8 of the group are left-handed i.e. $X = 8$. The proportion of left-handers in the group is $\hat{p} = X/n = 8/50 = 0.16$.

The sample proportion $\hat{p} = 0.16$ in the previous example is higher than the 0.1 for the population. Since X is a random variable (it is binomial) and $\hat{p}$ is based on X it is also a random variable. It is natural to wonder if 0.16 is too far away from 0.1 for the marketing majors to be representative of the population. Perhaps left-handers have a genetic predisposition to studying marketing? We will answer questions like these in

Chapter 9. In the meantime, let us look at the behaviour of this sample proportion. We use p for the population proportion with some characteristic of interest and $\hat{p}$ when we are discussing the proportion of a sample with the same characteristic. We note that if we collect many samples from the population, each will likely have a different sample proportion, as shown in Figure 7.2. This is because $\hat{p}$ is a random variable, and so like other random variables, it has a mean and a variance.

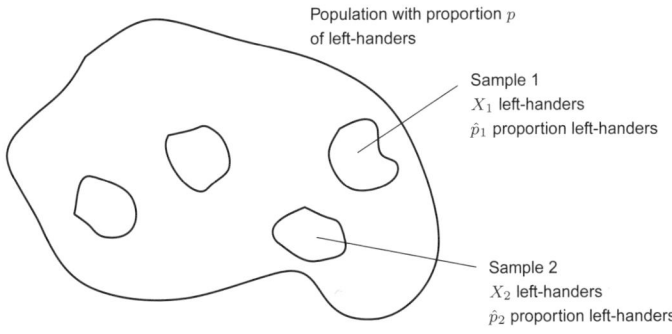

Figure 7.2 Sampling variation of $\hat{p}$

The sample proportion $\hat{p}$ has mean p and variance

$$\text{var}(\hat{p}) = \frac{p(1-p)}{n}.$$

On average, we know X will be $n \times p$, and so it is natural that $\hat{p} = X/n$ will be $np/n = p$ on average. The variance of $\hat{p}$ follows from some rules for variances, and so we just note that the n has moved from multiplying $p(1-p)$ to dividing it.

Example 7.11

Continuing Example 7.10, the sample proportion of left-handers in many samples of size 50 has mean $p = 0.1$ and variance $p(1-p)/n = 0.1 \times 0.9/50 = 0.0018$.

Exercises

Finding probabilities

7.1 Let $X \sim Bin(n, p)$. Find $P(X = 3)$ in each of the following instances:
(a) when $n = 3$ and $p = 0.4$;
(b) when $n = 10$ and $p = 0.2$;
(c) when $n = 6$ and $p = 0.01$;
(d) when $n = 10$ and $p = 0.75$;
(e) when $n = 15$ and $p = 0.65$.

7.2 Let $X \sim Bin(n, p)$. Find $P(X \leqslant 3)$ in each of the following instances:
(a) when $n = 3$ and $p = 0.4$;
(b) when $n = 10$ and $p = 0.2$;
(c) when $n = 6$ and $p = 0.01$;
(d) when $n = 10$ and $p = 0.75$;

(e) when $n = 15$ and $p = 0.65$.

7.3 Let $X \sim Bin(20, 0.25)$. Find:
  (a) $P(X = 0)$;
  (b) $P(X \geq 1)$;
  (c) the probability that X is odd;
  (d) the probability that X is even.
  (e) Why do the probabilities that X is odd and X is even not add up to one?

7.4 Let $X \sim Bin(12, 0.4)$. Find:
  (a) $P(2 < X < 5)$;
  (b) $P(2 \leq X \leq 5)$;
  (c) $P(X > 10)$;
  (d) $P(X < 10)$.

7.5 Let $X \sim Bin(15, 0.8)$. Find:
  (a) $P(X = 10)$;
  (b) $P(X > 10)$;
  (c) $P(X \leq 4)$;
  (d) $P(4 \leq X \leq 8)$.

Mean and variance

7.6 Calculate the mean and standard deviation of each of the binomial random variables in Exercise 7.1.

7.7 Calculate the mean and standard deviation of the binomial random variables in Exercises 7.3 to 7.5.

7.8 If X is binomial with 100 trials and probability of success 32%, what are the mean, variance and standard deviation of X?

7.9 If a fair coin is tossed 30 times, what are the mean and standard deviation of the number of heads?

Applications

7.10 Let $X \sim Bin(10, 0.4)$, and let X be the number of males in a sample containing ten people.
  (a) What is the expected number of males? What is the standard deviation of the number of males?
  (b) What is the expected number of females? What is the standard deviation of the number of females?

7.11 In 2009, almost exactly 20% of New Zealand's population was under 15 years of age. Consider a random sample of 20 New Zealanders collected at this time, and assume $p = 0.2$.
  (a) Is it reasonable to assume that the number in this sample under the age of 15 years old is binomial? Are the binomial assumptions satisfied?
  (b) What is the probability that no-one in the sample was under the age of 15?
  (c) What is the expected number of under-15s in the sample? What is the standard deviation of the number of under-15s?

7.12 Survey evidence indicates New Zealanders have a 15% chance of being scammed or tricked out of money. Let X be the number of people scammed or tricked out of money in a random sample of 12 New Zealanders.
  (a) Justify the use of the binomial distribution for X and give its parameters.
  (b) What is $P(X = 0)$?
  (c) What is $P(X \geq 1)$. Describe this probability and the event it applies to in words.
  See also Exercise 7.13.
  (*Data from* consumeraffairs.govt.nz)

7.13 Following Exercise 7.12, 60% of those scammed lose only a small amount (less than $1000). Suppose a random sample of 15 New Zealanders who had been scammed is obtained.
  (a) What is the probability that more than 10 of these individuals have lost in excess of $1000?
  (b) What is the probability that exactly half of them lost *less than* $1000?

(c) What is the probability that more than half of them lost *less than* $1000?

(*Data from* consumeraffairs.govt.nz)

7.14 Using a survey conducted in Winter 2009, it was estimated that only 1% of New Zealanders have solar panels on their houses to generate electricity. Consider a random sample of 20 New Zealand houses.
  (a) What is the probability that none of them have solar panels to generate electricity?
  (b) What is the probability that at least one of them has solar panels to generate electricity?

(*Data from* eeca.govt.nz)

7.15 Using a survey conducted in Winter 2009, it was estimated that 25% of New Zealand homes have complete or partial underfloor insulation. Consider a random sample of 20 New Zealand homes.
  (a) What is the expected number of these homes without complete or partial underfloor insulation?
  (b) What is the probability that none of these homes have complete or partial underfloor insulation?
  (c) What is the probability that half or more of these homes have complete or partial underfloor insulation? (*Data from* eeca.govt.nz)

7.16 40% of New Zealand adults receive instruction for a sport or recreation activity over a 12-month period. Consider a random sample of 15 New Zealand adults.
  (a) What is the probability that none of the sample received instruction for a sport or recreation activity?
  (b) What is the probability that over half of the sample received instruction for a sport or recreation activity?

(c) What are the mean and standard deviation of the number who received instruction?

(*Data from* sparc.org.nz)

7.17 Tail shedding in skinks is a defence mechanism. In a population, 80% of adults have incomplete tails, indicating recent tail shedding. In a random sample of 15 adult skinks:
  (a) What is the expected number with an incomplete tail?
  (b) What is the probability that all have an incomplete tail?
  (c) What is the probability that none have an incomplete tail?
  (d) Is it impossible that no skinks in this sample have an incomplete tail?

7.18 Of New Zealand Māori, close to 75% use English only, with the remaining 25% using Te Reo Māori and NZ Sign Language with or without English. In a random sample of 20 Māori, consider the number who use English only.
  (a) Justify use of the binomial distribution for this number.
  (b) What is the probability that all in the sample use English only?
  (c) What is the probability that at least 50% use English only?
  (d) What is the expected number who use English only?

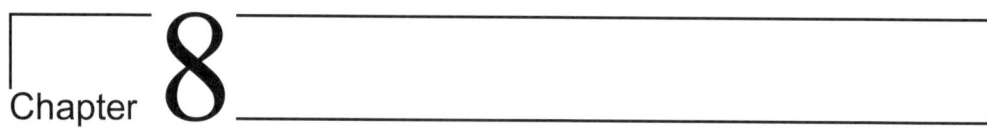

The normal distribution

8.1 The normal distribution and its uses

In Chapter 7 we looked at the properties of the binomial distribution – the probabilities attached to a binomial random variable X that *counts* the number of successes in a fixed number, n, of independent and identical trials, which each have two outcomes (commonly called "success" and "failure"), where the probability of success p is constant. We worked out the probabilities of X being 0, 1, 2, etc. This is a discrete random variable as we are counting occurrences and we can only observe whole numbers.

Many natural phenomena such as heights, weights, reaction times, flood levels, and pollution levels are continuous, i.e. they are the result of measuring something. These measurements can be to many decimal places, depending on the accuracy of our measuring instruments. A random variable that describes a surprising number of these continuous phenomena rather well is the so-called *normal distribution*. As far as we know, this distribution was first studied in depth by the German mathematician Gauss, and so it is sometimes called the Gaussian distribution.

Like the binomial, there are many normal distributions that all have the same basic description: that most observations are clumped around a middle value with a few very much bigger and a few very much smaller.

> The values of any normal random variable are:
> - symmetric either side of the mean
> - unimodal (one main hump)
> - have a "bell" shape
> - have no outliers.

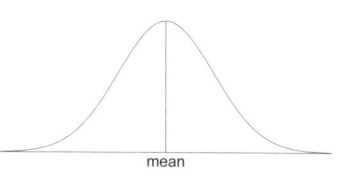

A variable X with a normal distribution must be a measurement, so an appropriate graph would be a histogram and the relative frequency polygon in this situation is the normal curve or distribution.

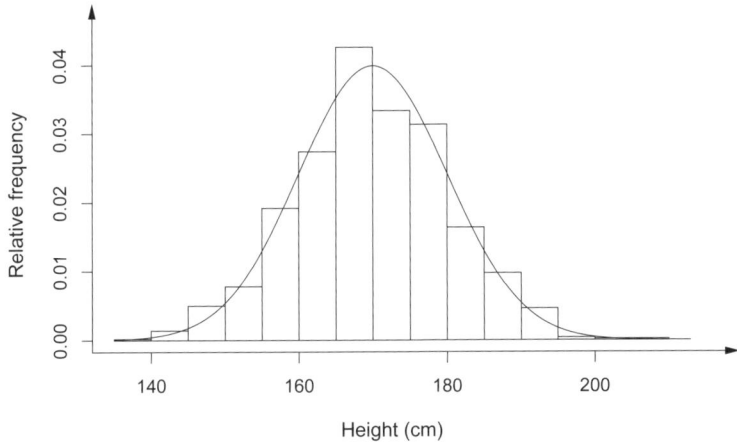

Figure 8.1 Normally distributed data, and corresponding normal curve

The total area of all the bars of the histogram must equal the sum of all the relative frequencies, which must be equal to 1. This means that the total area under the normal "bell" curve must also be 1.

We can use the curve to find $P(a < X < b)$ using Figure 8.2 as a guide. Shown in this figure are the bell curve, the mean of the distribution (at the mode), and vertical lines at a and b. The area of the shaded region is exactly equal to $P(a < X < b)$. While the curve we have plotted has a specific mathematical form, the formula for the area beneath the curve cannot be easily evaluated. Instead we need to use approximate values from a table, such as Table D.4 on page 329.

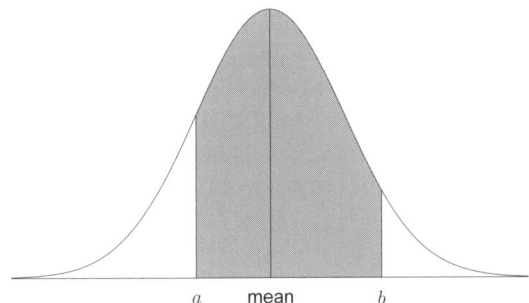

Figure 0.2 Shaded area showing $P(a < X < b)$

When we are dealing with continuous data it is not sensible to try to work out $P(X = 21)$. This is in contrast to the discrete binomial situation. This is because what we really mean by "$X = 21$" is "$X = 21.0000000\ldots$", and not $X = 21.000001$ or $X = 20.999999$, and in the continuous case, there is virtually no chance of getting exactly $X = 21$ (remember we can always improve our measurement). For this reason, $P(X = 21)$ is considered to be zero. What we do instead is consider ranges of X, and so probabilities are always of the form $P(a < X < b)$ for the normal distribution (and all other continuous distributions). In Figure 8.2, we represented this probability as an area under the normal curve.

Because $P(X = 21)$ or $P(X =$ any particular value$) = 0$, we do not need to worry about whether to use $<$ or $\leq$ in our probability statements, and the same is true for $>$ or $\geq$. It is true that $P(a < X < b) = P(a \leq X \leq b)$ and any other combination of $<$ and $\leq$. We usually use $<$ or $>$, and not $\leq$ or $\geq$, in our normal probability statements.

To describe a normal distribution properly, we need to give its mean and its standard deviation (or alternatively, its variance). Since this is quite a lot of information, statisticians have developed a shorthand.

> If X is a normally distributed random variable, with mean μ and variance σ^2, we write $X \sim N(\mu, \sigma^2)$.

The notation we have used is similar to that for the binomial distribution; however, the two *parameters* of the distribution are different. As before, we use X to represent the random variable (this is just shorthand, and it saves us from having to say exactly what X represents). The $\sim$ means "is distributed as" and the N stands for the normal distribution. In this case, the parameters we use are the mean μ and the variance σ^2. Although the standard deviation is a much more natural measure of spread, the variance is always used in the $N(\mu, \sigma^2)$ notation.

Some naturally occurring processes have normal distributions and many others are like truncated normals, i.e. normals with the extremes chopped off, e.g. babies' weights at birth, heights of 17-year-old males.

It is not sensible to have tables for all of the possible normal distributions for each possible pair of μ and σ^2. Indeed we do not need this, as we can scale *any* normal distribution so that it becomes the $N(0, 1)$, i.e. the normal with mean 0 and variance (and standard deviation) equal to 1. This normal distribution is conventionally labelled Z and is called the *standard normal distribution*. This property means we only need one set of tables rather than a whole bookful. To get from $X \sim N(\mu, \sigma^2)$ to $Z \sim N(0, 1)$ we must standardise X.

> If $X \sim N(\mu, \sigma^2)$, then
> $$Z = \frac{X - \mu}{\sigma} \sim N(0, 1)$$
> i.e. Z has a standard normal distribution. This transformation from X to Z is called *standardisation*.

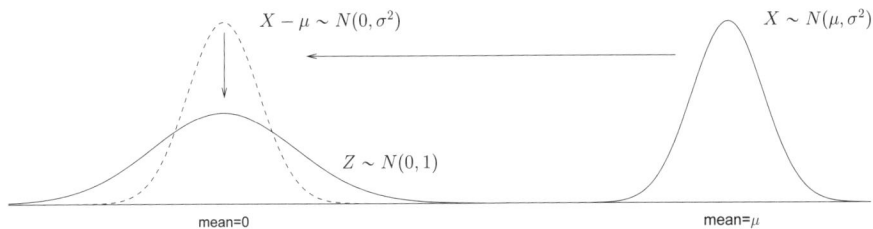

Figure 8.3 Standardisation of a normal random variable

That is, to scale a normal random variable $N(\mu, \sigma^2)$ so that it becomes a standard normal we subtract the mean and divide by the standard deviation. The result is commonly called a z-score. When we standardise any random variable X with mean μ and variance σ^2, we obtain a variable with mean 0 and variance 1. A particular (very useful) property of the normal distribution gives us a normal distribution *before and after* standardising. This process is illustrated in Figure 8.3. The distributions of X, $X - \mu$ (shown by the dotted line) and $Z = (X - \mu)/\sigma$ are all normal.

Notice that if we did the same thing to a binomial random variable, it would have mean 0 and variance 1, but it would no longer be binomial.

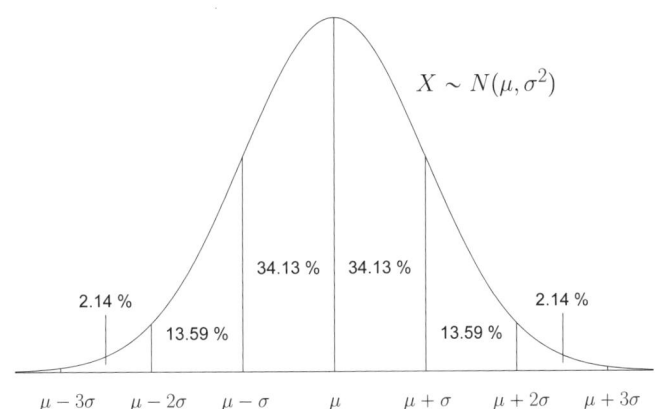

Figure 8.4 Probabilities under the normal curve

Other properties of any normal distribution, which are summarised in Figure 8.4, are

- 68.3% of all observations lie within one standard deviation either side of the mean
- 95.4% of values lie within two standard deviations of the mean
- 99.7% of values lie within three standard deviations either side of the mean.

We can make the rule of thumb that for any normal distribution, approximately $\frac{2}{3}$ of the observations lie within one standard deviation of the mean, 95% within two and 99% within three.

8.1.1 Using normal tables to compute probabilities

The symmetric nature of the standard normal curve is used extensively when working out probabilities of particular events in normal situations. As the area under the whole curve is equal to 1, then the area to the left or right of the mode (and median and mean) of the distribution equals $\frac{1}{2}$, or 50%.

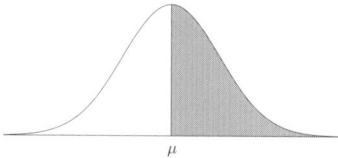

Figure 8.5 Shaded area $= 0.5 =$ unshaded area

The tables provided are always for the standard normal, and an example of such a table is given on page 329. We look up a positive number z, and the table gives us the probability $P(0 < Z < z)$. We notice that as z increases, the probability gets larger and larger, eventually reaching 0.5. Note that this is not exactly $\frac{1}{2}$, but rather 0.5000 rounded to four decimal places. When we use this table, we will have a z value to look up, and we will obtain a probability.

Example 8.1

Q: Calculate $P(0 < Z < 1.32)$; $P(0 < Z < 1)$; $P(0 < Z < 1.5)$.

A: The probability that Z lies between 0 and 1.32 is the area under the curve (shaded) between 0 and 1.32 as shown below:

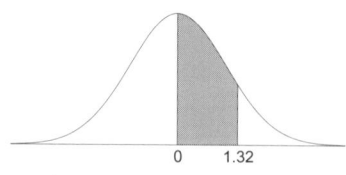

z	0	1	2	3	4
0.0	0.0000	0.0040	0.0080	0.0120	0.0160
0.1	0.0398	0.0438	0.0478	0.0517	0.0557
⋮	⋮	⋮	⋮	⋮	⋮
1.1	0.3643	0.3665	0.3686	0.3708	0.3729
1.2	0.3849	0.3869	0.3888	0.3907	0.3925
1.3	0.4032	0.4049	0.4066	0.4082	0.4099
1.4	0.4192	0.4207	0.4222	0.4236	0.4251
1.5	0.4332	0.4345	0.4357	0.4370	0.4382

The probability that Z lies between 0 and 1.32 is found by going across from 1.3 in the z column to the column headed 2, giving the value 0.4066, which is the required probability. From Table D.4 on page 329, $P(0 < Z < 1) = 0.3413$. Likewise, we see $P(0 < Z < 1.5) = 0.4332$.

NB: The diagram used in this example is intended to be a sketch only. It is designed to lessen the chance of you making a simple mistake. It is not to scale. Similar sketches are used throughout the remainder of this book.

Often, the probabilities we wish to look up are not of the form $P(0 < Z < z)$. For example, we might wish to find $P(-z < Z < 0)$. This is easily found by using the symmetry of Z as demonstrated in the following example.

Example 8.2

Q: Find $P(-2 < Z < 0)$ and $P(-2.39 < Z < 0)$.
A: These two problems are simple by the symmetry of Z.
$P(-2 < Z < 0) = P(0 < Z < 2)$ by symmetry, and $P(0 < Z < 2) = 0.4772$ from the table.

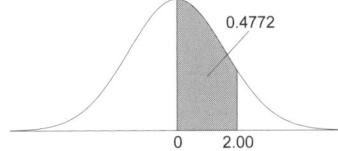 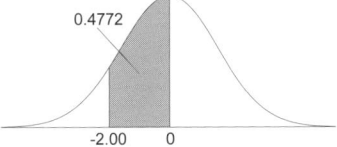

So $P(-2 < Z < 0) = 0.4772$. Secondly, $P(-2.39 < Z < 0) = P(0 < Z < 2.39) = 0.4916$.

To find probabilities like $P(Z < z)$, $P(Z > z)$, or $P(a < Z < b)$ (here a and b may be negative or positive numbers with $a < b$) is reasonably straightforward, provided you are careful. The problems can be made much simpler by using a small sketch such as the one in Figure 8.6.

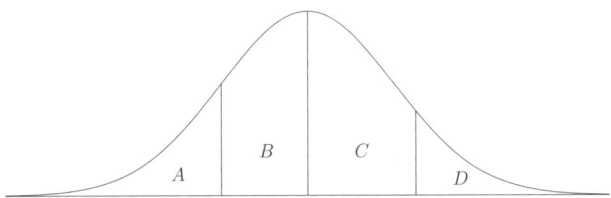

Figure 8.6 Regions under the normal curve

We make use of the symmetry of the normal distribution. Referring to Figure 8.6, we see that $A + B = \frac{1}{2}$ and $C + D = \frac{1}{2}$. We know that we can look up probability C directly, and probability B is found using symmetry as in Example 8.2. Finally, since $C + D = \frac{1}{2}$, D can be found using $D = \frac{1}{2} - C$ and, similarly, probability $A = \frac{1}{2} - B$.

We suggest the following general procedure:

- Draw a bell curve, and mark on it the values of z you are interested in.

- Shade the region you are interested in.

- Identify any probabilities of the sort $P(0 < Z < z)$ or $P(-z < Z < 0)$.

- Find the probability you are interested in by adding or subtracting the probabilities you have found to or from each other, or 0.5 as required.

This procedure is illustrated in the following example.

Example 8.3

Q: Find $P(-1 < Z < 2)$; $P(Z > 1.3)$; $P(Z < 1.57)$ and $P(1 < Z < 2)$.
A: For the first, we check with Figure 8.6, and see we are dealing with areas B and C. We have dealt with these individually before, and we find
$P(-1 < Z < 2) = 0.3413 + 0.4772 = 0.8185$.

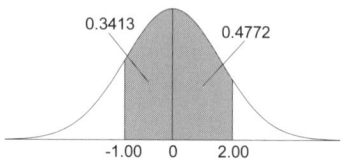

The next one we see corresponds to area D in Figure 8.6. Here we know we can look up C, which is $P(0 < Z < 1.3) = 0.4032$. Now we note that $C + D = \frac{1}{2}$ and so $P(Z > 1.3) = 0.5 - 0.4032 = 0.0968$.

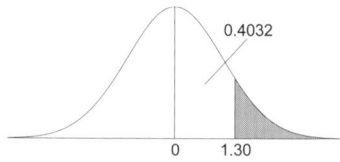

The third probability is all but area D, i.e. we want $A + B + C$. We have just seen $A + B = \frac{1}{2}$, and $P(0 < Z < 1.57) = 0.4418$ and so $P(Z < 1.57) = 0.5 + 0.4418 = 0.9418$.

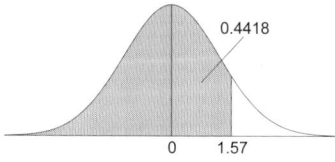

The final problem is a bit different. We have two probabilities like C and we need to subtract the smaller from the larger. We have $P(0 < Z < 2) = 0.4772$ and $P(0 < Z < 1) = 0.3413$.

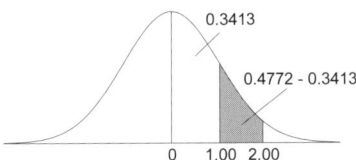

The answer is the difference between the two, $P(1 < Z < 2) = 0.4772 - 0.3413 = 0.1359$.

It is extremely unlikely that we will ever be interested in a random variable that actually has a normal distribution with mean 0 and variance 1. The reason that Z is so important is because of the standardisation properties we explained earlier. Suppose you are not dealing with a standard normal but some other normal, X (e.g. $X \sim N(12, 16)$, so that X is normal with mean 12 and standard deviation $4 = \sqrt{16}$). We might wish to find $P(X > 14)$ for example, and the first task is always to standardise the expression you are interested in. After doing this, we simply use the table for Z to find the probability.

Example 8.4

Q: Find $P(X > 14)$, where $X \sim N(12, 16)$.
A: We start with

$$P(X > 14) = P\left(\frac{X - 12}{4} > \frac{14 - 12}{4}\right) = P(Z > 0.5)$$

where in the first step, we standardise, and secondly, we recognise $Z = \frac{X-\mu}{\sigma}$ and that the original 14 has been rescaled to give a z-score of 0.5.

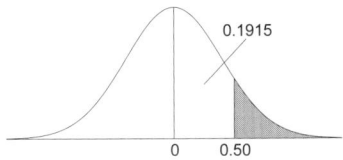

Finally, we note that $P(0 < Z < 0.5) = 0.1915$ and so $P(X > 14) = P(Z > 0.5) = 0.5 - 0.1915 = 0.3085$.

8.1.2 Using the inverse normal table

In the examples we have already seen, we have begun with a z value and used the standard normal table to find a probability. In some situations we actually want to go the other way, from a specified probability to a z value. While we can use the normal table on page 329, it is usually easier to use a special table called the inverse normal table. This can be found on page 330 (Table D.5).

We see that the way the inverse normal table is set out makes it easy to look up a probability p, and find the corresponding z value. We will do this a lot in Part III of this book, and in the following example, we use it to find the upper quartile of the standard normal distribution.

Example 8.5

Q: Find the upper quartile of Z.
A: We wish to find z so that $P(Z < z) = 0.75$ or $P(Z > z) = 0.25$. It is clear that we are dealing with the following identical situations

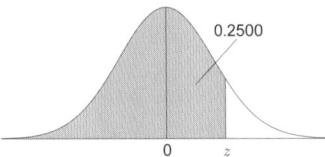 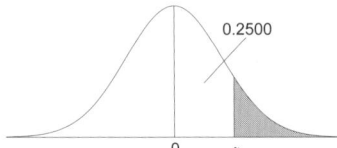

We could use the normal table on page 329, by looking in the probabilities for the figure 0.25. Doing so, we see that $P(0 < Z < 0.67) = 0.2486$ and $P(0 < Z < 0.68) = 0.2518$ so we know the upper quartile is between 0.67 and 0.68, but we are not sure exactly where in this range.

A better way to solve this problem is to look at the inverse normal probability table on page 330. Here, we have $p = 0.25$, and looking at this value in the inverse normal table, we are given $z = 0.6745$.

Example 8.6

Q: In children with "glue" ear (*otitis media*), the age at which a child is first affected is roughly normal with $\mu = 3.4$ years and standard deviation $\sigma = 1.2$ years. If untreated, this condition may lead to partial hearing loss and is detrimental to learning. Consider the following.
  (i) What is the probability that a randomly selected child with glue ear was first affected at age 4 or younger?
  (ii) What proportion of the population with glue ear are first affected by age 5 or older?
  (iii) By what age are 70% of the population with glue ear first affected?
  (iv) What are the possible implications of these results?

A: We start by labelling our variable of interest. Let's call X the age at which a child is first affected with glue ear. Now write down any information you have: $X \sim N(3.4, 1.2^2)$. Now turn the questions into mathematical statements.
  (i) $P(X < 4) = P(\frac{X-3.4}{1.2} < \frac{4-3.4}{1.2}) = P(Z < 0.5) = 0.5 + 0.1915 = 0.6915$.
  (ii) $P(X > 5) = P(\frac{X-3.4}{1.2} > \frac{5-3.4}{1.2}) = P(Z > 1.33) = 0.5 - 0.4082 = 0.0918$.
  (iii) This problem is more awkward. Let's call the age by which 70% of sufferers from glue ear are first affected k. We know $P(X < k) = 0.7$, and standardising, this is equivalent to $P(\frac{X-3.4}{1.2} < \frac{k-3.4}{1.2}) = 0.7$. From the inverse normal table on page 330, we see that $P(0 < Z < 0.524) = 0.2$, and so $P(Z < 0.524) = 0.7$.

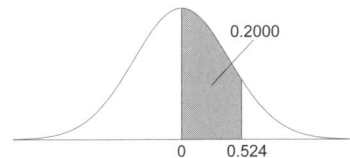

This means that $\frac{k-3.4}{1.2} = 0.524$ is required, giving $k = 3.4 + 1.2 \times 0.524 = 4.0288$.

(iv) From (ii), we see that about 90% of children with glue ear are first affected before age 5, so any screening programme needs to take place well before school entry.

8.1.3 Relative standing

The standard normal distribution can help us evaluate and compare probabilities from different normal distributions. Standard scores (z-scores) measure the relative standing of scores. If a score (or measurement) is X, then the standard score corresponding to X measures the number of standard deviations that X is away from its mean (above or below). From our discussion of the normal distribution, we know that the z-scores of most normal observations lie between -3 and $+3$ (i.e. the X values lie between $\mu - 3\sigma$ and $\mu + 3\sigma$).

Example 8.7

Q: One bed has become available in an old people's home specialising in the care of the elderly with impaired cognitive functioning. Two elderly women on the waiting list are tested for cognitive functioning to determine which of them is in the most urgent need of the place. They live in different towns and are tested for cognitive functioning using different tests. Woman A is tested with a test standardised on a mean of 58 and standard deviation 10. Woman B is tested using a test standardised on a mean of 55, and standard deviation of 12. Both women scored 45. Who is in greater need of the place?

A: Standardising the standard score of woman A gives $z_A = \frac{45-58}{10} = -1.3$. The standard score of woman B is $z_B = \frac{45-55}{12} = -0.83$. Woman A has the lower z-score, indicating she is further from the mean in a negative direction. This indicates that (all other things being equal, such as availability of care currently, etc.) woman A should be given the place.

These properties and the widespread occurrence of phenomena that follow a roughly normal distribution make the normal an important distribution to study. However, there is a curious property of the normal distribution that takes it out of the important category into the central focus of many statistics courses. We shall look at this in Section 8.2.

8.2 The central limit theorem

The normal distribution has a specific form, and many random variables seen in practice do not exhibit this behaviour – we commonly see skewed distributions and outliers. Why the normal distribution is so useful is largely due to the subject of this section: the central limit theorem.

While the central limit theorem is easy to describe, unfortunately it is not at all obvious. Consequently, it is something we will investigate at some length. This idea is generally attributed to Laplace (1749-1827), who first presented the result in 1810. Since saying "central limit theorem" is a bit of a mouthful, and it is something that statisticians use all the time, we will simply refer to it as "the CLT". Thankfully, this doesn't tell us any less about the result than the name "central limit theorem" itself, which is a fairly cryptic description.

A simple statement of the CLT is the following:

> The sum or average of a large number of independent and identically distributed random variables is approximately normal.

Let's look at the conditions in this statement:

- The CLT applies to the sum or average of a whole lot of random variables.
- These random variables need to be independent and identically distributed, i.e. we need to have a *random sample*.
- The sample size needs to be large – we'll discuss "how large is large?" a bit later.

Provided each of these conditions is met, the CLT says that the sum or average will be approximately normal! This is true no matter what the distribution of the individual random variables is. It doesn't matter if they are not normal, nor if they are some other distribution, nor if we don't even know what sort of distribution we have. So long as we have a large number of observations, the sample mean will be normal, and it is this that makes the CLT such a useful result.

While the CLT is pretty easy to state, and not too hard to interpret, it is very hard to *prove*. This is well beyond the scope of this book, and indeed, of most undergraduate statistics courses; however, we will have a closer look at this result nonetheless. We won't look at any mathematics, but it will be really useful to see the CLT in action, and we'll do this using some computer-generated samples.

8.2.1 A demonstration of the CLT

In this demonstration, we specify a very simple non-normal population. We imagine that each observation is a randomly selected (whole) number between 100 and 199. This situation might be equivalent to a lottery, where we have 100 numbered tickets available. If we were to draw the histogram of this population, it would be flat, reflecting that each observation has the same relative frequency, or probability. Although this probability "distribution" has very simple properties, we'll pretend that we do not know much about this situation, and rely on the CLT to provide approximate normality, which we can then use to calculate probabilities.

We now sample from this distribution: imagine a hat with 100 tickets labelled 100 through to 199. We select one ticket at random, record the number on it, replace it, and select another. In this way we obtain a sample of 30 observations. The first four samples are shown in the stemplots in Figure 8.7.

10	0000599	10	0189	10	688	10	12467	
11	0199	11	3567	11	178	11	56	
12	46	12	148	12	26889	12	22369	
13	125	13	8	13	28	13	224	
14		14	4	14	09	14	224	key
15	1568	15	18	15	78	15	34	11 | 5 = 115
16	47789	16	0029	16	012239	16	06788	
17	1	17	0779	17	18	17	01	
18	27	18	06	18	67	18	4	
19	02	19	55899	19	778	19	35	

Figure 8.7 Stemplots of four samples of 30 observations, randomly selected from the integers between 100 and 199

The four samples shown in Figure 8.7 have sample means $\bar{x}_1 = 140.23$, $\bar{x}_2 = 151.63$, $\bar{x}_3 = 148.87$ and $\bar{x}_4 = 141.6$ respectively (you should check at least one of these). These are each fairly close to the theoretical (population) mean of 149.5, which is the average of all the tickets. [Note that $(100+199)/2 = (101+198)/2 = \cdots = (149+150)/2 = 149.5$.] The population distribution contains an equal number of observations in each "stem" since every ticket has an equal probability of selection. Further, we notice that the stemplot of each sample is similar in shape to the population distribution. Suppose we sample over and over, collecting sample means as we go. Most will be close to 149.5, but a few will be a lot smaller or larger depending on the actual tickets drawn. In this way we can study the distribution of the sample mean. If $n = 30$ is large enough, the CLT predicts that the sample means will be approximately normally distributed.

The sample means of 1000 samples (like those in Figure 8.7) are shown in Figure 8.8. The shape of the histogram agrees very well with the normal curve that has been added. In addition, the boxplot is very symmetric, and only six outliers (out of 1000 values) are indicated. The behaviour that we observe is exactly what the CLT would predict.

This is the CLT at work: we sampled from a population that was not at all normal, yet our sample mean turned out to be roughly normally distributed.

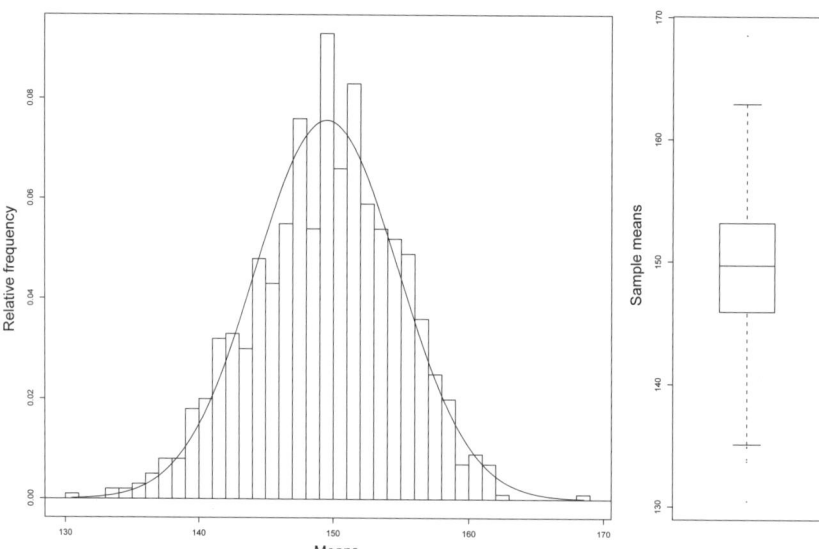

Figure 8.8 Histogram and boxplot of 1000 sample means, from samples like those in Figure 8.7, with the approximating normal distribution

8.2.2 How large is large?

The sample size we require in order to be able to use the CLT really depends on the sort of distribution we are sampling from. In Section 8.2.1, we saw that $n = 30$ was certainly large enough for the distribution considered there.

The two important features of the underlying distribution we are sampling from are its symmetry, and absence of outliers. As the distribution becomes more and more symmetric, then other things being equal, the required sample size will be smaller. On the other hand, if we increase the relative frequency of outliers in the distribution, other things being equal, we will need a larger sample size for the CLT to apply. Unfortunately, we cannot possibly know how large n needs to be in every possible situation, so we simply use a rough figure. In most cases $n = 30$ is sufficient for the CLT to hold, and this is what we use as our rule of thumb. We might restate the CLT as follows:

> The sum or average of at least 30 independent random variables with the same distribution is approximately normal.

If we know our underlying population is very symmetric, and has no outliers, we can apply the CLT to much a smaller sample size. Alternatively, if we know our underlying population is very skewed, or has many outliers, we would increase the sample size to compensate.

In Section 8.4 we consider the normal approximation to the binomial distribution. For the purposes of this section, assume that the CLT applies to the binomial situation, provided n is large enough. This is illustrated in Figure 8.9 for the $Bin(n, \frac{1}{4})$ distributions with $n = 1, 2, 4, 8, 16$ and 32. The plots feature the relative frequency polygon for the binomial distributions, as well as the normal curve with matching mean and variance. When n is very small, the normal curve does not match the binomial polygon at all well; however, when $n = 16$ and 32, the two are virtually indistinguishable. This is in spite of the fact that when $p = \frac{1}{4}$ the binomial distribution can be very asymmetric.

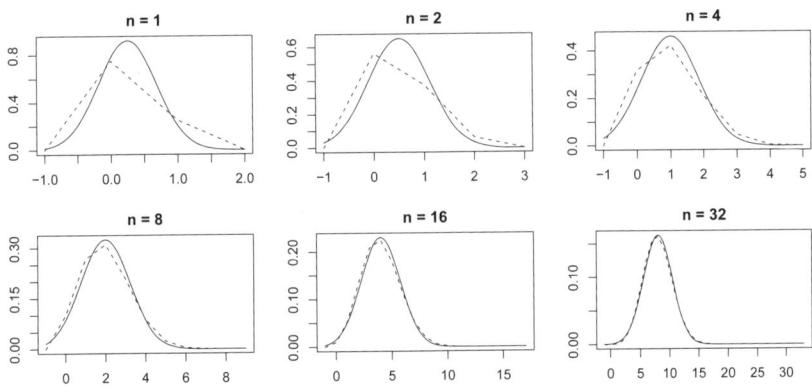

Figure 8.9 Relative frequency polygons for $Bin(n, \frac{1}{4})$ random variables, with $n = 1, 2, 4, 8, 16$ and 32, and corresponding normal curves

8.3 Sampling distributions

In the previous section we learned that provided the sample size n is large enough (at least 30) the sample mean is approximately normally distributed. We can combine this with a few facts about the sample mean to obtain the distribution of the sample mean – commonly called the sampling distribution.

Before looking at what the sampling distribution actually is, we stress that the sample mean is actually a random variable. If we look back at Figure 8.8, we notice that each of the 1000 samples gave a different sample mean. If each of us was to take another sample now, each of us would get a different sample, and consequently a different sample mean. It follows that the sample mean is a *random variable*, and it will have a mean, a standard deviation and a distribution. This sampling process is represented graphically in Figure 8.10.

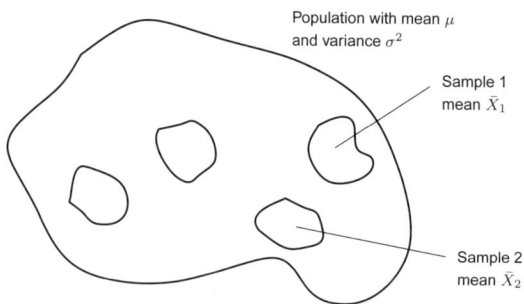

Figure 8.10 Sampling variation of $\bar{X}$

It should come as no surprise that the average of the sample mean is μ, since the sample mean is telling us about the middle of the sample, and the sample comes from a population with mean μ.

The standard deviation of $\bar{X}$ is a little more complicated though, and it is called the *standard error*.

> The sample mean $\bar{X}$ has mean μ and standard error $\sigma/\sqrt{n}$.

The standard error is actually just the standard deviation of the sample mean. All random variables have a standard deviation, which can get confusing when there is more than one random variable. The term *standard error* is used as a reminder that we are not just talking about the standard deviation of the individual observations σ. The way that the standard error depends on sample size n is a curious thing, but natural when you think about the averaging process. The sample mean is always in the middle of the sample, and as the sample size increases, the mean is forced to be closer to the middle of the population μ. Indeed, when n gets really large, the sample is roughly equal to the population, and $\bar{x}$ is almost exactly equal to μ.

We can remember to include n by thinking that as n increases, $\bar{x}$ gets closer and closer to μ. This closeness is measured by its standard deviation, which is given by $\sigma/\sqrt{n}$. As n gets larger, this quantity gets smaller, since n is in the denominator.

> - If the observations $X_1, \ldots, X_n$ are normally distributed with mean μ and variance σ^2, then the sample mean $\bar{X}$ is *exactly* normally distributed with mean μ and variance σ^2/n.
>
> - If the observations $X_1, \ldots, X_n$ are independent and identically distributed with mean μ and variance σ^2, *and the sample size is large* ($n \geq 30$), then the sample mean $\bar{X}$ is *approximately* normally distributed with mean μ and variance σ^2/n.

Note that *if the observations are normal, the CLT is not needed.* However, if the observations are not normal, provided the sample size is large enough, the CLT can be applied, and a normal distribution is appropriate for the sample mean. Since we now know the sample mean is normal, and since we also know the mean and standard deviation of the sample mean, we can standardise the sample mean, and obtain a standard normal random variable.

> If the X_i are normal, or the sample size n is large, then
> $$Z = \frac{\bar{X} - \mu}{\frac{\sigma}{\sqrt{n}}} \quad \text{where } Z \sim N(0, 1).$$

This equation is not new. In Section 8.1 we saw that for any normal random variable X with mean μ and variance σ^2, we can standardise to get the standard normal random variable $Z = (X - \mu)/\sigma$. Here, the variable we are interested in is $\bar{X}$, which has mean μ and variance σ^2/n, and we standardise $\bar{X}$ in the usual way.

Example 8.8

Q: Imagine a population with mean $\mu = 10$ and $\sigma = 5$. Take a random sample of $n = 100$ observations, and a boxplot of these casts doubt on the normality of the population. Can you calculate $P(9.5 < \bar{X} < 10.5)$ nonetheless?

A: The answer is yes, since $n = 100$ is large enough for the CLT to hold. In the usual way, we first standardise $\bar{X}$:

$$P(9.5 < \bar{X} < 10.5) = P\left(\frac{9.5 - 10}{5/\sqrt{100}} < \frac{\bar{X} - 10}{5/\sqrt{100}} < \frac{10.5 - 10}{5/\sqrt{100}}\right)$$

and the problem reduces to finding $P(-1 < Z < 1) = 0.3413 + 0.3413 = 0.6826$.

8.4 Normal approximation to the binomial

A very useful application of the CLT is to the binomial distribution. At first glance, this might seem a bit strange, since the binomial distribution is discrete, whereas the normal is continuous, and further, the binomial isn't naturally seen as a sum or average. If you have a look at the binomial table on pages 325 and 326 (Table D.2), you'll notice that probabilities are listed only for $n \leqslant 20$. If n was made much larger, the table would become very cumbersome indeed. On the other hand, the binomial situation is very common, and you often want probabilities for n much larger than 20.

Let's first convince ourselves that the CLT applies to a binomial situation. The binomial random variable X is a count: it is the number of successes in n trials. Suppose we think of each individual trial, and give each individual with a success a score of 1, and each failure a score of 0. If we add up these scores, the sum is exactly equal to the number of successes, which in turn is the binomial random variable X. Now, since X is a sum (we got it by adding the scores), and since each of the trials is independent and identically distributed, provided n is large enough, the CLT applies.

Usually the CLT only requires n to be large enough, but in the context of a binomial "sum", we need to impose additional conditions. This is because we know $0 \leq X \leq n$ is always satisfied by a binomial X, i.e. the number of successes is always at least zero, but never larger than the number of trials. In contrast, a normal random variable can be any value. A look at the normal table on page 329 tells us that $P(-3 < Z < 3) = 0.9974$ and so a normal random variable nearly always lies within three standard deviations of its mean. This leads us to the following result:

> The binomial random variable $X \sim Bin(n,p)$ has an approximate normal distribution with mean np and variance $np(1-p)$, provided $n \geq 30$ and $np \pm 3\sqrt{np(1-p)}$ both lie between 0 and n.

Let's look at the features of this statement.
- We know $X \sim Bin(n,p)$ and so X has mean np and variance npq, where $q = 1-p$.
- The approximate normal distribution also has mean np and variance npq, so the approximating distribution has the same mean and variance as the actual distribution.
- We require $n \geq 30$ which ensures the CLT applies, and gives us normality.
- We also require $0 < np \pm 3\sqrt{npq} < n$ so that "most" of the normal distribution is between 0 and n.

Before using the normal approximation to the binomial, we have to consider one further detail, this is the *continuity correction*, which corrects for the difference between the discrete binomial, and the continuous normal. This situation is shown in Figure 8.11, where the rectangles have width equal to 1, centred on x, and heights equal to $P(X = x)$.

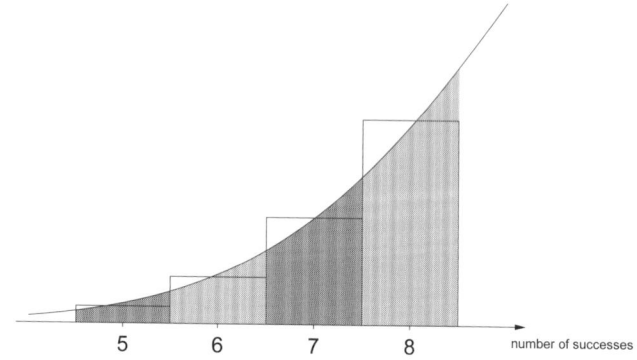

Figure 8.11 Binomial probabilities and approximating normal distribution

Consequently, the areas of the bars are equal to width times height, which is just height (since each width is equal one) and so probability may be thought of as the area of the rectangular bars. The histogram bar centred on $X = 5$ has area $P(X = 5)$ and extends from 4.5 to 5.5. The approximating normal curve is also shown, and probabilities are areas beneath this curve. To correspond to the histogram bars, the probabilities $P(x - \frac{1}{2} < X' < x + \frac{1}{2})$ have been shaded, where X' denotes the approximating normal random variable.

> The normal approximation to the binomial is
> $$P(X = x) \approx P(x - \tfrac{1}{2} < X' < x + \tfrac{1}{2})$$
> where $X \sim Bin(n, p)$ and $X' \sim N(np, npq)$. This approximation requires $n \geqslant 30$ and $0 < np \pm 3\sqrt{npq} < n$.

The factors $\frac{1}{2}$ in the normal approximation to the binomial are what is known as a continuity correction, and this applies when a discrete random variable is approximated by a continuous one.

Example 8.9

Q: A hotelier has 250 standard rooms in her hotel. Each of these has a probability of 0.75 of being occupied, and these are assumed to be occupied independently of one another. Define X = the number of occupied rooms. Consider the following:

(i) What is $P(X < 175)$?
(ii) What is $P(175 \leqslant X < 200)$?
(iii) What is $P(X \geqslant 200)$?

A: This situation is a binomial one, since X counts the number of successes in n trials, the trials have two outcomes, are independent and identical with a fixed P(success), and fixed n, i.e. $X \sim Bin(250, 0.75)$. There are no binomial tables for $n = 250$, so can we use the normal approximation? $n \geqslant 30$, $np = 187.5$ and $3\sqrt{npq} = 20.54$ so $np \pm 3\sqrt{npq} = 187.5 \pm 20.54 = 166.96, 208.04$ are both between 0 and $n = 250$. So, the approximating normal distribution is $X' \sim N(187.5, 46.875)$.

(i)
$$P(X < 175) = P(X = 0, 1, 2, \ldots, 173, 174)$$
$$\text{and } P(X = 174) = P(173.5 < X' < 174.5)$$
$$\text{So, we want } P(X' < 174.5) = P(\tfrac{X'-187.5}{\sqrt{46.875}} < \tfrac{174.5-187.5}{\sqrt{46.875}})$$
$$= P(Z < -1.90)$$
$$= 0.0287.$$

Computer software (e.g. a spreadsheet using instructions in Appendix C) gives the actual binomial probability as 0.0307.

(ii)
$$P(175 \leqslant X < 200) = P(X = 175, \ldots, 199) = P(174.5 < X' < 199.5)$$
$$\text{since } P(X = 175) = P(174.5 < X' < 175.5) \quad \text{Standardising, we have}$$

$$P(174.5 < X' < 199.5) = P(\tfrac{174.5-187.5}{\sqrt{46.875}} < \tfrac{X'-187.5}{\sqrt{46.875}} < \tfrac{199.5-187.5}{\sqrt{46.875}})$$
$$\text{giving } P(-1.90 < Z < 1.75) = 04713 + 0.4599$$
$$= 0.9314.$$

A spreadsheet gives the actual probability as 0.9319, showing how good the approximation is in this case.

(iii) For the final problem we do not need to go through the same procedure (though for practice, you might). Here we note that

$$P(X \geq 200) = 1 - P(X < 200)$$
$$= 1 - P(X < 175) - P(175 \leq X < 200)$$
$$= 0.0399.$$

The actual probability is 0.0374.

The final example gives an instance where the normal approximation to the binomial should not be used.

Example 8.10

Q: In Example 7.10, we observed that 16% of a sample of 50 marketing students were left-handers. If 10% of the general population are left-handers, what is the probability of getting 16% or more in a sample of 50 people?

A: Since $n = 50 \geq 30$, the first condition for use of the CLT applies. The second condition is that $np - 3\sqrt{np(1-p)} = 50 \times 0.1 - 3\sqrt{50 \times 0.1 \times 0.9} = -1.36$ should be greater than zero, which is not true. Unfortunately, the normal approximation does not apply here, since too much of the probability is outside the allowable range for a binomial count. In order to answer this question, we would need to use a computer to evaluate $P(X \geq 8)$ where $X \sim Bin(50, 0.1)$.

8.5 Summary

It is important to carefully distinguish the two uses of the normal distribution you have met in this chapter.

- When you are dealing with individual observations from a normal or roughly normal distribution and are trying to find probabilities of events, you use the standardised z-score

$$Z = \frac{X - \mu}{\sigma} \sim N(0, 1).$$

- When you are dealing with a random sample of 30 or more observations from *any* distribution and you are interested in the averages (means) you use the z-score

$$Z = \frac{\bar{X} - \mu}{\frac{\sigma}{\sqrt{n}}} \sim N(0, 1)$$

approximately (or exactly if the X_i are normal).

Instructions for how to compute normal probabilities in a spreadsheet, rather than manually using the normal table, are given in Appendix C.

Exercises

Finding probabilities

8.1 Find the following probabilities:
(a) $P(0 < Z < 1.45)$;
(b) $P(0 < Z < 2.97)$;
(c) $P(0 < Z < 4.39)$;
(d) $P(0 < Z < 0.62)$.

8.2 Find the following probabilities:
(a) $P(-2.12 < Z < 0)$;
(b) $P(-1.45 < Z < 0)$;
(c) $P(-3.1 < Z < 0)$;
(d) $P(-2.64 < Z < 0)$.

8.3 Find the following probabilities:
(a) $P(-1.7 < Z < 1.7)$;
(b) $P(-2.04 < Z < 1.80)$;
(c) $P(-4.1 < Z < 2.3)$;
(d) $P(-1.96 < Z < 1.96)$.

8.4 Find the following probabilities:
(a) $P(0.82 < Z < 1.40)$;
(b) $P(-2.04 < Z < -0.5)$;
(c) $P(-3.3 < Z < -0.32)$;
(d) $P(0.47 < Z < 2.06)$.

8.5 Find the following probabilities:
(a) $P(Z > 1.35)$;
(b) $P(Z > 2.18)$;
(c) $P(Z < -1.20)$;
(d) $P(Z < 0.95)$;
(e) $P(Z < -3.1)$;
(f) $P(Z > -1.14)$.

Standardising

8.6 Let $X \sim N(10, 5^2)$. Find:
(a) $P(X > 10)$;
(b) $P(10 < X < 20)$;
(c) $P(X > 20)$;
(d) $P(X < 0)$.

8.7 Let $X \sim N(12, 6)$. Find:
(a) $P(8 < X < 10)$;
(b) $P(X < 8)$;
(c) $P(X > 13)$;
(d) $P(X > 18)$.

8.8 Let $X \sim N(120, 40)$. Find:
(a) $P(X > 132)$;
(b) $P(X > 115)$.
(c) The probability that X is within 10 of its mean.
(d) The probability that X is at least 10 from its mean.

8.9 Let X be normal with mean 40.3 and standard deviation 11.7. Find:
(a) the probability that X exceeds 30;
(b) the probability that X exceeds 35;
(c) the probability that X exceeds 45.

8.10 Let X be normal with mean 5.34 and standard deviation 3.7. Find:
(a) the probability that X is negative;
(b) the probability that X is between zero and 10;
(c) the probability that X exceeds 10.

Sampling distribution and the CLT

The following questions involve the sample mean $\bar{X} = \frac{1}{n}\sum X_i$ where the X_i are a random sample from the distribution of X, and n is as given.

8.11 Let $X \sim N(10, 5^2)$ and $n = 9$. Find:
(a) $P(\bar{X} > 10)$;
(b) $P(10 < \bar{X} < 20)$;
(c) $P(\bar{X} > 20)$;
(d) $P(\bar{X} < 0)$.
(e) Compare your answers with those for Exercise 8.6 and explain any patterns.
(f) Is the CLT needed for normality of the sample mean in this case?

8.12 Let $X \sim N(12, 6)$ and $n = 5$. Find:
(a) $P(8 < \bar{X} < 10)$;
(b) $P(\bar{X} < 8)$;
(c) $P(\bar{X} > 13)$;
(d) $P(\bar{X} > 18)$.
(e) Compare your answers with those for Exercise 8.7 and explain any patterns.
(f) Is the CLT needed for normality of the sample mean in this case?

8.13 Let $X \sim N(20, 4)$. Find $P(\bar{X} > 22)$ when:
(a) $n = 1$;
(b) $n = 2$;
(c) $n = 4$;
(d) $n = 8$;
(e) $n = 16$.
(f) Explain the effect n has on these probabilities.

8.14 Let X have mean 32.6. Can you determine $P(X > 32.6)$? Under what conditions will this be 0.5?

8.15 Let X have mean 20.7 and standard deviation 6.9.
(a) Can $P(X > 23)$ be calculated using this information alone? Draw sketches of three distributions for X and label $P(X > 23)$ on these plots. Are the areas the same?
(b) Let $n = 15$. Can $P(X > 23)$ be calculated now? What about $P(\bar{X} > 23)$? If so, perform the calculation.
(c) Let $n = 40$. Can $P(X > 23)$ be calculated now? What about $P(\bar{X} > 23)$? If so, perform the calculation.

Inverse normal

8.16 Find the values of z which solve the following equations:
(a) $P(0 < Z < z) = 0.45$;
(b) $P(-z < Z < 0) = 0.2$;
(c) $P(-z < Z < z) = 0.8$;
(d) $P(Z < z) = 0.75$;
(e) $P(Z > z) = 0.1$;

8.17 Find the values of z which solve the following equations:
(a) $P(0 < Z < z) = 0.3$;
(b) $P(-z < Z < 0) = 0.49$;
(c) $P(-z < Z < z) = 0.7$;
(d) $P(Z > -z) = 0.6$;
(e) $P(Z > z) = 0.05$.

8.18 Find the value that 25% of observations of Z lie above. What is the name given to this number?

8.19 Quintiles are the 20th, 40th, 60th and 80th percentiles of a random variable. Find the quintiles of Z.

8.20 Let $X \sim N(100, 20^2)$. Find the values of x which solve the following equations:
(a) $P(X > x) = 0.75$;
(b) $P(X > x) = 0.5$;
(c) $P(X > x) = 0.25$.
(d) What are the names given to the numbers you found in (a), (b) and (c)?

8.21 Let $X \sim N(3.2, 4.6)$. Find the values of x which solve the following equations:
(a) $P(X > x) = 0.3$;
(b) $P(X > x) = 0.7$;
(c) $P(X > x) = 0.9$.

8.22 Let $X \sim N(50, 100)$. Find:
(a) $P(X > 72)$;
(b) the 5th percentile of X, i.e. find x that solves $P(X > x) = 0.95$;
(c) the 95th percentile of X, i.e. find x that solves $P(X < x) = 0.95$.

8.23 Let $X \sim N(124, 28)$ and let $\bar{X}$ be the sample mean of 25 observations drawn from X.
(a) Find the 60th percentile of X.
(b) Find the probability that the sample mean exceeds the 60th percentile of X.
(c) Find the 60th percentile of the sample mean.

8.24 Let X have mean $\mu = 18$ and standard deviation $\sigma = 5$.
(a) What is $P(\bar{X} > 20)$ when $n = 30$?
(b) What is the 95th percentile of $\bar{X}$ when $n = 50$?

Applications

8.25 The beak of a fledgling male kea (the native parrot *Nestor notabilis*) is approximately normally distributed with mean 47.5 millimetres (mm) and standard deviation 2.5 mm. Consider

a random sample of n fledgling male kea.
(a) If $n = 1$, what is the probability that the selected kea has a beak longer than 50 mm? Longer than 45 mm?
(b) If $n = 8$, what is the probability that the average beak length is longer than 50 mm? Why is this probability smaller than that in (a)?
(c) If $n = 20$, what is the probability that the average beak length is longer than 50 mm? Why is this so unlikely?
(d) Why is the CLT not necessary to answer (a)-(c)?

(Thanks to Clio Reid for the data)

8.26 The firms listed on a large stock exchange have annual returns that are normally distributed with mean 15.1% per annum and standard deviation 9.4% per annum. The risk-free interest rate in this market is 6% per annum. Consider a random sample of n firms.
(a) If $n = 1$, what is the probability that the selected firm has a negative return, i.e. its value goes down?
(b) If $n = 1$, what is the probability that the selected firm has a return that exceeds this risk-free rate?
(c) What proportion of firms earn a return at least three times the risk-free rate?
(d) If $n = 20$, what is the probability that the average return is at least double the risk-free rate?

8.27 House prices in a city have a highly skewed distribution with the mean much greater than the median. Assume the mean price is $320,000 and the standard deviation is $240,000.

(a) Draw a sketch of a possible distribution for the population of house prices.
(b) What is the z-score for a price of $10,000? Label this score z^* and find $P(Z < z^*)$. Is this a problem given the likely probability of a house worth less than $10,000?
(c) Consider a random sample of $n = 100$ houses. What is the probability their average price is within $30,000 of the mean?
(d) Consider a random sample of $n = 100$ houses collected from one suburb in this city. Does you answer to (c) apply to this sample? Why, or why not?

8.28 Social support scores for a large number of 13-year-old boys are distributed with a mean of 20 and a standard deviation of 5.2. The scores are not symmetrically distributed.
(a) Is the normal distribution suitable for finding the probability of a selected boy having a large score? Why or why not?
(b) Is the normal distribution suitable for finding the probability that the mean of a sample of randomly selected boys is large? Why or why not?
(c) If possible, calculate $P(X > 24)$, $P(\bar{X} > 24)$ when $n = 20$, and $P(\bar{X} > 24)$ when $n = 40$. If the calculation is not possible, explain why not.

8.29 A population of mussel larvae *Mytilus galloprovincialis* was provided an abundant food source over a 20-day period followed by a limited food source over an additional 9 days. Their growth rates after "settlement" were recorded, and were found to be approximately normally distributed with mean 25 microns per day and standard deviation 6 microns per day.

Consider a randomly selected juvenile mussel.
(a) What is the probability its growth rate exceeds 30 microns per day?
(b) What is the probability its growth rate puts it in the lower quartile of the population?
(c) What is the lower quartile of the population growth rate?

(*Thanks to Nicole Phillips for access to raw data, from Phillips, Variable timing of larval food has consequences for early juvenile performance in a marine mussel, Ecology, 2004*)

Normal approximation to the binomial

Optional: use a spreadsheet to calculate the actual binomial probabilities in these questions. Instructions for doing so are given in Appendix C.

8.30 Let X be binomial with $n = 240$ and $p = 0.32$. Estimate $P(60 \leq X \leq 80)$ and $P(60 < X < 80)$.

8.31 Let X be binomial with $n = 600$ and $p = 0.85$. What are the mean and variance of X? What is the probability that X exceeds 500?

8.32 A hotel has 125 rooms, and has over time had an occupancy rate of 62% per night.
(a) What are the conditions for the number of occupied rooms on a randomly selected night to have a binomial distribution?
(b) Assuming the binomial conditions hold, what are the mean and standard deviation of the distribution?
(c) Estimate the probability that at least half of the rooms are occupied on a randomly selected night.

8.33 Following Exercise 7.14, suppose a random sample of 1500 houses is considered.
(a) Show that the conditions for the normal approximation to the binomial are met in this instance.
(b) What is the expected value and standard deviation of the number of homes with solar panels for electricity generation?
(c) Using the normal approximation to the binomial, estimate the probability that more than 10 homes have solar panels.

(*Data from* eeca.govt.nz)

8.34 Using a survey conducted in Winter 2009, it was estimated that 17% of New Zealand homes have complete underfloor insulation. Consider a random sample of 250 homes. Estimate the probability that between 20 and 40 of these homes (inclusive) have complete underfloor insulation.

(*Data from* eeca.govt.nz)

Part III

Estimation and testing

Chapter 9

Single population

In Part III of this book, we consider two different, but related, topics.

The first is estimation:

> *estimation* is when we provide our best guess for an unknown population parameter based on sample evidence

and the second is hypothesis testing:

> *hypothesis testing* is when we check a claim about an unknown population parameter (property) based on sample evidence.

We have had a hint of these two ideas already. In Chapter 3 we discussed how we might use sample statistics to estimate population parameters. For example, if we were wondering what the population mean μ might be for a particular population, we might collect a sample, and approximate it by the sample mean $\bar{x}$.

Example 9.1

We wish to estimate the average value of residential apartments in Central Wellington. Sample data are collected on recent sales, and these data yield a mean of $\bar{x} = \$328{,}700$. This was a *point* estimate of the true average value of Wellington Central apartments, μ, at the time that the data were collected.

The problem with point estimation as in Example 9.1 is that we have no idea how good our estimate is. The user of our estimate has no idea how close this is likely to be to the true value. Mostly, if we have no further information about the true population value, it is preferable to present an *interval estimate* which is usually called a *confidence interval.* If we have a preconceived idea about the population, e.g. from the most recent national census, or from some recently published research, we might check this by asking "how close is the sample (point) estimate to this value?" This is done formally using a *hypothesis test.* These techniques are the subject of Part III of this book.

9.1 Large sample inference for a mean

When we have a large sample, i.e. $n \geq 30$, the CLT holds, and *regardless of the actual population distribution*, the sample mean is approximately normal. This means that when we standardise the sample mean, the resulting statistic is $N(0, 1)$. Typically, we won't have σ, but when n is large, we can replace this by the sample standard deviation s without changing things too much. Written symbolically, we have

$$\frac{\bar{X} - \mu}{\frac{\sigma}{\sqrt{n}}} \sim N(0, 1) \quad \text{and} \quad \frac{\bar{X} - \mu}{\frac{s}{\sqrt{n}}} \sim N(0, 1).$$

The second expression becomes our starting point for both confidence intervals and tests in this chapter.

9.1.1 Confidence intervals for μ

A confidence interval (CI) gives the likely range for a population parameter based on sample information. We form this interval using the appropriate probability distribution – in this case, the standard normal. For instance, in Chapter 8, we discovered that 95% of the time, Z lies between ± 1.96, i.e. $+1.96$ and -1.96. Since we know $(\bar{X} - \mu)/(\sigma/\sqrt{n})$ has the same distribution as Z (by the CLT), 95% of the time, $(\bar{X} - \mu)/(\sigma/\sqrt{n})$ should lie between ± 1.96 (see Exercise 8.3d). We unravel this statement to get a likely range for μ as follows:

$$
\begin{aligned}
0.95 &= P(-1.96 < Z < 1.96) && \text{(by the properties of } Z) \\
&= P\left(-1.96 < \frac{\bar{X} - \mu}{\sigma/\sqrt{n}} < 1.96\right) && \text{(standardised } \bar{X}) \\
&= P\left(-1.96 \frac{\sigma}{\sqrt{n}} < \bar{X} - \mu < 1.96 \frac{\sigma}{\sqrt{n}}\right) && \text{(multiplying by } \frac{\sigma}{\sqrt{n}}) \\
&= P\left(-1.96 \frac{\sigma}{\sqrt{n}} < \mu - \bar{X} < 1.96 \frac{\sigma}{\sqrt{n}}\right) && \text{(multiplying by } -1) \\
&= P\left(\bar{X} - 1.96 \frac{\sigma}{\sqrt{n}} < \mu < \bar{X} + 1.96 \frac{\sigma}{\sqrt{n}}\right) && \text{(adding } \bar{X}).
\end{aligned}
$$

The final statement can be interpreted "the probability that μ lies between $\bar{X} \pm 1.96 \frac{\sigma}{\sqrt{n}}$ is 95%"; however, care is needed, because at this point, the random variable in the expression is $\bar{X}$. The object of interest is μ, and this is fixed, but unknown. The thing we are unsure about (the *random variable*) in this context is $\bar{X}$, the sample mean, and this determines the position of the two end-points $\bar{X} - 1.96\frac{\sigma}{\sqrt{n}}$ and $\bar{X} + 1.96\frac{\sigma}{\sqrt{n}}$, and whether or not the CI will actually contain μ.

In order to calculate the confidence interval using sample data we must

- calculate the sample statistics $\bar{x}$ and s
- use these in in the equation above (using s in place of σ) to form the confidence interval.

> For $n \geqslant 30$
>
> $$\bar{x} \pm 1.96 \frac{s}{\sqrt{n}} \quad \text{or equivalently} \quad \bar{x} - 1.96\frac{s}{\sqrt{n}}, \bar{x} + 1.96\frac{s}{\sqrt{n}}$$
>
> is a 95% confidence interval for μ.

The notation $\pm$ is common when we are dealing with confidence intervals, and in this context, it means we add and subtract (+ and −) the term $1.96\frac{s}{\sqrt{n}}$ from $\bar{x}$. By ensuring $n \geqslant 30$, we know that the CLT holds, and replacing σ by s is reasonable. So we can safely use the normal distribution, and ± 1.96 are the appropriate z-scores for a 95% confidence interval.

Note that the form of the confidence interval is (sample value) $\pm$ (value from the normal distribution) × (standard error of the sample value). This is a common structure that will recur in subsequent situations.

Example 9.2

Q: "Glue" ear is a medical condition which affects children. In a sample of 35 children aged under ten from Canterbury with a known "glue" ear condition, the mean age at which they were first affected was 3.3 years with a sample standard deviation of 1.2 years. Calculate a 95% confidence interval for μ, the population mean age of affliction.

A: Since $n = 35 > 30$, the CLT applies, and the form of the confidence interval for μ is $\bar{x} \pm 1.96\frac{s}{\sqrt{n}}$. Substituting the values $\bar{x} = 3.3$ and $s = 1.2$, the end points of the interval are

$$\bar{x} - 1.96\frac{s}{\sqrt{n}} = 3.3 - 1.96\frac{1.2}{\sqrt{35}} = 3.3 - 0.40 = 2.90$$

$$\bar{x} + 1.96\frac{s}{\sqrt{n}} = 3.3 + 1.96\frac{1.2}{\sqrt{35}} = 3.3 + 0.40 = 3.70$$

or alternatively, we could write 3.3 ± 0.40. The 95% confidence interval for μ is $(2.90, 3.70)$ and this gives us the likely range for μ.

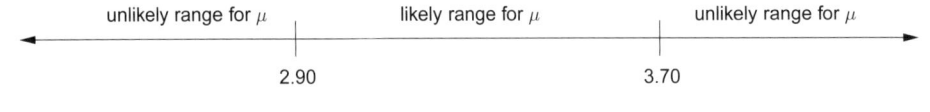

Having calculated a confidence interval (using sample data) we must *not* say "the probability μ lies in this interval is 95%." Once we have replaced the random variable $\bar{X}$ by its realisation $\bar{x}$, we can no longer use a probability statement. Either the true value μ lies in the interval or it does not, we just do not know which. So, we must make an alternative statement, e.g. "we are confident that μ lies in this interval."

A more precise statement, based on the interval in Example 9.2, would be, "a 95% confidence interval for μ is $(2.90, 3.70)$, and 95% of such intervals will contain μ." This is demonstrated in Figure 9.1 where 20 independent random samples are taken from a population with mean μ, and from each sample a 95% confidence interval is calculated. These intervals are drawn and the true mean μ shown. This figure was based on simulated data, and we see that most of the confidence intervals do include the true value (in this instance only two from the 20 exclude μ). Due to the fact that we have shown only 20 samples, it is fine that the actual proportion of intervals that contains μ is not equal to the true probability 95%.

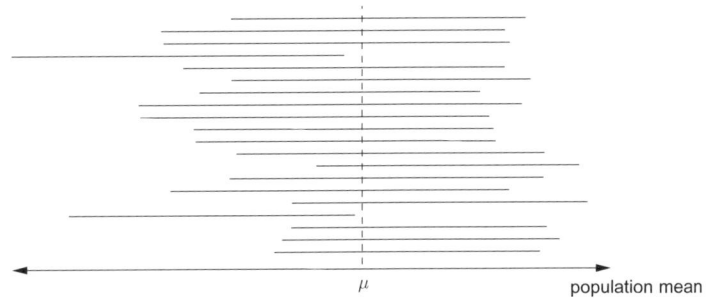

Figure 9.1   95% confidence intervals for μ from 20 independent random samples

A confidence interval gives us the likely range for μ, and so it also tells us the *unlikely* range(s) for μ. In Example 9.2 the likely range for μ was $(2.90, 3.70)$, and so μ is unlikely to be less than 2.90 or greater than 3.70. We can use this to answer questions like "is μ equal to μ_0?" (where μ_0 is a number like 2.75) by seeing whether or not the proposed μ_0 lies in the confidence interval. An alternative procedure is to use a hypothesis test.

9.1.2   Hypothesis tests for μ

Although a confidence interval can be used to answer the question "is μ_0 a likely value for μ?" (where μ_0 is some number) we typically prefer to use a technique called a *hypothesis test*. The basic idea behind this technique is to measure the observed difference between sample and population values (in appropriate units) and ask whether or not this observed difference is too large for μ to actually equal μ_0.

> The hypothesis test consists of four steps:
>
> 1. forming the hypotheses
> 2. calculating the test statistic
> 3. determining the rejection region
> 4. drawing a conclusion.

Hypotheses

For a hypothesis test, we must always have two hypotheses. The first is called the *null hypothesis* and is denoted H_0. The second hypothesis is called the *alternative hypothesis* and is denoted H_a. The notations H_A and H_1 are also commonly used for the alternative; however, we will stick with H_a throughout this book. We form both hypotheses *before we collect any data*.

The null hypothesis will typically have an equals sign '=' in it, and what the population parameter μ equals will be based on our past belief. An example is $H_0 : \mu = 26$, i.e. we assume that the population from which our sample is drawn has a mean of 26. We will have some prior reason to think that 26 is an appropriate and interesting value for μ.

> The alternative hypothesis takes one of three forms:
>
> - the two-sided alternative is $H_a : \mu \neq \mu_0$, i.e. μ is not equal to the value suggested by H_0; and
> - the one-sided alternatives are:
>   - upper tail $H_a : \mu > \mu_0$, i.e. μ is greater than the value suggested by H_0
>   - lower tail $H_a : \mu < \mu_0$, i.e. μ is less than the value suggested by H_0.

If we have no reason to suspect an increase or decrease in the population parameter from the past belief, we choose the two-sided alternative. This is our default alternative hypothesis. If we suspect that the population values have increased on average, e.g. we might think that birth weights of babies are getting larger, we would specify the one-sided alternative $H_a : \mu > \mu_0$ indicating the increase in μ. Conversely, if we have some reason to believe a decrease in μ has occurred, we use the one-sided alternative $H_a : \mu < \mu_0$. It is important that we make this choice *before* analysing the sample information – it should be based on our knowledge of the situation, *not* on what we observe in the sample.

> **Example 9.3**
>
> Q: Formulate null and alternative hypotheses in the following instances and specify the population from which we should sample.
>   1. Normal 70-year-olds take on average 30 minutes to complete a test of cognitive skills. Patients with early Alzheimer's are expected to take longer.
>   2. It is wondered if frogs on Maud Island have a different life span from those on Stephen's Island, which is on average 30 years.
>   3. A new medication is hoped to reduce high blood pressure in people on medication for this condition.
>   4. The terrorist events of 9/11 are expected to have an effect on New Zealand visitor arrivals.
>   5. The life of automatic washing machines is thought to be longer than the two-year warranty period.
>
> A: For each of the above examples, we must determine an appropriate value μ_0 for the null hypothesis, and choose between $\neq$, $<$ and $>$ for H_a.
>   1. $H_0 : \mu = 30$, $H_a : \mu > 30$. We must sample from 70-year-old Alzheimer's patients.
>   2. $H_0 : \mu = 30$, $H_a : \mu \neq 30$. We need to collect data from Maud Island frogs.
>   3. $H_0 : \mu = \mu_0$, $H_a : \mu < \mu_0$, where μ_0 is the existing population average. We must sample from individuals on the new medication.
>   4. $H_0 : \mu = \mu_0$, $H_a : \mu \neq \mu_0$ since the expected effect is not specified. We should sample visitor arrivals since 9/11.
>   5. $H_0 : \mu = 2$, $H_a : \mu > 2$. We should sample lifetimes of washing machines with two-year warranties.

Test statistic

The second stage of the hypothesis test is to calculate the test statistic. This stage is always based on sample data, and so a precursor to forming the test statistic is collection of the sample. In this section we assume that the sample $X_1, \ldots, X_n$ is large, so $n \geq 30$ is assumed. To form the test statistic, we standardise the sample mean. This allows us to use the tabulated $N(0, 1)$ probabilities.

> The test statistic is
> $$Z = \frac{\bar{X} - \mu}{\frac{s}{\sqrt{n}}}$$
> and this has an approximate standard normal distribution.

Here s is used instead of σ because when n is large, we can be confident that s is very close to σ. We denote the test statistic by Z in this case, because the test statistic is approximately $N(0, 1)$ due to the CLT.

The formula for the test statistic contains μ. When we were forming confidence intervals, this population parameter was considered unknown, and indeed, here it is unknown too. Unlike interval estimation, here we do have a *proposed* value for μ: it appears in H_0. It is this proposed value we use to standardise $\bar{X}$.

In order to complete this stage of the hypothesis test, we must use sample data to calculate the sample mean $\bar{x}$ and the sample standard deviation s, and with them and the value of μ from H_0, we evaluate the formula for the test statistic. Consequently, the test statistic is just a single number whose value is based on the assumed value of μ given by H_0.

Example 9.4

For a sample of $n = 35$ observations, with $\bar{x} = 3.3$ and $s = 1.2$, and for a null hypothesis $H_0 : \mu = 2.6$, the test statistic is

$$Z = \frac{\bar{x} - \mu}{s/\sqrt{n}} = \frac{3.3 - 2.6}{1.2/\sqrt{35}} = 3.45$$

and since $n = 35 > 30$, this has an approximate $N(0, 1)$ distribution.

Rejection region

In this stage, we identify the range of values of the test statistic that is unlikely to arise if H_0 is true.

> The rejection region depends on three things:
>
> 1. the distribution of the test statistic
> 2. which tails of the distribution we will use
> 3. the probability we will allocate to the appropriate tails.

The distribution of the test statistic is related to how we have calculated the test statistic, and consequently, to the situation we are dealing with. To get the test statistic in this case we standardised the sample mean $\bar{X}$ which is a random variable. Because we have a large sample, the CLT applies to $\bar{X}$ and so the test statistic is approximately $N(0, 1)$. It is this distribution we use to calculate the rejection region.

The second consideration is made by looking at H_a. We know there are three different possibilities: $\mu \neq \mu_0$, $\mu > \mu_0$ and $\mu < \mu_0$. Suppose $\mu > \mu_0$ is the alternative we are interested in. If it is true (rather than H_0 being true), the μ we used to form the test statistic is too small, and so $\bar{x} - \mu$ will be too large. Consequently, we would expect the test statistic to be too large. This means the rejection region will be in the upper tail of the distribution. The mathematical form of the rejection region will be $Z > z^*$ but we can more easily represent it using a diagram, as shown in Figure 9.2, where it is the part of the axis beneath the shaded area.

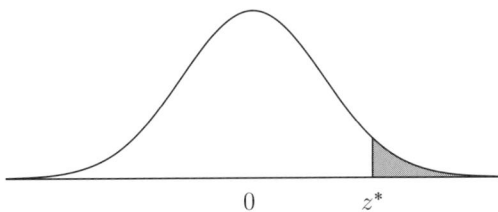

Figure 9.2 Upper-tail rejection region (below shaded area)

The opposite holds for $\mu < \mu_0$, and we end up using the lower tail of the normal distribution, and a rejection region of the form $Z < -z^*$. In the case that we have a two-sided alternative it doesn't matter whether μ is above or below μ_0; too much less than μ_0 or too much more than μ_0 will both cause us to reject H_0. Consequently, we need a rejection region at both ends of the normal distribution, and the rejection region is given by $Z < -z^*$ and $Z > z^*$ (or equivalently $|Z| > z^*$ where $|Z|$ is the absolute value of Z).

In order to pin down the value for z^*, we need to allocate a probability to the (shaded) rejection region. This probability is denoted α and is usually chosen from 10%, 5% and 1%, the most common choice being 5%. We call α the *level of the test*, and in the case of a two-sided test, the quantity $100(1-\alpha)\%$ corresponds to the confidence level of the equivalent confidence interval, where α is treated as a decimal, i.e. 0.05 for the 5% level. The level α is also commonly described as the *significance level* or the *level of significance*.

We might write the rejection region as $Z > z^*$ for a one-sided upper tail test, $Z < -z^*$ for a one-sided lower tail test, or, in the case of a two-sided test, either $Z > z^*$ or $Z < -z^*$ for the appropriate values of z^* taken from the normal table. Various rejection regions are shown in Figure 9.3 with the values z^* indexed by the level of the test α.

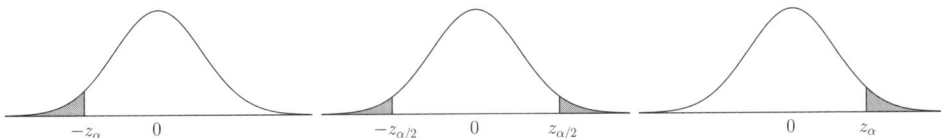

Figure 9.3 Rejection regions: lower tail, two-sided and upper tail

The value at the cut-off point of the rejection region is called the *critical value*, as this will be what we compare the test statistic to. We obtain the critical values by looking at the inverse normal probability table. Remember that the standard normal table is laid out to allow us to look up a z-score and read off a probability, whereas here we have a probability (α, or more precisely $0.5 - \alpha$) and wish to look up a z-score. This is most easily done using the inverse normal table as shown in the following example.

Example 9.5

Q: Determine the rejection regions for the following tests.
1. $H_0 : \mu = 2.6$ vs $H_a : \mu > 2.6$ at the 5% level.
2. $H_0 : \mu = 2.6$ vs $H_a : \mu < 2.6$ at the 5% level.

3. $H_0: \mu = 2.6$ vs $H_a: \mu \neq 2.6$ at the 5% level.
A: We assume the test statistic is normally distributed for these examples.
1. This is an upper-tail test, with 5% in the upper tail, and so we have the situation

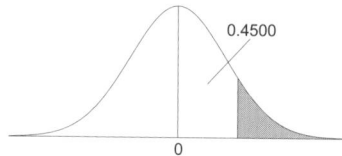

and looking up $0.5 - 0.05 = 0.45$ in the inverse normal table gives us a z-score of 1.645, and so this is our critical value. We reject H_0 if $Z > 1.645$.
2. In this situation, following the previous example, we take the critical value to be -1.645, and we reject H_0 if $Z < -1.645$.
3. For the two-tail test we need to halve α, and subtract this from 0.5. In this case, we have $\alpha/2 = 0.025$ and so looking up $0.5 - 0.025 = 0.475$ in the inverse normal table we read off the critical value $z = 1.96$. The rejection region is $Z < -1.96$ and $Z > 1.96$, or equivalently $|Z| > 1.96$. Notice that since we have reduced the probability in each tail, a larger test statistic is needed before we can reject H_0.

Conclusion

The final stage of the hypothesis test is to draw a conclusion.

> The conclusion is one of the following two statements:
>
> - "Since the test statistic lies in the rejection region, we reject the null hypothesis in favour of the alternative at the $100\alpha\%$ level."
>
> - "Since the test statistic does not lie in the rejection region, we cannot reject the null hypothesis in favour of the alternative at the $100\alpha\%$ level."

We choose between our two conclusions depending on whether the test statistic lies in the rejection region or not. If it does, it is very unlikely that the value of μ given by H_0 is correct, and we reject H_0. Notice that we either *reject the null* or *fail to reject the null*. If we fail to reject the null, we have not shown it to be true, but rather, we have not found evidence against it. For this reason, we prefer not to accept H_0 if we cannot reject it. We are not proving anything, just reporting which hypothesis the data favour. We state the level of the test ($100\alpha\%$, usually 5%) as part of the conclusion to remind ourselves that if we change α we might change the conclusion we draw.

Example 9.6

Q: Referring to Example 9.2, "glue" ear affects young children in New Zealand. It is thought that children who get "glue" ear in less humid parts of the country are first affected at a later age than the population in general. It is thought that the mean age nationally at which children are first affected by "glue" ear is 2.6 years. The sample from Canterbury (a less humid area) is collected to test this claim.

A: In order to complete the test, we perform the four steps of the hypothesis test.

- We wish to test hypotheses about a population mean μ. The null hypothesis comes from the belief about the general population of children and is $H_0 : \mu = 2.6$. The claim is that the mean age should be higher in Canterbury, since it is less humid and the children are affected at a later age, giving the alternative $H_a : \mu > 2.6$.
- The sample data give $\bar{x} = 3.3$, $s = 1.2$ with $n = 35$. We use the value in H_0 to standardise the sample mean. The test statistic is

$$Z = \frac{\bar{x} - \mu}{s/\sqrt{n}} = \frac{3.3 - 2.6}{1.2/\sqrt{35}} = 3.45$$

and since $n = 35 > 30$, this has an approximate $N(0,1)$ distribution.

- The alternative is $H_a : \mu > 2.6$ and so we have an upper-tail (one-sided) test. The rejection region will be of the form $Z > z$, with $\alpha = 5\%$. From Example 9.5, we look up $0.5 - 0.05 = 0.45$ in the inverse normal table, giving a critical value of 1.645 and a rejection region of the form $Z > 1.645$.
- We compare the test statistic $Z = 3.45$ to the rejection region $Z > 1.645$.

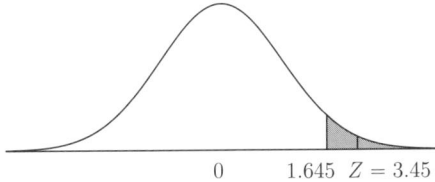

The test statistic clearly lies in the rejection region, and so we reject the null hypothesis in favour of the alternative at the 5% level. It appears that the children in Canterbury are afflicted with "glue" ear at an older age than the national average. Note that we cannot definitely attribute this to the low humidity levels in Canterbury.

Example 9.7

Q: A second group of researchers believes there is no reason why Canterbury should be any different from the rest of the country for age of affliction of "glue" ear. In addition, they feel that dietary changes indicate that the current national average is more likely to be 3.1 years. Test their claim using the Canterbury sample.

A: Once again, we complete the four steps of the hypothesis test.

- This time, the null hypothesis is $H_0 : \mu = 3.1$ against the two-sided alternative $H_a : \mu \neq 3.1$. This is based on the statement that "there is no reason why Canterbury should be any different from the rest of the country" indicating no preference for an increase or decrease.
- The test statistic is

$$Z = \frac{\bar{x} - \mu}{s/\sqrt{n}} = \frac{3.3 - 3.1}{1.2/\sqrt{35}} = 0.99$$

and since $n = 35 > 30$, this has an approximate $N(0, 1)$ distribution.
- The alternative is $H_a : \mu \neq 3.1$ and so we have a two-sided test. The rejection region will be of the form $Z > z$ and $Z < -z$, with $\alpha = 5\%$. From Example 9.5, we look up $0.5 - 0.05/2 = 0.475$ in the inverse normal table, giving a critical value of 1.96 and a rejection region of the form $Z > 1.96$ and $Z < -1.96$.
- We compare the test statistic $Z = 0.99$ with the rejection region $Z > 1.96$ and $Z < -1.96$.

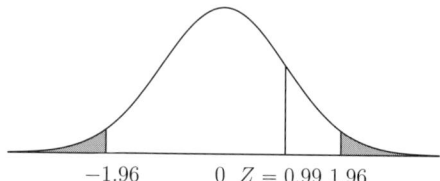

The test statistic does not lie in the rejection region, and so we cannot reject the null hypothesis in favour of the alternative at the 5% level. In this case, the Canterbury data are consistent with the claim made by the research team.

9.1.3 *p*-value approach to hypothesis testing

An alternative testing procedure uses a *p*-value, also known as the *observed significance level*. Here, we alter the rejection region step of the hypothesis test.

The *p*-value is a specific probability that gives you an indication of whether the null hypothesis is likely to be true.

> The *p*-value is a probability that reflects how likely our sample was, based on a null hypothesis. If the *p*-value is small, we conclude that the sample was unlikely to arise if the null hypothesis is true, and we conclude the null hypothesis must be incorrect, i.e. we trust the data over our prior beliefs.

The *p*-value is defined to be the probability of getting a sample result as extreme or more extreme than the one we observed, if H_0 is assumed to be true. If this probability is large, we can conclude our observed test statistic was a likely one to obtain if H_0 were true, and we do not reject H_0. Conversely, a small *p*-value indicates that the sample was an unlikely one to get if H_0 were true, and so we reject H_0.

> If a *p*-value is smaller than α, we reject H_0 in favour of H_a. If not, we fail to reject H_0.

This indicates the usefulness of the *p*-value approach. If we have available the *p*-value for a hypothesis test, then we can immediately draw a conclusion for any value of α,

9.1 Large sample inference for a mean

simply by comparing α with the p-value. If we were to use the rejection region approach, each time we consider a different α, we would have to use the tables to find the new rejection region.

We calculate the p-value using the following technique.

- Draw the normal curve, and add to it the value of the test statistic.

- If the test is two-sided, add to the diagram the negative of the value of the test statistic.

- Mark on the curve the end(s) corresponding to the alternative hypothesis H_a (left for $<$, right for $>$ and both for $\neq$; you can think of the $<$ as an arrow left $\leftarrow$, and the $>$ as an arrow right $\rightarrow$).

- Shade from the marked end(s) to the marked value(s). If H_a is two-sided, your sketch should look like this:

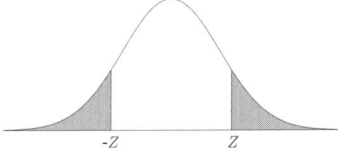

- Calculate the probability of the shaded region using the normal table. This area is the p-value.

Example 9.8

Q: We repeat the test in Example 9.6 using the p-value approach.
A: The hypotheses and test statistic are unchanged, i.e. we are testing $H_0 : \mu = 2.6$ against the alternative $H_a : \mu > 2.6$. The test statistic is

$$Z = \frac{\bar{x} - \mu}{s/\sqrt{n}} = \frac{3.3 - 2.6}{1.2/\sqrt{35}} = 3.45$$

and since $n = 35 > 30$, this has an approximate $N(0,1)$ distribution. We now calculate the p-value.
- We first draw the normal curve, and add to it the test statistic.

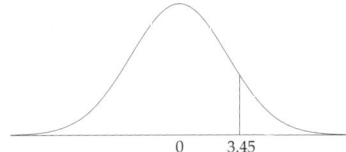

- Since the test is an upper-tail test, we mark the upper tail, and shade from that end toward the test statistic.

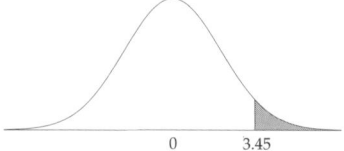

- From the normal table, the probability Z lies between 0 and 3.45 is $P(0 < Z < 3.45) = 0.4997$, and so the probability of the shaded area is p-value $= P(Z > 3.45) = 0.5 - 0.4997 = 0.0003 = 0.03\%$. The p-value is much smaller than $\alpha = 5\%$ and so we reject the null hypothesis in favour of the alternative at the 5% level.

Note that the first two steps of the hypothesis test are unchanged, and the conclusion we draw is the same. Only the third step, and the way we obtain the conclusion, alters.

To see why the p-value approach always gives the same conclusion as the test, it can be useful to draw a diagram. Referring to Figures 9.4 and 9.5, the test statistic can only lie in the rejection region when the p-value is smaller than α. When the test statistic lies outside the rejection region, the p-value must be larger than α. This leads us to reject H_0 only when the p-value is less than α.

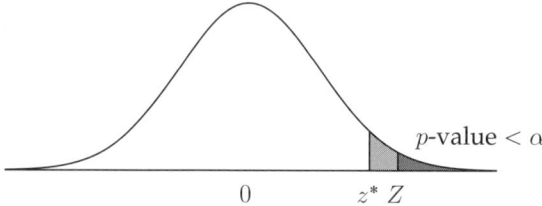

Figure 9.4 The test statistic Z lies in the rejection region, and the p-value (shaded in dark grey) is less than α

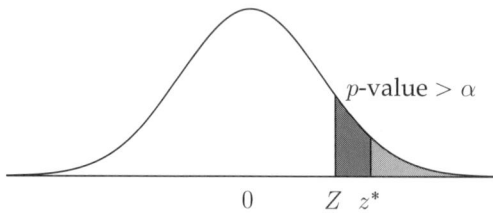

Figure 9.5 The test statistic Z does not lie in the rejection region, and the p-value is greater than α (shaded in light grey)

Statistical computer programs typically report p-values for any hypothesis tests they conduct on our behalf, leaving the conclusion up to the user. Users with different test levels (values of α) can easily draw their own conclusion based on the p-value. The rejection region approach requires users to know the test statistic, and be able to calculate the rejection region corresponding to their own personal choice of α.

In the situation where the test statistic is $N(0,1)$, we can use the standard normal table to determine the p-value (as in Example 9.8); however, in other situations, the tables are typically not comprehensive enough and we will not be able to calculate the p-value. In such cases, we use the rejection region approach to hypothesis testing.

There are two special situations that are worth noting due to the confusion they can cause.

Very large test statistic: When the test statistic is very large (positive or negative), this is strong evidence against H_0, since the observed result is many standard errors away from the null value of μ. Values of the test statistic larger than 3.99 do not appear in Table D.4 on page 329. However, we can see from the table that a probability of 0.5000 is reached at $z = 3.90$. Values of z larger than this will have larger probabilities since we are adding additional area. The total area is not exactly one half, but *it is* to 4 decimal places. The tail area (p-value) is therefore $0.5 - 0.5000 = 0.0000 = 0$ (4dp). For a two-sided test, we double this and still get zero.

Test statistic equal to zero: When the sample mean is exactly equal to the value of μ in H_0, both the numerator ($\bar{x} - \mu$) and the test statistic equal zero. The instructions above work, even though the diagram can look a bit strange. For a one-sided test, your sketch would look like this:

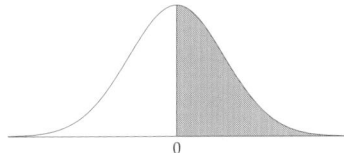

and your p-value would be exactly 0.5. This large probability indicates an entirely likely result. In other words, a z-score of 0 supports the claim that the mean as given in H_0 – in fact, you can't get better support of it. If the alternative is two-sided, we would shade from both ends of the curve, which would lead to the whole curve being shaded, and a p-value of 1. This is always the case: a z-score of 0 means your evidence totally supports the claim and is, in a sense, your target score.

9.1.4 Type I and type II errors

When we conduct a hypothesis test, we do not prove anything; rather, we identify support for a particular statement (either H_0 or H_a). There are four possible outcomes of a test, and these are shown in Table 9.1.

Test outcome	True situation	
	H_0 true	H_a true
reject H_0	Type I error	correct conclusion
fail to reject H_0	correct conclusion	Type II error

Table 9.1 Test conclusions

When the null hypothesis H_0 is true, we either draw the correct conclusion (which is exactly what we want to do), or we incorrectly reject it. This mistaken rejection of H_0 is called a Type I error. The probability of this occurring is exactly the level we choose for the test α. If we conduct a test at the 5% level, and if H_0 is true, there is a 5% chance we will reject H_0 anyway. We can reduce this error rate by reducing α, but this makes it harder for us to detect a shift from H_0 if H_0 is actually false.

If H_a is true, the correct conclusion is to reject H_0. If this is not done, we have made a Type II error. The probability of this occurring is usually denoted β, and we would like β to be as small as possible.

In practice, there is a trade-off between the two types of error. Each time we reduce the probability of one type of error, it is usually the case that the probability of the other type increases. The most effective way of making the probabilities as small as possible is through good test design; however, this is not an elementary issue, and we leave it for more advanced books.

9.2 Small and normal sample inference for a mean

When the sample size is no longer large, i.e. $n < 30$, this has two effects on the distribution of the test statistic. The first is that the CLT no longer applies, and we cannot presume the sample mean $\bar{X}$ is normal. The second effect is that replacing the population standard deviation σ by the sample estimate s is no longer without consequence.

If we want to draw valid conclusions based on a small sample, we can use very similar techniques to the large-sample ones *if* we can assume that the population from which the sample is drawn is normally distributed. Recall that when the sample was large, the CLT gave us normality for $\bar{X}$ no matter what the underlying population looked like. Without the CLT, $\bar{X}$ will only be normal if the population is itself normal, and we assume that this is the case in this section. In general, this assumption will need to be checked before we can proceed. In particular, the sample data should be

- symmetric
- unimodal
- and have no outliers.

9.2 Small and normal sample inference for a mean

If these conditions do not hold, we will be unable to assume the sample mean is normal, and we will have to use the technique in Section 9.3 rather than that which follows.

> When we assume that the population is normal, we must first *check* the validity of this assumption.

The common ways to check for normality are to draw a histogram or stemplot and look for the "bell" shape, or draw a boxplot and check for symmetry and absence of outliers.

If the population is normal, then it follows that $\bar{X}$ is normal, and that

$$\frac{\bar{X} - \mu}{\frac{\sigma}{\sqrt{n}}} \sim N(0, 1)$$

exactly. We will not usually be able to evaluate this test statistic because σ will be unknown. For a large sample, replacing σ by s does not unduly affect the distribution of the test statistic, and we can continue to assume that $(\bar{X} - \mu)/(s/\sqrt{n})$ is approximately $N(0, 1)$. In contrast, when the sample is *small*, we must replace the normal distribution with one called *Student's t-distribution*.

> When $n < 30$ and the population is normally distributed
>
> $$\frac{\bar{X} - \mu}{\frac{s}{\sqrt{n}}} \sim t_{n-1}$$
>
> where t_{n-1} is the t-distribution with $n - 1$ degrees of freedom.

The t-distribution requires a parameter called the *degrees of freedom parameter* and denoted ν, to identify it. For each value of ν there is a different t-distribution. Like the normal distribution, t-distributions are symmetric and unimodal; however, depending on the size of ν, a t-distribution may have values much further from the mean than we'd expect from a normal distribution, i.e. outliers.

The degrees of freedom ν are whole numbers $1, 2, \ldots$, and if the sample size is n, we have $n - 1$ degrees of freedom. For small ν, the t-distribution has very heavy tails (and extreme values are relatively likely), but as ν increases, the t-distribution becomes more and more like the normal. In fact, the t-distribution with $\nu = \infty$ is the standard normal distribution $N(0, 1)$. When $\nu \geq 30$, there is so little difference between the actual distribution (t_ν) and the $N(0, 1)$, that we can safely use the standard normal distribution as the distribution of the test statistic rather than the t-distribution. Figure 9.6 compares various t-distributions with the standard normal distribution. In that figure the tail behaviour of the t-distribution described above is apparent.

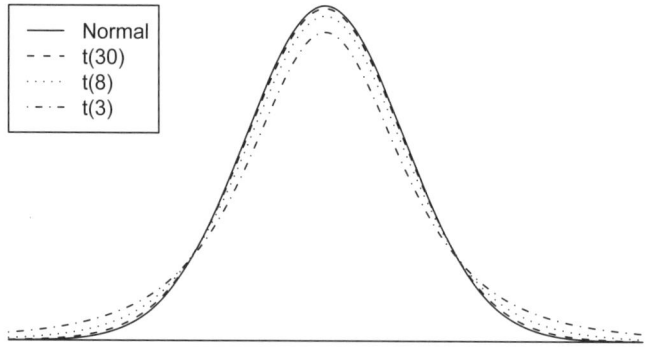

Figure 9.6 The normal distribution, and the t-distributions with 30, 8 and 3 degrees of freedom

9.2.1 Confidence intervals for μ

Calculation of a confidence interval for μ when we have only a small sample is very similar to the method used in Section 9.1.1. The only change is to use values from the appropriate t-distribution in place of values from the normal distribution. The values will come from the t-distribution table found on page 331 (Table D.6). We need to know the degrees of freedom, which will be $n - 1$ in this situation, and also the confidence level for the interval $100(1 - \alpha)\%$, usually 95%.

We need to choose the correct row of the t-distribution table according to the degrees of freedom that we have. Since a confidence interval is always two-sided, the correct column of the table is headed by α, and for a 95% confidence interval we choose the column designated by the "two-tail" probability value of $5\% = 0.05$. The final row of the table is for the normal distribution (with $\nu = \infty$ degrees of freedom), and we see the familiar 1.96 figure in the 5% two-sided column.

Example 9.9

Q: It is suspected that a bay has been polluted by an unknown pollutant. Some decrease in average cortex thickness of kina is suspected. To check this, a sample of 12 kina were collected from the bay and the thickness of each cortex was measured. The following measurements were obtained (in mm): 4.4, 5.4, 3.9, 5.7, 4.4, 3.2, 4.0, 3.7, 3.8, 4.8, 4.7, 4.3. Give a 95% confidence interval for the population mean thickness of the cortex.

A: The sample mean and standard deviation of these data are $\bar{x} = 4.51$ and $s = 1.10$ (both rounded to 2dp), with the sample size $n = 12$. Since $n < 30$ we must assume the data are normally distributed, and use a t-distribution with $n - 1 = 11$ degrees of freedom. From the t-table, the appropriate value for a 95% with 11 degrees of freedom is 2.201, and so the confidence interval is

$$\bar{x} \pm 2.201 \frac{s}{\sqrt{n}} = 4.51 \pm 2.201 \frac{1.10}{\sqrt{12}} = 4.51 \pm 0.70 = (3.81, 5.21).$$

9.2.2 Hypothesis tests for μ

The only aspect of the small sample hypothesis test that differs from the large sample technique in terms of *what we do* is the rejection region step. The formula for the test statistic remains unchanged; however, the distribution of the test statistic is different.

> The test statistic is
> $$T = \frac{\bar{X} - \mu}{\frac{s}{\sqrt{n}}}$$
> and this has a t-distribution with $n - 1$ degrees of freedom.

The t-tables are particularly easy to use for finding the rejection region (for this reason we may choose to use the t_∞ entry to find the rejection region in the large sample case).

This test is commonly referred to simply as a t-test.

Example 9.10

Q: The average (population) cortex thickness of healthy kina is 5.68 mm. Using the data from Example 9.9, test whether there is any evidence of a decrease in kina cortex thicknesses at this site.

A: We follow the four steps of a hypothesis test in order to check this claim.
- The hypotheses are $H_0 : \mu = 5.68$ against $H_a : \mu < 5.68$.
- The test statistic is

$$T = \frac{\bar{x} - \mu}{s/\sqrt{n}} = \frac{4.51 - 5.68}{1.10/\sqrt{12}} = \frac{-1.17}{0.32} = -3.67$$

and this has an approximate t-distribution with $n - 1 = 11$ degrees of freedom.
- The test is a lower tail (one-sided) test, and we choose $\alpha = 5\%$. Using the t-table with a one-tail probability of 0.05 and 11 degrees of freedom, we read off 1.796, and so the rejection region is $T < -1.796$.
- We compare the test statistic $T = -3.67$ with the rejection region $T < -1.796$.

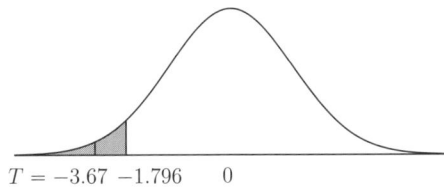

$T = -3.67 \quad -1.796 \quad 0$

The test statistic clearly lies in the rejection region, and so we reject the null hypothesis in favour of the alternative at the 5% level. It appears that the kina at the beach have lower cortex thicknesses.

9.3 Small and non-normal sample inference for a median

If we discover that our sample is not normal, for example by observing an asymmetric sample distribution or outliers, we risk drawing the wrong conclusion if we use a t-test. For non-normal data, the t-test is not at all suitable, and we can do much better using a non-parametric test called the *sign test*. This test is called non-parametric because we do not need to estimate any population parameters, e.g. $\bar{x}$ for μ, or s for σ. It is based on the binomial distribution.

The test is for a population median $\mathcal{M}$ rather than the population mean μ. If we assume the population is symmetric, then $\mu = \mathcal{M}$ and the sign test can be used to test a hypothesis about the mean. If the population is not symmetric, we focus on the population median instead. The test works by observing that the probability that a randomly selected observation is above the median is $\frac{1}{2}$, i.e. we expect half our observations to fall below the median and half above. Assuming that we can't get observations exactly equal to the population median, the number of observations in a sample that are above the population median has a binomial distribution with number of trials equal to the sample size, and probability of success equal to 0.5.

Hypotheses

We test the null hypothesis $H_0 : \mathcal{M} = m_0$ against the alternative $H_a : \mathcal{M} < m_0$ or $H_a : \mathcal{M} > m_0$ or $H_a : \mathcal{M} \neq m_0$. If the null hypothesis is true, then the number of observations in the sample above m_0 has a binomial distribution with probability of success $p = 0.5$.

Test statistic

We assume that the population is continuous, and then, theoretically, the probability that an observation equals m_0 is zero. In practice, we *can* get observations equal to m_0, and because the binomial distribution allows only two possible outcomes for each trial, we must discard these observations. Since we can only use observations above or below m_0, we may find our sample size is reduced.

> Call the reduced sample size n, then for a one-sided test the test statistic is
>
> $$X = \text{the number of observations greater than } m_0.$$
>
> For a two-sided test, the test statistic is
>
> $$X = \text{the smaller of} \begin{cases} \text{the number of observations greater than } m_0 \\ \text{the number of observations less than } m_0. \end{cases}$$
>
> The test statistic has a binomial distribution with parameters n and $p = 0.5$.

p-value

Because the test statistic is binomial, we use the *p*-value approach. As usual, the *p*-value depends on the test statistic, and the alternative hypothesis, and in this case, the binomial distribution. The *p*-value is

$$p\text{-value} = \begin{cases} P(Bin(n, 0.5) \geqslant X) & \text{if } H_a : \mathcal{M} > m_0 \\ P(Bin(n, 0.5) \leqslant X) & \text{if } H_a : \mathcal{M} < m_0 \\ 2 \times P(Bin(n, 0.5) \leqslant X) & \text{if } H_a : \mathcal{M} \neq m_0 \end{cases}$$

where $Bin(n, 0.5)$ is a binomial random variable with parameters n and $p = 0.5$.

Notice that, for the one-sided tests, the sign in the *p*-value calculation matches the sign in H_a, i.e. $\geqslant$ with $>$ and $\leqslant$ with $<$. For the two-sided test we have to collect probability from both ends of the distribution. Since the binomial distribution with $p = 0.5$ is symmetric, we can simply multiply one of the tail probabilities by two.

In each case, the *p*-value is calculated from values which signal a move away from H_0 toward H_a, where the starting point is determined by the test statistic. For example, if H_a has $\mathcal{M} < m_0$ and the test statistic is $X = $ the number of observations above m_0, then if H_a is actually true, we would expect the test statistic to be too small (the true median is lower, so the probability of being greater than m_0 is less than 0.5). The *p*-value is the probability of the test statistic and smaller values, i.e. *p*-value $= P(Bin(n, 0.5) \leqslant X)$.

Conclusion

As with other tests using the *p*-value, we draw conclusions by comparing the *p*-value with the level of the test α. Usually $\alpha = 5\%$, and if the *p*-value is smaller than this, we are unlikely to observe the sample result, given H_0 is true. For small *p*-values, we reject H_0 in favour of H_a. Conversely, if the *p*-value is larger than α, the probability of

Example 9.11

Q: Closer examination of the kina data in Example 9.9 indicates the presence of an outlier. The stem plot for these data is

```
3 | 2 7 8 9
4 | 0 3 4 4 7 8      Key: 3 | 2 = 3.2mm
5 | 4
6 |
7 | 5
```

with an obvious outlier. Thus, we should not have assumed normality for the cortex thickness. Re-test the earlier hypotheses using the sign test.

A: Since the data contains an outlier, the earlier test is invalid.

- We now have the null hypothesis $H_0 : \mathcal{M} = 5.68$, which we wish to test against the alternative $H_a : \mathcal{M} < 5.68$.
- To find the test statistic, we compare the observations to $m_0 = 5.68$, as in the following table, where we have recorded a plus (+) if the observation is above m_0 and a minus (−) if it is below.

4.4	5.4	3.9	7.5	4.4	3.2	4.0	3.7	3.8	4.8	4.7	4.3
−	−	−	+	−	−	−	−	−	−	−	−

Since the test is one-sided, counting the number of observations above $m_0 = 5.68$, the test statistic is $X = 1$ and this has a binomial distribution with $n = 12$ and $p = 0.5$.

- Since $H_a : \mathcal{M} < 5.68$, the p-value is

$$P(Bin(12, 0.5) \leqslant 1) = 0.0002 + 0.0029 = 0.0031 = 0.31\%$$

where we add the probabilities of getting 0 and 1 from the binomial table.

- The p-value 0.31% is less than $\alpha = 5\%$, so as for the t-test performed earlier, we reject the null hypothesis in favour of the alternative at the 5% level.

Example 9.12

Q: Redo Example 9.11 for a two-sided alternative.

A: Here, the hypotheses become $H_0 : \mathcal{M} = 5.68$ against $H_a : \mathcal{M} \neq 5.68$. The number of observations above 5.68 is 1, and the number below is 11, and the test statistic is the smaller of these, i.e. $X = 1$. The p-value is

$$2 \times P(Bin(12, 0.5) \leqslant 1) = 2 \times 0.0031 = 0.62\%$$

and since this is smaller than 5%, we again reject the null hypothesis in favour of the alternative.

9.4 Inference for a proportion

In Sections 9.1 and 9.2 we have discussed estimation and testing for a population mean. A related issue is estimation and testing for a population proportion. As with means, proportions are widely used in many disciplines, and it is natural to gather sample information to better understand them. For the time being, we will assume that we have a large sample, with $n \geqslant 30$.

In Section 8.4 we explained why the CLT applies to a binomial random variable. Essentially, counting successes is equivalent to summing them, and so the number of successes in a binomial situation can be thought of as a sum, and since the CLT applies to sums, the CLT applies to the binomial random variable.

The proportion, p, of a population with some characteristic, is naturally estimated using the proportion in a sample with that characteristic, $\hat{p}$, defined to be

$$\hat{p} = \frac{X}{n} = \frac{\text{the number of successes in the sample}}{\text{the number of observations in the sample}}$$

where "success" is defined to be having the characteristic of interest, and where X is the binomial random variable. We know from Section 7.4 that the sample proportion has mean p and standard deviation $\sqrt{p(1-p)/n}$. As a consequence of the CLT, provided n is large enough, $\hat{p} \sim N(p, p(1-p)/n)$ and so standardising $\hat{p}$ we have

$$\frac{\hat{p} - p}{\sqrt{\frac{p(1-p)}{n}}} \sim N(0, 1).$$

We use this statement to construct the form of a confidence interval for p, and a hypothesis test for p, parallel with their development for the mean.

Since p is a proportion, it will always lie between 0 and 1, and the same holds for $\hat{p}$. A normal random variable does not have any constraints on it, so we need to check that "most" of the probability is for the region between 0 and 1. Alternatively, we require the probability of $\hat{p} < 0$ or $\hat{p} > 1$ to be approximately zero, i.e. very small. Conditions which ensure the normal approximation is reasonable are

$$0 < \hat{p} \pm 3\sqrt{\hat{p}(1-\hat{p})/n} < 1 \quad \text{and} \quad 0 < p \pm 3\sqrt{p(1-p)/n} < 1.$$

These conditions ensure neither p nor $\hat{p}$ is too close to 0 or 1. If we are estimating p, we will not be able to check the second set of conditions, but in this case, checking the conditions based on $\hat{p}$ is good enough.

9.4.1 Confidence Intervals for p

In Section 9.1, the standardised score $Z = (\bar{X} - \mu)/(\sigma/\sqrt{n})$ gave a confidence interval for μ of the form $\bar{X} \pm Z\sigma/\sqrt{n}$. Remembering the development of the confidence interval, we began with the standardised score in-between $\pm Z$ and rearranged the expression to leave μ between the confidence interval limits $\bar{X} \pm Z\frac{\sigma}{\sqrt{n}}$. Comparing, we have a standardised score equal to

$$\frac{\hat{p} - p}{\sqrt{\frac{p(1-p)}{n}}} \qquad \text{instead of} \qquad \frac{\bar{X} - \mu}{\frac{\sigma}{\sqrt{n}}}$$

and it follows that the confidence interval for p is given by

$$\hat{p} \pm Z\sqrt{\frac{p(1-p)}{n}} \qquad \text{instead of} \qquad \bar{X} \pm Z\frac{\sigma}{\sqrt{n}},$$

i.e. the confidence interval is once again of the form: (sample value) $\pm$ (value from the normal distribution) $\times$ (standard error of the sample value).

A careful look at the form of the confidence interval for p indicates an interesting problem: to calculate the interval, we need to use p, the parameter we are trying to estimate, in the standard error. Clearly, we will not have p (otherwise we wouldn't be bothering to estimate it with the confidence interval). A sensible thing to do would be to replace the population proportion with its sample estimate, giving an approximate confidence interval.

> An approximate 95% confidence interval for p is given by
>
> $$\hat{p} \pm 1.96\sqrt{\frac{\hat{p}(1-\hat{p})}{n}}.$$

Strictly speaking, when we use $\hat{p}$ in place of p in the standard error, we should replace n in the standard error by $n - 1$. Since n is large (and the numerator of the standard error is small), this doesn't have a very large effect on the confidence interval limits, so we ignore this correction. Different confidence levels are accommodated by changing 1.96 to the appropriate value from the normal table.

Instead of replacing p by $\hat{p}$ in the standard error, we could take a conservative approach, and this is to replace p by 0.5. We note that $0 \leqslant p \leqslant 1$, and in this range, the graph of $p(1-p)$ is shown in Figure 9.7.

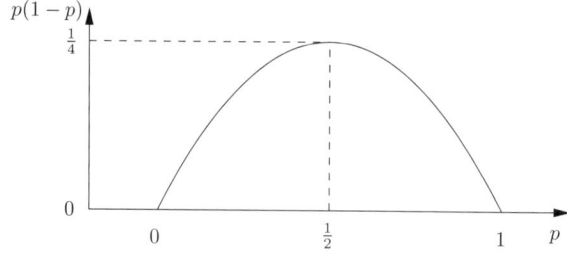

Figure 9.7   The function $p(1-p)$ for $0 \leqslant p \leqslant 1$

There we see that the value of $p(1-p)$ ranges from a low of 0 (at $p = 0$ and $p = 1$) to a high of 0.25 (at $p = 0.5$). If we use $p = 0.5$ in the standard error, $p(1-p)$ takes its largest possible value, as does $p(1-p)/n$ and as does the standard error. Since the standard error is at its maximum, the confidence interval is as wide as it can be. Since it is possibly wider than it should be, it will contain more than $100(1-\alpha)\%$ of the probability, and so it is regarded as a conservative interval. This will be fine when the actual (unknown) p is close to 0.5, but might be way too wide when p is actually close to 0 or 1.

Example 9.13

Q: A programme is implemented to support small businesses in their first years of operation. A sample of 60 firms is monitored until the end of their fifth year, with nine firms leaving the sample due to failure. Calculate a 90% confidence interval for p, the proportion of firms which fail within the first five years of operation, using $\hat{p}$ in the standard error. Compare this with a conservative confidence interval using the maximum standard error.

A: Since $n = 60 \geq 30$ is large, and since $\hat{p} = 9/60 = 0.15$ is not too close to zero ($\hat{p} - 3\sqrt{\hat{p}(1-\hat{p})/n} > 0$), we can use the normal approximation for $\hat{p}$. Using $\hat{p}$ in the standard error, the 90% confidence interval for p is

$$\hat{p} \pm 1.645\sqrt{\frac{\hat{p}(1-\hat{p})}{n}} = 0.15 \pm 1.645\sqrt{\frac{0.15 \times 0.85}{60}} = 0.15 \pm 0.076 = 0.074, 0.226.$$

Using a more conservative $p = 0.5$ in the standard error, the CI for p is

$$\hat{p} \pm 1.645\sqrt{\frac{0.5 \times 0.5}{n}} = 0.15 \pm 1.645\sqrt{\frac{0.25}{60}} = 0.15 \pm 0.106 = 0.044, 0.256.$$

Because $\hat{p}$ is not close to 0.5, the conservative interval is quite a bit wider than the one based on $\hat{p}$.

9.4.2 Large sample hypothesis testing for p

The basic procedure involved in hypothesis testing for a proportion is very similar to the other hypothesis tests we have performed. Again, we rely on standardising the sample estimate, and in this case, the CLT, to provide normality for this standardised score. We follow the same four steps as in the previous cases.

Hypotheses

We test the null hypothesis $H_0 : p = p_0$ against the alternative $H_a : p < p_0$ or $H_a : p > p_0$ or $H_a : p \neq p_0$. Which of these three alternatives we choose will be determined by prior knowledge of the situation, and we should complete this stage before analysing the sample information.

Test statistic

Prior to calculating the test statistic we need to collect the sample information. We need to determine the number of "successes" out of the n observations, X, and from this determine the sample proportion $\hat{p} = X/n$.

> The test statistic is
> $$Z = \frac{\hat{p} - p}{\sqrt{\frac{p(1-p)}{n}}}$$
> and this has a standard normal distribution.

In order to calculate the test statistic, we use $\hat{p}$, and everywhere we see p in the formula for the test statistic, we use p_0 from the null hypothesis. This is because H_0 gives us some idea of the likely value of p, and since we have it, we use it! Notice that this is quite different from the standard error in the confidence interval, where we were forced to use either $\hat{p}$ or the conservative value $p = 0.5$.

Rejection region or p-value

As with the large sample test of a mean, the third stage of the hypothesis test may consist of forming a rejection region, or calculating a p-value. To form the rejection region, as before, we need to take into account the alternative hypothesis H_a, and the level of the test α, and use the normal distribution to find the critical values.

If we use the p-value approach, we use the method outlined on page 181. Once we have calculated the test statistic, this has a standard normal distribution, and it is immaterial that it is based on proportions. The p-value is calculated in *exactly* the same way as before.

Conclusion

The conclusion is drawn by comparing the test statistic with the rejection region, or comparing the p-value with α. If the test statistic lies in the rejection region, or if the p-value is smaller than α we reject the null hypothesis in favour of the alternative at the $100(1 - \alpha)\%$ level. Conversely, if the test statistic does not lie in the rejection region, or the p-value exceeds α, we cannot reject H_0.

Example 9.14

Q: Following from Example 9.13, the support agency would like to use the sample data to judge the success of the programme. Prior to its implementation, approximately 23% of small businesses would fail within the first five years of business. Do the data show a significant improvement in this failure rate?

A: The hypotheses we must test are $H_0 : p = 0.23$ against $H_a : p < 0.23$, since the agency is looking for an improvement, i.e. a smaller proportion of firms failing. The test statistic is

$$Z = \frac{\hat{p} - p}{\sqrt{p(1-p)/n}} = \frac{0.15 - 0.23}{\sqrt{0.23 \times 0.77/60}} = \frac{-0.08}{0.054} = -1.47$$

where we have used $p = 0.23$ from H_0 in the standard error. The p-value for this one-sided test is $P(Z < -1.47) = 0.5 - 0.4292 = 0.0708 = 7.08\%$. We would reject H_0 at the 10% level, but not at the 5% level (since $5\% < 7.08\% < 10\%$). The data do not show a significant reduction in failures at the 5% level.

9.4.3 Small sample hypothesis testing for p

When the sample size is small, we cannot use the CLT to provide approximate normality for $\hat{p}$. In this case, we can appeal directly to the fact that the number of "successes" in the sample has a binomial distribution. The test we use is similar to the small sample test of a median described in Section 9.3, except that instead of the $Bin(n, 0.5)$ distribution, we use the $Bin(n, p_0)$ where p_0 comes from H_0.

Hypotheses

Writing the hypotheses is exactly the same as for a large sample test, i.e. we test the null hypothesis $H_0 : p = p_0$ against the alternative $H_a : p < p_0$ or $H_a : p > p_0$ or $H_a : p \neq p_0$.

Test statistic

The test statistic is similar to the small sample test of a median.

> The test statistic is
>
> $$X = \text{the number of successes in the sample.}$$
>
> For a two-sided test, the test statistic is
>
> $$X = \text{the smaller of} \begin{cases} \text{the number of successes} \\ \text{the number of failures} \end{cases}$$
>
> and this has a binomial distribution with parameters n and probability of success p_0.

The distribution of the test statistic depends on the probability p_0 given in the null hypothesis.

p-value

The p-value is calculated using the binomial distribution with n equal to the number of observations in the sample, and probability of success, p_0 given by the null hypothesis. As with the test for the mean, we need to take into account the alternative hypothesis, and the p-value is

$$\text{p-value} = \begin{cases} P(Bin(n, p_0) \geq X) & \text{if } H_a : p > p_0 \\ P(Bin(n, p_0) \leq X) & \text{if } H_a : p < p_0 \\ P(Bin(n, p_0) \leq X) + P(Bin(n, p_0) \geq n - X) & \text{if } H_a : p \neq p_0 \end{cases}$$

where $Bin(n, p_0)$ is a binomial random variable with parameters n and p_0. Notice that the p-value for the two-sided test is slightly different from that in the previous situation. This is because unless $p_0 = 0.5$, the binomial distribution is not symmetric, and if this is the case, we cannot simply double a convenient tail probability. Note that $n - X$ can be thought of as the number of failures we have observed.

A difficulty remains if the p_0 in our null hypothesis is not available in the tables. In this case, we should use the closest available probability of success, and report an approximate p-value, or use computer software to evaluate it for us exactly.

Conclusion

The conclusion is drawn by comparing the p-value with α. If the p-value is smaller than α, the chance of getting such an extreme sample result was slim (if H_0 is actually true), and so we reject the null hypothesis in favour of the alternative at the $100(1 - \alpha)\%$ level. Conversely, if the p-value exceeds α, we cannot reject H_0.

Example 9.15

Q: When John's daughter was born, she was the only daughter born to the eight couples who attended a series of ante-natal classes together. Does this sample provide evidence that the probability of having a girl in New Zealand is not 50%?

A: Since the sample is small, with $n = 8$ we must use the binomial distribution for the test. The hypotheses are $H_0 : p = 0.5$ against the two-sided alternative $H_a : p \neq 0.5$. In advance of collecting the sample data, we have no reason to assume that girls are more likely than boys, or vice versa. Since the test is two-sided, we take the smaller of the number of boys and girls (failures and successes) and so the test statistic is $X = 1$. The p-value is

$$P(Bin(8, 0.5) \leq 1) + P(Bin(8, 0.5) \geq 7) = 0.0039 + 0.0313 + 0.0313 + 0.0039 = 0.0704$$

and this is greater than $\alpha = 0.05$ and so we cannot reject the null hypothesis in favour of the alternative at the 5% level. Even though the result seems surprising, the small sample does not offer enough evidence to reject H_0.

9.5 Finite populations

The confidence intervals and test statistics we have calculated in previous sections of this chapter had an implicit assumption that the population from which sample data were drawn is very, very large (actually, infinite). If the population size is not large, since the sampling we do is typically *without replacement*, i.e. we do not include an individual more than once in the sample, we can actually end up including a large proportion of the population in our sample. As a consequence, the sample mean $\bar{X}$ ends up being a lot "closer" to the true value μ (or $\hat{p}$ to p) than we would expect from an infinite population. Since closeness is measured by standard deviation, we must amend the standard errors by including a *finite population correction factor* (FPCF).

> The finite population correction factor for a population of size N is
> $$\sqrt{\frac{N-n}{N-1}}$$
> and this applies to the standard error of $\bar{X}$ or $\hat{p}$.

For a finite population, the standard error of the sample mean becomes

$$se(\bar{X}) = \frac{\sigma}{\sqrt{n}}\sqrt{\frac{N-n}{N-1}}$$

and so the 95% confidence interval for μ becomes

$$\bar{X} \pm 1.96 \frac{\sigma}{\sqrt{n}}\sqrt{\frac{N-n}{N-1}}$$

and the test statistic becomes

$$\frac{\bar{X} - \mu}{\frac{\sigma}{\sqrt{n}}\sqrt{\frac{N-n}{N-1}}}$$

instead of the earlier versions without the FPCF. Similarly, when dealing with proportions, the standard error becomes

$$se(\hat{p}) = \sqrt{\frac{p(1-p)}{n}}\sqrt{\frac{N-n}{N-1}}$$

and we have

$$\hat{p} + 1.96\sqrt{\frac{p(1-p)}{n}}\sqrt{\frac{N-n}{N-1}} \quad \text{and} \quad \frac{\hat{p} - p}{\sqrt{\frac{p(1-p)}{n}}\sqrt{\frac{N-n}{N-1}}}$$

in the same way. No other changes need to be made to either confidence intervals or hypothesis tests.

The FPCF $\sqrt{(N-n)/(N-1)}$ is a number less than 1. To see this, we note that $N-n$ will always be less than $N-1$, since $n > 1$ is a sample size. When the sample size is large relative to the population size, $N - n$ will be small, and the FPCF will be quite a small number. As a result, the standard error of either $\bar{X}$ or $\hat{p}$ will be much smaller than the infinite population case. This is the behaviour we would hope for, since the sample becomes very close to the population, and the sample estimates become very close to the population parameters.

On the other hand, if the sample size is very small relative to the population size, $N - n$ is not that different from $N - 1$, and the FPCF is very close to 1. In that case the standard errors are not very different numerically from the standard, infinite population cases.

Example 9.16

Q: The ages of $n = 50$ individuals from a professional society with $N = 912$ members are collected. They have $\bar{x} = 43.7$ and $s = 9.2$ (both measured in years). Are the society members older than the population at large, which has a mean age of 31.6 years?

A: Since this is a finite population, we must use the FPCF for our test statistic. The hypotheses are: $H_0 : \mu = 31.6$ and $H_a : \mu > 31.6$. The test statistic is

$$Z = \frac{\bar{X} - \mu}{\frac{\sigma}{\sqrt{n}}\sqrt{\frac{N-n}{N-1}}} = \frac{43.7 - 31.6}{\frac{9.2}{\sqrt{50}}\sqrt{\frac{912-50}{912-1}}} = \frac{12.1}{1.301 \times 0.973} = 9.56.$$

This test statistic is much greater than the 1% critical value of $z = 2.3263$, and so we reject the null hypothesis in favour of the alternative at the 1% level. The society members are significantly older than the general population.

Example 9.17

Q: The sample of $n = 50$ professional society members described in Example 9.16 contained $X = 31$ females. Are females more highly represented in this society than in the population at large, where they constitute approximately 50%?

A: In this case the hypotheses are $H_0 : p = 0.5$ against $H_a : p > 0.5$. The sample proportion is $\hat{p} = X/n = 31/50 = 0.62$, and the test statistic is

$$\frac{\hat{p} - p}{\sqrt{\frac{p(1-p)}{n}}\sqrt{\frac{N-n}{N-1}}} = \frac{0.62 - 0.5}{\sqrt{\frac{0.5 \times 0.5}{50}}\sqrt{\frac{912-50}{912-1}}} = \frac{0.12}{0.071 \times 0.973} = 1.74.$$

The p-value for this test is $P(Z > 1.74) = 0.5 - 0.4591 = 0.0409$, so we would reject the null hypothesis at the 5% level, but not at the 1% level. There is fairly strong evidence that the society contains a higher proportion of women than the general population.

9.6 Margin of error and sample size

When we produce a confidence interval for a population parameter rather than simply a point estimate, we are taking into account what is called the *margin of error*, or *accuracy* of the sample estimate.

> A large sample confidence interval is of the form:
>
> point estimate $\pm Z \times$ standard error of the point estimate
>
> point estimate $\pm$ margin of error
>
> point estimate $\pm$ accuracy.

From the second representation it is clear that the margin of error depends on the Z-score (which in turn depends on α) and the standard error of our point estimate. The two standard errors we have encountered are

$$\frac{\sigma}{\sqrt{n}} \quad \text{for a mean, and} \quad \sqrt{\frac{p(1-p)}{n}} \quad \text{for a proportion.}$$

Since the standard error of the sample mean can be rewritten

$$se(\bar{X}) = \frac{\sigma}{\sqrt{n}} = \sqrt{\frac{\sigma^2}{n}}$$

the two standard errors are very similar, and in particular, they both feature $\sqrt{n}$ in the denominator. In the situation where we fix the margin of error, provided we have a prior estimate of σ^2 or $p(1-p)$, for any given α, we will be able to determine the sample size that is required to ensure the required margin of error.

Example 9.18

Q: Returning to Example 9.2 on children with "glue" ear, how large a sample would be needed to estimate the population mean age of affliction to within one month, at the 95% confidence level?

A: Assume that the population standard deviation is equal to the sample standard deviation $s = 1.2$ years. One month is $1/12 = 0.0833$ years, so $m = 0.0833$, where $m =$ the required margin of error. We must solve

$$\text{confidence interval margin of error} \leq m$$

$$\text{i.e.} \quad Z\frac{\sigma}{\sqrt{n}} \leq m$$

$$\text{or, using the sample data:} \quad 1.96\frac{1.2}{\sqrt{n}} \leq 0.0833.$$

The only unknown value in this expression is n, so we can rearrange for this value. Replacing $\leq$ by $=$, and using symbols, we have:

$$Z \times \sigma = \sqrt{n} \times m \quad \text{giving} \quad \sqrt{n} = \frac{Z}{m}\sigma \quad \text{and} \quad n = \left(\frac{Z}{m}\right)^2 \sigma^2.$$

So we need at least

$$n = \left(\frac{Z}{m}\right)^2 \sigma^2 = \left(\frac{1.96}{0.0833}\right)^2 \times 1.2^2 = 797.23$$

individuals. We cannot have a decimal number of children in the sample. If we round this down to 797, the margin of error is

$$Z\frac{\sigma}{\sqrt{n}} = 1.96 \times \frac{1.2}{\sqrt{797}} = 0.08331 > 0.0833$$

i.e. slightly larger than what we wanted. Rounding *up* to 798, we obtain a margin of error of

$$Z\frac{\sigma}{\sqrt{n}} = 1.96 \times \frac{1.2}{\sqrt{798}} = 0.08326 < 0.0833$$

which is within the margin of error we have specified. We need to sample 798 children to estimate μ to within one month.

You can perform the necessary algebra each time you compute the required sample size for a given margin of error, or you can learn a formula for the sample size.

> For a given margin of error m, the sample size must satisfy
>
> $$n \geqslant \left(\frac{Z}{m}\right)^2 \sigma^2$$
>
> for a mean. If we are estimating a population proportion, we replace σ^2 by $p(1-p)$.

Typically, when we calculate the required n, this will be a decimal number. Obviously, sample size must be a whole number, so to ensure the margin of error, we must *round our answer up*, as demonstrated in Example 9.18. It is tempting to use proper rounding (i.e. round down if the decimal is less than 0.5) but if we do this we will find that our margin of error is slightly above what was specified.

In the case of a finite population, the standard error contains n in two places, and the algebra is a little more messy.

> For a finite population of size N, and for a given margin of error m, the required sample size is approximately
>
> $$n \geqslant \frac{n^*}{\left(1 + \frac{n^*}{N}\right)} \quad \text{where} \quad n^* = \left(\frac{Z}{m}\right)^2 \sigma^2$$
>
> for a mean. If we are estimating a population proportion, we replace σ^2 by $p(1-p)$.

We note that n^* for a finite population is exactly the required sample size for an infinite population. We correct this size on account of the finite population, by performing the additional calculation. As before, we round the sample size up to the next whole number.

Example 9.19

Q: Referring back to Example 9.17, how large a sample is needed to estimate the proportion of women in the population to within $\pm 2\%$, at the 5% level?

A: The sample estimate was $\hat{p} = 0.62$ from a population of $N = 912$ people. We first calculate n^*, the required sample size in the case of an infinite population. To do so, we use the conservative value of $p = 0.5$. Using $m = 2\% = 0.02$, we find

$$n^* = \left(\frac{Z}{m}\right)^2 p(1-p) = \left(\frac{1.96}{0.02}\right)^2 \times 0.5 \times 0.5 = 2401.$$

Clearly this figure is a lot bigger than the whole population, and we need to correct it. The corrected sample size is

$$n = \frac{n^*}{\left(1 + \frac{n^*}{N}\right)} = \frac{2401}{1 + \frac{2401}{912}} = \frac{2401}{1 + 2.633} = 660.95$$

so we need to sample at least 661 of the society members. We can check the margin of error, by calculating

$$1.96\sqrt{\frac{p(1-p)}{n}}\sqrt{\frac{N-n}{N-1}} = 1.96 \times \sqrt{\frac{0.5 \times 0.5}{661}} \times \sqrt{\frac{912-661}{912-1}} = 1.96 \times 0.0194 \times 0.5249$$

$= 0.01996$, which is within the required 2% figure.

Exercises

Large sample inference for a mean

We suggest you use $\alpha = 5\%$ for these tests.

9.1 Researchers conducting a study of the semi-aquatic skink *Oligosoma suteri* collected various morphological measures before their experiments. 62 adult males were weighed, with sample mean $\bar{x} = 14.6$ grams and standard deviation $s = 2.4$ grams.
  (a) Form a 95% confidence interval for the population mean weight of an adult male skink.
  (b) Give a careful interpretation of this interval.
  (Data from Miller, Hare & Nelson, Do alternate escape tactics provide a means of compensation for impaired performance ability? *Biological Journal of the Linnean Society*, 2010)

9.2 Following Exercise 6.16, the average number of ambiguous words recalled from a list of nine words for the 100 students was 6.63, with a standard deviation of 1.33 words.
  (a) Justify use of the binomial distribution for the number of words

recalled by each student. What are the parameters of this distribution?
(b) Why does the central limit theorem apply to the average number of words recalled by the students, even though there were only nine words in the original list?
(c) Give a 95% confidence interval for the population mean number of words recalled.

(Data from Takarangi, Polaschek, Hignett & Garry, Chronic and temporary agression causes hostile false memories for ambiguous information, Applied Cognitive Psychology, 2008)

9.3 The television watching habits of a sample of 40 overweight and obese adults were monitored. The subjects were asked to self-report TV viewing time in response to the question "How many hours do you watch TV per day, on average", while electronic monitors were used to calculate their actual viewing. Subjects underestimated their actual viewing by 0.6 hours per day on average, with a standard deviation of 2.3 hours per day.
(a) Form a 95% confidence interval for the population mean understatement.
(b) Is there evidence of understatement *on average*?

(Research published in Obesity, 2009)

9.4 The 2006 New Zealand census reports a median income for the 15-24 year age group of $11,525. A random sample of 150 university students has a mean of $6250 and a standard deviation of $3100.
(a) What assumption is necessary in order to conduct a test with $H_a : \mu < 11525$?
(b) Conduct the test mentioned in (a).

(c) What conclusions can you draw from your test? What is the population?

9.5 My 45 most recent Facebook status updates have been "Liked" by on average 2.3 friends, with a standard deviation of 2.5 friends. Is this number significantly greater than one?

9.6 A random sample of 110 New Zealanders were asked about their domestic holiday patterns in the previous year. The following table summarises the number of holidays taken:

Holidays	Frequency
0	13
1	30
2	29
3	19
4+	16

(a) Assuming a suitable representative value for the "4+" category is 4.5, estimate the sample mean and standard deviation of the number of trips.
(b) Form a 95% confidence interval for the population mean number of domestic holidays.
(c) Do New Zealanders take more than one domestic holiday per annum, on average?
(d) How might the year of the survey impact the conclusions of the test?

(Data from Schänzel, Family time and own time on holiday: generation, gender, and group dynamic perspectives from New Zealand, VUW thesis, 2010)

9.7 During a study monitoring a donation behaviour at City Gallery Wellington, a prominent donation box was emptied by gallery staff after every donation. In total 104 donations were observed, and these had average size $1.97, with standard deviation $1.60.
(a) The donation distribution is not normal: it is discrete, and very

asymmetric. Why does this not preclude us using the normal distribution for $\bar{x}$?

(b) Form a 95% confidence interval for μ, the population average donation.

(c) When the donation box is preloaded with a large number of coins, with total value $100, the average donation is approximately $1.70. Is the average per donor higher in the empty box?

(d) When the donation box is preloaded with a number of notes, with total value $100, the average donation is approximately $2.30. Is the average per donor higher in the empty box?

(e) Why can we not make a recommendation to the gallery on the basis of the information given?

(From Martin & Randal, How is donation behaviour affected by the donations of others? Journal of Economic Behavior and Organization, 2008)

9.8 The average weight of the shells of 81 *Mytilus californianus* mussels at Piedras on the coast of California, USA is 7.01 grams, with a standard deviation of 2.15 grams.

(a) Form a 95% confidence interval for the average weight of the mussel shells at this location.

(b) Assuming a population mean weight of 7.9 grams at another location, does it appear that the mussel shells are lighter at Piedras?

(Thanks to Nicole Phillips for the data)

Small sample intervals and t-tests

We suggest you use $\alpha = 5\%$ for these tests.

9.9 The heights of the 2009 All Black backs were given in Exercise 3.22. The average height of a New Zealand adult male (25-34 years old) is given as 177.3cm. Is the average All Black back significantly taller than the average New Zealand adult male?

(Data from moh.govt.nz)

9.10 The following observations are the estimated payback periods (in years) for installation of aluminium-framed double glazing in a random sample of ten homes in Christchurch. These periods reflect the costs of installation, and energy savings from the insulation.

| 12.5 | 20.3 | 22.3 | 19.7 | 19.5 |
| 24.5 | 26.9 | 13.4 | 18.4 | 18.3 |

(a) Draw a boxplot of these data. Do you think a normality assumption is reasonable?

(b) Calculate a 95% confidence interval for the mean payback period for this sort of retro-fitted double glazing.

(c) How large a sample would be needed to estimate the mean payback period to within one year?

(Data from Smith, A cost benefit analysis of secondary glazing as a retrofit alternative for New Zealand homes, VUW thesis, 2009)

9.11 The growth rate in gross domestic product (GDP) describes the change in economic output of an economy. When this is negative, the economy is said to be in recession. The following quarterly time series data (in %) are New Zealand's GDP growth rates over a period which contained the "Global Financial Crisis" of 2007-2010.

	Q1	Q2	Q3	Q4
2006		0.00	0.10	0.30
2007	1.30	0.80	0.70	0.90
2008	−0.30	−0.60	−0.60	−1.10
2009	−0.80	0.10	0.30	0.90
2010	0.60			

(a) Draw a stemplot of these data. Comment on any relevant features.
(b) Is there evidence that the average growth rate over this period is positive? Use a t-test.
(c) What assumptions are necessary for the t-test to be valid? Do these appear to be met?

(Data from stats.govt.nz*)*

9.12 The following table contains the weight (in grams) of a random sample of 15 fledgling male kea from Aoraki/Mount Cook.

860	860	880	880	800
840	900	880	800	880
840	840	830	780	940

(a) Is there evidence that the mean weight of a fledgling male kea is more than 830 grams?
(b) Calculate a 95% confidence interval for the mean weight of a fledgling male kea.
(c) Are your hypothesis test result and confidence interval consistent regarding the mean weight of a fledgling male kea?

(Thanks to Clio Reid for the data)

9.13 Persistent coughing is debilitating and impairs quality of life. A study investigating the use of theobromine (a chemical present in cocoa) as a cough supressant measured the concentration of a cough-inducing chemical to induce five coughs in ten healthy adult males. The average (log)-concentration was $\bar{x} = 1.86$ with standard deviation $s = 0.58$.

(a) Form a 95% confidence interval for the population (log)-concentration.
(b) What assumptions are necessary for the confidence interval in (a) to be valid?
(c) Assume that extensive studies have shown the mean (log)-concentation needed to induce five coughs in someone untreated is $\mu = 1.43$. Is the average concentration for someone "treated" with theobromine significantly higher than for someone untreated?
(d) Codeine is an alternative cough supressant. Assume that extensive studies have shown the mean (log)-concentration needed in someone using codeine is $\mu = 1.59$. Is the average concentration for someone "treated" with theobromine significantly higher than for someone using codeine?
(e) Summarise your findings.

(Adapted from Usmani et al., Theobromine inhibits sensory nerve activation and cough, FASEB, 2004)

9.14 A random sample of 14 children with Autism Spectrum Disorder (ASD) were monitored while at play. The children were scored on a variety of measures, including imaginative drawing, at which they had an average score of 0.41 and a standard deviation of 0.23. A typical figure for typically developing children of the same age is 0.65.

(a) Calculate a 95% confidence interval for the mean score of children with ASD. Carefully interpret this interval.
(b) Does the interval in (a) contain the typically developing children's average score? What implications does this have?
(c) Formally test whether or not the ASD children's scores are significantly lower on average than the typically developing children's average.
(d) What care must be taken in using a confidence interval to test a one-

sided alternative hypothesis as in (c)?

(*Data from Den-Kaat, Cognitive underpinnings of symbolic pretend play and impossible entity drawings: imaginative ability in typical development and autism spectrum disorder, VUW thesis, 2008*)

9.15 The survivorship of populations of the native frog, *Leiopelma pakeka* transplanted from Maud Island to the Karori Wildlife Sanctuary, was studied. While the mainland sanctuary is predominantly pest-free, the transfer is not without risk. The following data are weights of individual frogs recaptured within a mouse-proof enclosure in the sanctuary.

```
4.1   4.3   4.4   4.8   5.0
3.9   4.9   4.8   4.9   5.1
5.7   5.1   5.7   6.8   6.0
```

(a) Give a 95% confidence interval for the population mean weight of frogs based on this sample. To what population does this apply?

(b) Assuming a mean weight of 5.7 g for adult frogs on Maud Island, are these translocated frogs significantly different in size from those in the parent population?

(*Data based on Lukis, Returning an endemic frog to the New Zealand mainland: Transfer and adaptive management of* Leiopelma pakeka *at Karori Sanctuary, Wellington, VUW thesis, 2009*)

9.16 Like reef-building corals, some temperate anemones harbour symbiotic microalgae called "zooxanthellae". Algal photosynthetic products are absorbed by the host. At 15°C, the peak photosynthetic functioning is 600. In a study of the effects of climate change on this temperate symbiosis, five anemones were exposed to a constant water temperature of 25°C. The mean peak photosynthetic functioning of these anemones was 427, and the standard deviation was 85.

(a) Form a 95% confidence interval for the mean peak photosynthetic function of anenomes in 25°C water.

(b) What assumptions are necessary for your confidence interval to be valid?

(c) Is there sufficient evidence to indicate lower photosynthetic function at the warmer water temperature? Conduct a formal hypothesis test.

9.17 A study into tax evasion behaviour in New Zealand adults invited Flybuys members to play an online 'game', in which they were given fictitious income and asked to declare, none, some, or all of it for tax purposes. At the end of the game, total net income was calculated, being gross income less tax paid on declared income, less any fines on unpaid tax identified at random audits throughout the game. The following summary statistics are from 23 participants who shared the same gross incomes, tax rate, audit probability and audit fine level but who may have had different audit sequences. The average amount of income declared over eight rounds of the game was $26,915 with a standard deviation of $15,195.

(a) Assuming the amount of *declared* income is normally distributed, calculate a 95% confidence interval for the underlying population mean.

(b) Given the minimum declared income for these particpants was $0, and the maximum was $40,000, comment on the normal distribution assumption. Why is it necessary in this case?

(*Data thanks to Kevin Holmes, Lisa Marriott and John Randal*)

Sign test

We suggest you use $\alpha = 5\%$ for these tests.

9.18 The median age of women having a baby in New Zealand in 2009 was 29.9 years old, down from 30.2 in 2007 and 30.0 in 2008. The ages of a random sample of 10 mothers in 2010 were:

23.1	27.4	29.2	25.8	30.0
41.2	31.3	28.4	29.5	19.4

(a) Does this sample provide support to a further decrease in median age?
(b) Why is a t-test not appropriate for these data?

9.19 Below is a list of multiple EBIT figures for a random sample of firms in the Property sector:

-4.7	28.0	78.2	-36.1
-11.1	-19.7	-10.6	-15.4

(a) Use a sign test to comment on whether or not the average multiple EBIT for firms in the property sector is negative.
(b) Why do you think a sign test is appropriate in this instance?
(*From* pwc.com/nz/en/cost-of-capital)

9.20 On a virtual treasure hunt, subjects were given a choice to "dig" either on the left region of the screen, or the right. Reinforcing messages would appear more often on one of these choices than the other. Results were analysed to see if the subjects' behaviour was altered by the reinforcement. The following sensitivity estimates quantify the response of the 18 subjects, with zero indicating no sensitivity. Positive values are expected if the subjects respond to the messages.

0.53	2.17	0.41	0.34	0.15	-0.02
0.10	0.70	0.25	0.80	0.14	0.30
0.23	0.41	0.85	0.35	1.72	0.01

(a) Draw a stemplot of the sensitivity estimates. Show that there are outliers in the sample. Identify which observations these are.
(b) Perform a t-test if the population mean sensitivity is positive with the outliers present.
(c) Repeat the t-test with the outliers removed. Do they affect the conclusions?
(d) Finally, perform a sign test on the entire sample.
(e) Summarise the evidence presented by this sample.
(*Data from Lie, Harper & Hunt, Human performance on a two-alternative rapid-acquisition choice task, Behavioural Processes, 2009*)

9.21 A boutique cheese-maker sells "kilogram" blocks of cheese. A random sample of 12 blocks yields the following weights, rounded to the nearest 10 grams (1000 grams = 1 kilogram).

1060	1100	1070	1000
1070	1030	1000	990
1050	860	1030	960

(a) Does the average weight exceed 1kg? Use a sign test
(b) Repeat (a) using a t-test.
(c) Why might the sign test be more reliable in this instance?

9.22 The 2006 New Zealand census reports a median income for the 15-24 year age group of 11.5 (in $ thousands). Evidence is sought for an increase in population median income level for this age group prior to the 2011 census. A random sample of 16 individuals across New Zealand, in this age group has the following incomes (in $ thousands):

9.8	5.7	12.4	16.1
15.4	37.6	1.9	57.5
1.9	16.0	18.1	13.7
1.8	7.0	9.9	11.5

(a) Give two reasons why a *t*-test is inappropriate for the purpose.
(b) Conduct a sign test using these data.
(c) What conclusions can be drawn?

Inference for a proportion

We suggest you use $\alpha = 5\%$ for these tests.

9.23 In 2006, 29% of New Zealanders made at least one online purchase. Three years later, in a random sample of 250 New Zealanders, 108 had made at least one online purchase in the 2009 calendar year.
(a) Find a 95% confidence interval for the proportion of New Zealanders who made an online purchase in 2009.
(b) Test the hypothesis that the population proportion has increased between 2006 and 2009.
(Data from stats.govt.nz*)*

9.24 Teaspoons disappearing from communal kitchens is a worldwide phenomenon. A study published by the *British Medical Journal* monitored 54 generic stainless steel teaspoons over a period of 150 days. At the end of the study period, only 10 of these spoons remained. 25% of more expensive teaspoons are known to disappear over an equivalent period. Is the "theft" proportion significantly different for the generic spoons?
(Data from Lim, Hellard & Aitken, The case of the disappearing teaspoons: longitudinal cohort study of the displacement of teaspoons in an Australian research institute, BMJ, 2005)

9.25 Despite a "large latte" and a "large flat white" being identical (and what the French would call a *café au lait*), Wellingtonians use both terms. In a random sample of 50 orders of this drink, 32 of them were for a latte. Is the observed proportion significantly greater than 50%?
(Thanks to Claire at Sweet Fanny-Anne's for the data, and the latte)

9.26 A survey of New Zealanders' understanding of their consumer rights proposed "When you buy an extended warrantee with your new television, you'll have more protection than provided by the law." 340 out of the 1000 respondents correctly answered that this statement is false. Is this evidence of an increase in awareness since a (population) proportion of $p = 30\%$ was established?
(Data from consumeraffairs.govt.nz*)*

9.27 The 2009 "swine-flu" pandemic influenza A (H1N1) virus affected populations globally. In New Zealand there were 3211 confirmed (notified) cases including 1122 hospitalisations and 35 deaths. A report prepared for the Ministry of Health sheds light on patterns of infection and response to the virus. Infected people without symptoms have implications with regard to spread of the virus. A survey was conducted on 1696 New Zealanders. 347 survey respondents tested immune to the virus, and of these, 130 had observed no symptoms. Give a 95% confidence interval for the proportion of immune New Zealanders who show no symptoms of the virus. Is there evidence that the population proportion is less than 50%?
(Data from Bandaranayake, Bissielo, Huang & Wood, Seroprevalence of the 2009 influenza A (H1N1) pandemic in New Zealand, 2010)

9.28 In 2006, 75.2% of all New Zealand Māori iwi used English only, while 1% used Te Reo Māori only. In the same year, a random sample of 4500 Ngati Porou found 67.1% used English only. The others spoke Te Reo only, used NZ Sign Language only, or used more

than one of these three official languages.
(a) Form a 95% confidence interval for the proportion of all Ngati Porou who use English only.
(b) Conduct a formal hypothesis test to see whether fewer Ngati Porou use English only than the proportion for all Māori.
(*Data from* stats.govt.nz)

9.29 An experiment on anger levels in adolescents required participants to have experience playing *Quake II*. While 50% of the applicants were female, all but 13 females were excluded on account of inexperience with the game. The final sample consisted of these 13 females, and 94 males. Does this sample provide sufficient evidence that females make up a smaller proportion of gamers than in the general population?
(*Data from Unsworth, Devilly & Ward, The effect of playing violent video games on adolescents: should parents be quaking in their boots?, Psychology, Crime & Law, 2007*)

9.30 In the ten-year period 1994-2003, there were 175 deaths or serious injuries to trespassers on New Zealand's rail corridor. Of these, only 18 occurred in the evening period of 6 pm to 10 pm. This four-hour period is one-sixth of the day. Test whether the sample proportion is significantly less than this fraction. (*Data from* ltsa.govt.nz)

9.31 A study of two species of limpets (*Benhamina obliquata* and *Siphonaria australis*) examined their position relative to the high tide mark. While all of the 40 specimens of *B. obliquata* were above the high tide mark, only 21 out of 60 specimens of *S. australis* were in the low zone. 13 were submerged and 26 were in a tidal pool. Form a 95% confidence interval for the population proportion of *S. australis* living above the high tide mark.
(*Data from Russell & Phillips, Synergistic effects of ultraviolet radiation and conditions at low tide on egg masses of limpets* (Benhamina obliquata *and* Siphonaria australis) *in New Zealand, Marine Biology, 2009*)

9.32 In 2008 and 2009, the Reserve Bank of New Zealand conducted quality surveys of circulating bank notes. In 2008, a random sample of 400 $10 notes was collected, and in 2009, 400 $5 notes and 400 $20 notes were collected. These were analysed for wear and tear. 62% of the $5 notes were found to be in "good" quality (not poor, or very good). 71% of the $10 were in good quality, while 86% of the $20 were in good quality.
(a) Is it valid to use the normal approximation for each of these three proportions? Support your claim.
(b) If suitable, calculate 95% confidence intervals for the three true proportions of "good" quality bills, and draw these on a number line.
(c) Comment on your findings.
(*From Langwasser, Recent trends and developments in currency – 2009, RBNZ, 2010*)

9.33 Following from Exercise 9.3, 23 of the 40 subjects underestimated their TV viewing. Is the sample proportion significantly below 50%? Use a 5% level for the test.

9.34 In the 2006 Census, individuals are asked "which ethnic group do you belong to?", and given a list which includes "New Zealand European", and "Māori" among other choices. An increasing number of respondents select "Other" and report "New Zealander", and in 2006 the number was about three times as many as expected. Of the 429,429 people who gave this response, 222,702 were male. In contrast, in the entire population, there were 1,965,618 males out of 4,027,947 people. Test whether or not the proportion of people describing themselves as "New Zealanders" who are male differs from the population proportion of males.

(*Data from* stats.govt.nz)

9.35 In a 2008 survey of arts participation in New Zealanders aged 15 years and over, 48% of the 2099 individuals questioned had participated in some aspect of the arts. Is this proportion significantly less than 50%?

(*Data from* creativenz.govt.nz)

9.36 On census day 2006, approximately 91% of the 1.4 million New Zealanders who travelled to work on the roads (i.e. in private or work vehicles, public buses, motorcycle or bicycles) travelled in private vehicles. A random sample of 65 New Zealand car drivers was asked to describe the last incident which frustrated them on the roads, and what other vehicle was involved. Only 21 respondents were frustrated by other cars, with the remainder of the sample complaining about buses, cyclists and goods trucks. On the basis of this sample and the 2006 census figure, test whether car drivers' frustrations are disproportionately directed away from their own mode of transport.

9.37 Prior to the completion of the FIFA 2010 World Cup, "Paul the Octopus" gained notoriety for correctly picking the outcome of Germany's seven matches at the tournament, and the eventual winners, Spain.
(a) Is the sample evidence sufficient to suggest that the octopus, also known as the "Oracle of Oberhausen", actually has an ability to predict the outcome of football matches, or could it plausibly be guessing?
(b) Actually, Paul only rose to fame after correctly picking the first six games. Is the *additional* sample evidence (picking the next two games correctly) sufficient to suggest that the octopus actually has an ability to predict the outcome of football matches, or could it plausibly be guessing?
(c) Which of the two tests above is more reasonable in this instance, and why? (*Data from* stuff.co.nz)

9.38 Researchers investigating predation of the Pacific Sand Dollar *Dendraster excentricus* conducted a feeding trial, whereby a sample of starving Red Rock Crabs (*Cancer productus*) were presented with five each of small, medium and large sand dollars. In a random sample of ten crabs, five consumed a large sand dollar first, and five a medium sand dollar. Let p be the proportion of small sand dollars eaten by this species of crab when a variety of sizes are available. Test $H_0 : p = \frac{1}{3}$ against a suitable alternative. Is there evidence of selective predation?
(*Thanks to Jeff Radford for the data*)

Margin of error

9.39 In a small sample of the semi-aquatic skink, approximately 80% of adults had incomplete tails, indicating recent shedding. How large a sample would be needed to estimate the population proportion with recently shed tails to within $\pm 2\%$ at the 95% level?
(*Data from Miller, Hare & Nelson, Do alternate escape tactics provide a means of compensation for impaired performance ability? Biological Journal of the Linnean Society, 2010*)

9.40 Are video games a modern art-form? With developments in the hardware and operating systems of personal computers and game consoles, game designers have had an increasing ability to develop visually sophisticated virtual worlds, and both characters and plots within modern games are capable of causing emotional reactions in players akin to what has been experienced with more traditional art forms. A random sample of 30 non-gamers were shown an excerpt from *BioShock* demonstrating its "sumptuous world of fallen splendour." They were asked "is this art?" and 21 of the sample responded "yes".
  (a) Calculate a 95% confidence interval for the population proportion of non-gamers who would view this excerpt from *BioShock* as art.
  (b) What is the margin of error of your confidence interval in (a)?
  (c) How is your margin of error affected by: an increase in the sample size? An increase in the level of significance? A decrease in the observed sample proportion?
  (d) How many would be needed in the sample to estimate the population proportion to within 2% at the 95% level?
  (From Tavinor, The Art of Videogames, 2009)

9.41 "The Great Morality Debate" – an online survey of New Zealanders – asked "is it wrong for men and women to have sex before marriage?" A staggering 35% of the 9000 responses were "yes" although there were strong regional differences (only of 23% of respondents from Wellington answered in the affirmative).
  (a) Calculate a 95% confidence interval for the population proportion who believe it is wrong to have sex before marriage.
  (b) Assume that the sample data exclude Wellingtonians, and that the population proportion for Wellingtonians is $p = 23\%$. Is the sample proportion significantly higher?
  (c) The sample actually does contain responses from Wellington, but the data do not enable them to be removed. How do you think your result in (b) would change if you could disaggregate the sample data and exclude the Wellingtonian sub-sample?

9.42 *Te Rau Hinengaro: The New Zealand Mental Health Survey* assessed the mental health of 12,992 New Zealanders using the Composite International Diagnostic Interview (CIDI 3.0). The survey found 39.5% of the sample had met criteria for a DSM-IV mental disorder at some time in their life before interview.
  (a) Give a 95% confidence interval for the population proportion of New Zealanders meeting the criteria for a DSM-IV mental disorder at some time in their life.
  (b) How large a sample would be needed to estimate the proportion to within $\pm 0.1\%$ (approximately 4000 New Zealanders)?
  See also Exercise 10.43.
  (Data from moh.govt.nz)

9.43 You wish to estimate the population proportion of a reasonably uncommon condition to within $\pm \frac{1}{2}\%$. You take a pilot random sample of 20 individuals and none of them has the condition.
  (a) Why is the sample proportion not a value of p for the sample size calculation?
  (b) Why is using $p = 0.5$ in the sample size calculation not a good idea?

(c) Choose a suitable value of p, and determine the required sample size.

9.44 Tail shedding (caudal autotomy) is a defence mechanism against predation used by lizards when other methods do not work. Generally autotomy occurs within three vertebrae of where the tail is grasped, which aids in minimising the length of tail lost. A sample of 44 adult *Hoplodactylus maculatus* geckos was collected from Mana Island and one mainland site. The accuracy of tail loss (in millimetres, mm) for these geckos has sample mean 11.2 mm and standard deviation 9.3 mm.
(a) What assumptions are necessary to apply to the central limit theorem to this sample mean?
(b) Give a 95% confidence interval for the mean accuracy of autotomy for this gecko species.
(c) How many geckos would be required to estimate the mean accuracy of autotomy to within $\pm$ 1 mm?
See also Exercise 11.13.
(*Thanks to Kelly Hare for the data. From Hare & Miller, Frequency of tail loss does not reflect innate predisposition in temperate New Zealand lizards, Naturwissenschaften, 2010*)

9.45 In a sample of 600 young people 6 to 13 years old, only 16 had no rules regarding their television viewing.
(a) Is it reasonable to assume the sample proportion is normally distributed? Why or why not?
(b) Calculate a 95% CI interval for the population proportion of 6-13 year olds with no television restrictions.
(c) What sample size would be required to estimate the population proportion to within ± 0.01 at the 95% level? (*Data from* bsa.govt.nz)

9.46 Following Exercise 9.28, how many Ngati Porou would need to be sampled to estimate the proportion who use English only to within $\pm 1\%$? How would this change if the total number of Ngati Porou, 71,910, were taken into account?

9.47 Using the sample standard deviation from Exercise 9.15:
(a) How many frogs would be needed to estimate the population mean to within ± 0.1 grams?
(b) The total population size of *Leiopelma pakeka* is estimated to be 34,499. What effect, if any, does this have on your answer in (a)?
(*From Le Roux, The remnant* Leiopelma pakeka *population on Maud Island: population size, distribution and morphology, VUW thesis, 2008*)

Chapter 10

Two populations

In Chapter 9, the concepts of confidence intervals and hypothesis testing in the context of a single population were introduced. Confidence intervals were found for an unknown population characteristic known as a *parameter* (a mean or a proportion), and these gave us a likely range for that parameter. When we conducted hypothesis tests, we formed the null and alternative hypotheses on the basis of some background knowledge of the population from which we sampled. In this chapter, we consider *comparing* two populations, by comparing their means, proportions, or variances.

The framework we use is the same as in Chapter 9: when we have large samples we rely on the central limit theorem to provide normality, and when we have small samples, we assume normality if we can.

10.1 Large sample inference for means

In order to see whether two population means, μ_1 and μ_2, are the same, we would like to calculate the difference between them, $\mu_1 - \mu_2$, which will equal zero, if in fact they are the same. We will rarely be able to do this in practice, so we must sample. We collect data from each population, and as in Chapter 9, the sample mean is used as a point estimate for the population mean. If $\bar{X}_1$ is the sample mean for data from the first population, this estimates μ_1. Likewise, $\bar{X}_2$, the sample mean for data from the second population, estimates μ_2. A point estimator of the difference in the population means, $\mu_1 - \mu_2$, is the difference in the sample means, $\bar{X}_1 - \bar{X}_2$.

Provided the first sample is a large one, with $n_1 \geqslant 30$, we know that

$$\bar{X}_1 \sim N\left(\mu_1, \frac{\sigma_1^2}{n_1}\right)$$

where the normality follows from the CLT (this is approximate unless the X_i are exactly normal). Similarly, provided the second sample is large, with $n_2 \geqslant 30$, we know that

$$\bar{X}_2 \sim N\left(\mu_2, \frac{\sigma_2^2}{n_2}\right)$$

where again, approximate normality follows from the CLT. We are interested in the distribution of $\bar{X}_1 - \bar{X}_2$.

> Provided *both* n_1 and n_2 are large, then
> $$\bar{X}_1 - \bar{X}_2 \sim N\left(\mu_1 - \mu_2, \frac{\sigma_1^2}{n_1} + \frac{\sigma_2^2}{n_2}\right)$$
> approximately.

This result follows from the CLT, and is probably worth gaining some intuition for. In particular, we notice that while the mean of the normal distribution is the *difference* of the population means, the variance is the *sum* of the individual variances.

Instead of differencing $\bar{X}_1$ and $\bar{X}_2$, let's imagine adding $\bar{X}_1$ and $-\bar{X}_2$. For the moment, we shall think of a single normal random variable $X \sim N(\mu, \sigma^2)$. Since the normal distribution is symmetric, multiplying every value by -1 has the effect of reflecting the distribution in the vertical line at $x = 0$. This is demonstrated in Figure 10.1. We can see from this picture that while the mean of the new distribution is $-\mu$, the variance is still exactly the same, σ^2. We can be confident that the distribution of $-X$ is normal, with mean $-\mu$ and variance σ^2.

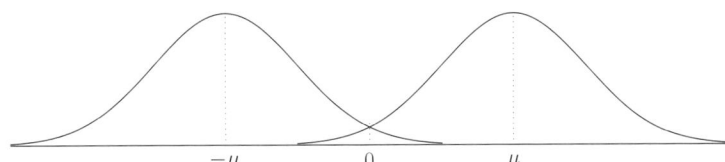

Figure 10.1 Distributions of X and $-X$

Now we need to think about what happens when we add two independent normal random variables. Ultimately, these will be the sample means from our two (independent) random samples. We learned earlier that if the population is normal, the sample mean is exactly normal, without the need for the CLT. The same is true for sums of independent normal random variables: the sum is also normal, with mean equal to the sum of the means, and variance equal to the sum of the variances. Applying this to $\bar{X}_1 - \bar{X}_2 = \bar{X}_1 + (-\bar{X}_2)$, we see that this is normal with mean $\mu_1 + (-\mu_2) = \mu_1 - \mu_2$ and variance $\sigma_1^2/n_1 + \sigma_2^2/n_2$, confirming our result.

Now that we know the distribution of $\bar{X}_1 - \bar{X}_2$, in developing the confidence intervals and test statistics as in the previous chapter, we make heavy use of the standardised random variable

$$Z = \frac{\bar{X}_1 - \bar{X}_2 - (\mu_1 - \mu_2)}{\sqrt{\frac{\sigma_1^2}{n_1} + \frac{\sigma_2^2}{n_2}}}$$

which is approximately $N(0, 1)$.

10.1.1 Confidence intervals for $\mu_1 - \mu_2$

In this case, we are interested in estimating the difference between the population means $\mu_1 - \mu_2$. The form of the confidence interval follows from the standardised score in exactly the same way as the confidence interval in the single population case did. In addition, its interpretation is the same. Comparing the standardised scores in the two population case with the single population case, we have

$$\frac{\bar{X}_1 - \bar{X}_2 - (\mu_1 - \mu_2)}{\sqrt{\frac{\sigma_1^2}{n_1} + \frac{\sigma_2^2}{n_2}}} \quad \text{instead of} \quad \frac{\bar{X} - \mu}{\frac{\sigma}{\sqrt{n}}}.$$

It follows that the confidence interval for $\mu_1 - \mu_2$ is given by

$$\bar{X}_1 - \bar{X}_2 \pm Z \sqrt{\frac{\sigma_1^2}{n_1} + \frac{\sigma_2^2}{n_2}} \quad \text{instead of} \quad \bar{X} \pm Z \frac{\sigma}{\sqrt{n}}.$$

To use this interval, we will typically need to assume that $\sigma_1 \approx s_1$ and $\sigma_2 \approx s_2$. Making this replacement, we have the confidence interval for $\mu_1 - \mu_2$.

> For n_1 and $n_2 \geqslant 30$
>
> $$\bar{x}_1 - \bar{x}_2 \pm 1.96 \sqrt{\frac{s_1^2}{n_1} + \frac{s_2^2}{n_2}}$$
>
> is a 95% confidence interval for $\mu_1 - \mu_2$.

Since the confidence interval gives us the likely range for $\mu_1 - \mu_2$, we can see whether or not specific values of $\mu_1 - \mu_2$ are likely by checking whether they lie in the confidence interval. The most common value we will be interested in is $\mu_1 - \mu_2 = 0$, which suggests that $\mu_1 = \mu_2$, i.e. the two populations have the same mean.

Example 10.1

Q: It is claimed that males committed for trial for minor offences are spending more time in prison on remand than females committed for trial for similar offences. A sample of 40 females and 49 males awaiting trial gave the following information where $X =$ the time on remand (in days):

	Sample mean	Sample standard deviation	Sample size
Females	16.3	14.6	40
Males	29.5	17.2	49

Use these data to estimate the difference in the population mean times for males and females.

A: Defining μ_1 = the population mean remand time for males, and μ_2 = the population mean remand time for females, a 95% confidence interval for $\mu_1 - \mu_2$ is:

$$\bar{x}_1 - \bar{x}_2 \pm 1.96\sqrt{\frac{s_1^2}{n_1} + \frac{s_2^2}{n_2}} = 29.5 - 16.3 \pm 1.96\sqrt{\frac{17.2^2}{49} + \frac{14.6^2}{40}} = 13.2 \pm 6.61 = 6.59, 19.81.$$

The population male remand time is likely to be within approximately 6.6 and 19.8 days longer than that for females.

10.1.2 Hypothesis tests for $\mu_1 - \mu_2$

While we can use a confidence interval to check whether individual values of the difference between the two population means are likely, it is much better – particularly when we have a one-sided alternative – to use a formal hypothesis test. As with the single population case, the hypothesis test has four stages: the hypotheses, the test statistic, the rejection region (or p-value) and the conclusion.

Hypotheses

The most commonly tested hypothesis is $H_0 : \mu_1 = \mu_2$ against one of the alternatives: $H_a : \mu_1 \neq \mu_2$, $H_a : \mu_1 < \mu_2$ or $H_a : \mu_1 > \mu_2$. It is convenient to write these in a different form, since the standardised score on which we will base our test statistic has $\mu_1 - \mu_2$ in it. Consequently, we write the null hypothesis as $H_0 : \mu_1 - \mu_2 = 0$. Note that we are not restricted to testing a zero difference, and we could replace "0" in the null hypothesis by any other value, e.g. we could test $H_0 : \mu_1 - \mu_2 = 12.2$, if this was well motivated by the situation.

Test statistic

Once again, the test statistic is based on standardising a normal random variable. In this case, that variable is the difference in the sample means. Since we assume both n_1 and n_2 are large, the CLT applies to both sample means and their difference is approximately normal. Further, for large samples, replacing σ^2 by s^2 is reasonable, and we do not need to change to a t-distribution.

For $n_1, n_2 \geq 30$, the test statistic is

$$Z = \frac{\bar{X}_1 - \bar{X}_2 - (\mu_1 - \mu_2)}{\sqrt{\dfrac{s_1^2}{n_1} + \dfrac{s_2^2}{n_2}}}$$

and this has an approximate standard normal distribution.

Rejection region and conclusion

Since the test statistic has an approximate standard normal distribution, the remaining stages are identical to the single large sample case. We obtain the rejection region from the $N(0,1)$ table, according to the alternative hypothesis H_a and the level of the test α. If the test statistic is in the rejection region, we reject the null hypothesis in favour of the alternative at the $100\alpha\%$ level. If the test statistic is not in the rejection region, we cannot reject H_0.

Since the normal distribution applies here, an alternative is to calculate the p-value for the test using the procedure described on page 181. If the p-value is smaller than α, then if H_0 is true, our sample is unlikely, and so we reject H_0. If the p-value is large (i.e. larger than α), then if H_0 is true the chance of getting the sample data is high, and so we cannot reject H_0.

Example 10.2

Q: Using the data from Example 10.1, test the claim that males and females committed for trial for minor offences are not spending the same amount of time in prison on remand on average. Find the p-value of this result, if in fact there is no overall difference across the country.

A: We wish to test $H_0 : \mu_1 - \mu_2 = 0$ against $H_a : \mu_1 - \mu_2 \neq 0$.

The test statistic is

$$Z = \frac{29.5 - 16.3 - 0}{\sqrt{\dfrac{17.2^2}{49} + \dfrac{14.6^2}{40}}} = \frac{13.2}{3.371} = 3.92.$$

To work out the p-value we sketch an $N(0,1)$ curve and mark the value of 3.92 on it. Since the test is two-sided, we add -3.92 to the plot, shade both ends, and calculate the shaded area. The diagram (clearly not to scale in this case) is as follows:

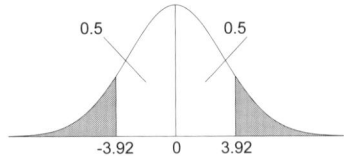

where $P(0 < Z < 3.92) = 0.5000 = 0.5$ from the table. There are two things to note here: the probability that Z lies between 0 and 3.92 is not actually equal to 0.5, but it is

0.5 to the 4dp that the tables report (a computer gives it as 0.4999557...). The p-value is $2 \times (0.5 - 0.5) = 0$ (4dp), and this is not exactly zero, but very close to it. What does it mean? There is almost no chance (0, to 4 decimal places) of getting this result, if it is true that the two means are equal. So, we reject the null hypothesis in favour of the alternative. This result is consistent with zero not falling in the confidence interval calculated in Example 10.1.

10.2 Small and normal sample inference for means

When either one or both of the sample sizes are small (i.e. less than 30), the CLT no longer holds, and we cannot guarantee normality for both sample means. As a consequence, we cannot assume that the *difference* in sample means is normal.

> For any small sample, we need to assume that the population from which it is drawn is normally distributed.

This assumption should be checked using plots of the sample data.

If populations for any small samples are normal, then

$$\frac{\bar{X}_1 - \bar{X}_2 - (\mu_1 - \mu_2)}{\sqrt{\frac{\sigma_1^2}{n_1} + \frac{\sigma_2^2}{n_2}}} \sim N(0, 1)$$

regardless of the sample sizes. However, if we have to replace the population variances by sample variances, in the case of small samples, this affects the distribution of the standardised score.

> In addition to assuming normality, we assume that the two populations have equal variances.

We provide a formal hypothesis test for this assumption in Section 10.3.

Having assumed that the two population variances are equal, we now wish to find the best estimate of the common value $\sigma^2 = \sigma_1^2 = \sigma_2^2$. We cannot simply combine the samples, because it is not certain that the population means are the same. If the means were the same, we would use

$$s^2 = \frac{\sum (X_i - \bar{X})^2}{n-1} = \frac{\sum_{i=1}^{n_1}(X_i - \bar{X})^2 + \sum_{j=1}^{n_2}(X_j - \bar{X})^2}{n_1 + n_2 - 1}$$

where we have split the combined sample back into its two parts, and $\bar{X}$ is the sample mean from the combined sample. Estimating a single sample mean is not a good idea, because the two populations might have different means, so instead, we use the individual sample means $\bar{X}_1$ and $\bar{X}_2$. In doing so, we need to reduce the degrees of freedom in the denominator by 1, because we have now estimated *two* sample means, and our remaining available degrees of freedom are $n_1 + n_2 - 2$. The resulting estimate of variance is

$$s_p^2 = \frac{\sum_{i=1}^{n_1}(X_i - \bar{X}_1)^2 + \sum_{j=1}^{n_2}(X_j - \bar{X}_2)^2}{n_1 + n_2 - 2}$$

where the X_i are from the first sample and the X_j are from the second sample. Going back to the original definition of the individual sample variances we see

$$\sum_{i=1}^{n_1}(X_i - \bar{X}_1)^2 = (n_1 - 1)s_1^2 \quad \text{and} \quad \sum_{i=1}^{n_2}(X_j - \bar{X}_2)^2 = (n_2 - 1)s_2^2$$

and so we can calculate the estimate using the sample variances we already have.

The *pooled variance* is given by

$$s_p^2 = \frac{(n_1 - 1)s_1^2 + (n_2 - 1)s_2^2}{n_1 + n_2 - 2}$$

and this estimates the common variance $\sigma^2 = \sigma_1^2 = \sigma_2^2$.

The pooled variance is a *weighted average* of the individual sample variances. The weight on the first variance is $(n_1 - 1)/(n_1 + n_2 - 2)$ and the weight on the second is $(n_2 - 1)/(n_1 + n_2 - 2)$ and these weights add to 1. Because s_p^2 is a weighted average of s_1^2 and s_2^2, it follows that s_p^2 always lies between s_1^2 and s_2^2. Also, it is always closer to the variance from the *larger* sample, since the larger the sample, the more accurate the point estimate, and the larger the weight we give to this estimate in our calculation.

Using the pooled variance, it follows that

$$T = \frac{\bar{X}_1 - \bar{X}_2 - (\mu_1 - \mu_2)}{\sqrt{\frac{s_p^2}{n_1} + \frac{s_p^2}{n_2}}} \sim t_{n_1 + n_2 - 2}$$

i.e. the standardised difference in sample means has a t-distribution with $n_1 + n_2 - 2$ degrees of freedom. The degrees of freedom are exactly the denominator of the estimate of σ^2. This is the same as the single sample situation where we had $n - 1$ degrees of freedom, also the denominator of the estimate of σ^2 in that case.

The standard error has a number of different forms, due to the variety of ways that we can represent it algebraically. The possibilities are:

$$\sqrt{\frac{s_p^2}{n_1} + \frac{s_p^2}{n_2}} = \sqrt{s_p^2 \left(\frac{1}{n_1} + \frac{1}{n_2} \right)} = s_p \sqrt{\frac{1}{n_1} + \frac{1}{n_2}}$$

and these all give the same numerical value. You should practise using each one, and decide on the formula you prefer to use.

10.2.1 Confidence intervals for $\mu_1 - \mu_2$

As with the shift from large to small samples in the single population case, there are very few changes that need to be made in the two population case. We must remember to state (and check) our assumptions:

- that the populations are normally distributed (this can be relaxed for any population that has a large sample)
- that the population variances are the same.

We replace the large sample z-score by the corresponding score from the t-distribution with $n_1 + n_2 - 2$ degrees of freedom, and we use s_p^2 for both population variances in the standard error term.

> The $100(1-\alpha)\%$ confidence interval for $\mu_1 - \mu_2$ is given by
>
> $$\bar{X}_1 - \bar{X}_2 \pm t\sqrt{\frac{s_p^2}{n_1} + \frac{s_p^2}{n_2}}$$
>
> where t is the $100(1-\frac{\alpha}{2})$th percentile of the t-distribution with $\nu = n_1 + n_2 - 2$, and both populations are assumed normal, with common variance.

Example 10.3

Q: A native lichen (*Pseudocyhellaria homoeophylla*) was sampled at two locations on the east coast of the North Island in order to establish which location suited it better. The amount of lichen was measured by the ratio of dry weight to wet weight (a common measure in physiology studies). The data, collected from seven samples in location 1 and ten samples in location 2, were:

$$\text{Location 1} \quad \bar{x}_1 = 0.73 \quad s_1 = 0.23 \quad n_1 = 7$$
$$\text{Location 2} \quad \bar{x}_2 = 0.84 \quad s_2 = 0.41 \quad n_2 = 10.$$

Estimate the difference in mean weight ratios for the two lichen populations, using a 95% level of confidence.

A: Since the samples are both small, we must assume that both populations are normally distributed. Further, we assume that both population variances are equal, an assumption that will be tested later in Example 10.5. The common population variance is estimated by the pooled variance

$$s_p^2 = \frac{(n_1-1)s_1^2 + (n_2-1)s_2^2}{n_1+n_2-2} = \frac{6 \times 0.23^2 + 9 \times 0.41^2}{7+10-2} = \frac{1.8303}{15} = 0.12202.$$

The estimate of σ is $s_p = \sqrt{0.12202} = 0.349$, which lies between the two sample standard deviations and closer to s_2, since it was from the larger sample. We have a t-distribution with $n_1+n_2-2 = 15$ degrees of freedom, and from the table the 95% t-value is $t = 2.131$. The 95% confidence interval for $\mu_1 - \mu_2$ is

$$\bar{x}_1 - \bar{x}_2 \pm 2.131\sqrt{\frac{s_p^2}{n_1} + \frac{s_p^2}{n_2}} = 0.73 - 0.84 \pm 2.131\sqrt{\frac{0.122}{7} + \frac{0.122}{10}} = -0.11 \pm 0.37$$

$= -0.48, 0.26$. This interval contains zero, so there is no significant difference indicated by this sample.

10.2.2 Hypothesis tests for $\mu_1 - \mu_2$

The hypotheses we test using two small samples are exactly the same as those we test using large samples. However, we change the form of the test statistic slightly, and use the t-distribution to find the rejection region.

Assuming the populations are normal, with common variance, the test statistic is

$$T = \frac{\bar{X}_1 - \bar{X}_2 - (\mu_1 - \mu_2)}{\sqrt{\frac{s_p^2}{n_1} + \frac{s_p^2}{n_2}}}$$

and this has a t-distribution with $n_1 + n_2 - 2$ degrees of freedom.

This test is commonly referred to as a two-sample t-test.

We obtain the rejection region from the t-table, and reject the null hypothesis in favour of the alternative if the test statistic lies in the rejection region. If the test statistic does not lie in the rejection region, we fail to reject H_0.

Example 10.4

Q: Assuming that the two lichen populations in Example 10.3 have equal variances, test whether the % dry weight to wet weight is significantly higher in the second location. Use $\alpha = 0.05$.

A: The hypotheses we wish to test are $H_0 : \mu_1 - \mu_2 = 0$ against $H_0 : \mu_1 - \mu_2 < 0$. The test statistic is

$$T = \frac{\bar{x}_1 - \bar{x}_2 - (\mu_1 - \mu_2)}{\sqrt{\frac{s_p^2}{n_1} + \frac{s_p^2}{n_2}}} = \frac{0.73 - 0.84 - 0}{\sqrt{\frac{0.122}{7} + \frac{0.122}{10}}} = \frac{-0.11}{0.172} = -0.64$$

and this has a t-distribution with $n_1 + n_2 - 2 = 15$ degrees of freedom. The critical value from the t-table is $t = -1.753$, and the rejection region is $T < -1.753$.

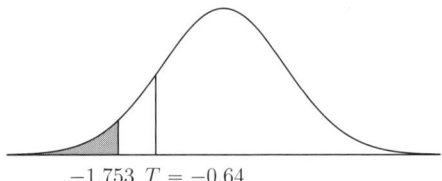

The test statistic does not lie in the rejection region, and so we cannot reject H_0 in favour of H_a at the 5% level. The location-2 ratios are not significantly higher than the location-1 ratios.

> Technical note: *In the case where the two populations have very different variances, and we cannot safely assume that $\sigma_1^2 = \sigma_2^2$, we can use a slightly different test. In this case, the test statistic is the same as in the large sample case, i.e.*
> $$T = \frac{\bar{X}_1 - \bar{X}_2 - (\mu_1 - \mu_2)}{\sqrt{\frac{s_1^2}{n_1} + \frac{s_2^2}{n_2}}}$$
> *and this has a t-distribution with degrees of freedom equal to the smaller of $n_1 - 1$ and $n_2 - 1$.*

10.3 Tests for variances

In the previous section, we needed to assume that two population variances were equal. That assumption allowed us to estimate a pooled sample variance, and conduct a *t*-test for the difference in means. Under the assumption that the populations are normally distributed, we can conduct a formal hypothesis test of that assumption. This is based on a distribution we have not yet met, called the *F*-distribution.

The *F*-distribution describes what happens when you divide one variance estimator (an average of some squared deviations, also called a mean square) by another. It has two associated parameters: the degrees of freedom of the numerator mean square, and the degrees of freedom of the denominator mean square, e.g. s_1^2/s_2^2 has an *F*-distribution with $n_1 - 1$ and $n_2 - 1$ degrees of freedom. This distribution is only appropriate if the underlying populations are normal.

Hypotheses

We test the null hypothesis $H_0 : \sigma_1^2 = \sigma_2^2$ against the alternative $H_a : \sigma_1^2 \neq \sigma_2^2$. We restrict ourselves to the two-sided alternative here, since we are most likely to use this test to test the assumption that two populations have equal variance. Note that an equivalent null hypothesis would be $H_0 : \sigma_1 = \sigma_2$, but it is standard to write H_0 in terms of variances rather than standard deviations. If H_0 is true, then s_1^2 and s_2^2 should be approximately equal, and so $s_1^2/s_2^2 \approx 1$. In other words, if H_0 is true, s_1^2/s_2^2 should be close to 1.

Test statistic

The test statistic is based on the ratio of the sample variances; however, we complicate things slightly to make it easier for ourselves at the rejection region stage. We know that the sample variances are always going to be positive, and so their ratio will always be positive. If $s_1^2 < s_2^2$, then $s_1^2/s_2^2 < 1$, whereas if $s_1^2 > s_2^2$, then $s_1^2/s_2^2 > 1$. Since the labelling of the populations is arbitrary, we label the populations so that we always get a test statistic greater than 1.

> The test statistic is
> $$F = \frac{s_\ell^2}{s_s^2}$$
> where $s_\ell^2 > s_s^2$, and, if the populations are normally distributed, this has an F-distribution with $n_\ell - 1$ and $n_s - 1$ degrees of freedom.

We have used the subscript notation ℓ and s (for *l*arge and *s*mall *variances*), in place of the usual 1 and 2, just to remind ourselves that we may need to relabel the populations.

Rejection region and conclusion

If the two sample variances are quite different, then we would tend to favour H_a over H_0. Since we have chosen the test statistic so that it is always greater than 1, we will reject the null hypothesis if the test statistic is *large*. Consequently, our rejection region is based on the upper tail of the F-distribution only, even though we are doing a two-sided test. The complication is that we assign probability $\alpha/2$ to this upper tail. The situation is shown in Figure 10.2, where we can see that the test statistic can never fall in the lower part of the rejection region (since that part of the rejection region is always below 1), so we need not bother calculating it. [Note that if we chose $F = s_1^2/s_2^2$ regardless of the relative size of the variances, we *would* need to calculate the lower rejection region.]

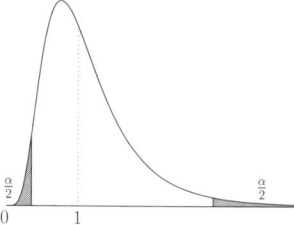

Figure 10.2 The rejection region for an F-test

We obtain the critical value from the $100\frac{\alpha}{2}\%$ F-table, F_{crit}, and reject H_0 in favour of H_a at the $100\alpha\%$ level if $F > F_{crit}$. If the test statistic is not in the rejection region, the data support H_0 and we fail to reject it.

Example 10.5

Q: In Examples 10.3 and 10.4, an assumption of equal population variances was made. Test this assumption at the 5% level.

A: The hypotheses are $H_0 : \sigma_1^2 = \sigma_2^2$ against the two-sided alternative $H_a : \sigma_1^2 \neq \sigma_2^2$. Since $n_2 > n_1$, the test statistic is

$$F = \frac{s_2^2}{s_1^2} = \frac{0.41^2}{0.23^2} = 3.178$$

and this has an F-distribution with $n_2 - 1 = 9$ and $n_1 - 1 = 6$ degrees of freedom. Using the $2\frac{1}{2}\%$ F-table, the critical value is $F_{crit} = 5.5234$, and we reject H_0 if the test statistic exceeds this value. Since $3.178 < 5.5234$, the test statistic does not lie in the rejection region and so we do not reject H_0 against H_a at the 5% level. This assumption was important for the validity of the CI and the test.

- If the F-test is two-sided, and $\alpha = 5\%$, use the $2\frac{1}{2}\%$ F-table.
- If the test is one-sided:
  - for $\alpha = 5\%$, use the 5% table
  - for $\alpha = 1\%$, use the 1% table.

10.4 Small and non-normal sample inference for medians

As in the case of the single sample, when the data are not normal, we must use a non-parametric test. In the context of two independent populations, we must use this test if at least one of our samples is small and not reasonably described by a normal distribution. Here, we use a test called the Mann-Whitney U-test, named after two statisticians and their chosen notation for the test statistic. The U-test is based on ranks, and as such it is robust to outlying observations (like the median or Spearman's correlation). The test can be used most generally to test the null hypothesis that the two samples are drawn from identical distributions. If we wish to focus on the means of these distributions, we will need to make additional assumptions about the populations; in particular, they will need to be symmetric, and have identical variability.

10.4.1 The Mann-Whitney U-test

If the two distributions are the same, then if the two equal-sized samples are combined, we would expect the observations from one sample to be fairly randomly spread out among the observations from the other sample. If we were to give the observations ranks, as we did with Spearman's correlation in Chapter 4, in the case of identical populations, we would expect the sums of each individual sample's ranks to be very similar to one another.

If the two distributions are not the same, we would expect to see clusters of observations from the different samples. Suppose, for example, that one population had a lower mean than the other. In this case, we would expect to see many observations from that population lying below the observations from the other population. When we allocate ranks, and add those ranks for each sample, we would expect to see a reasonable difference between those sums. The U-test gives us a formal procedure for determining whether the difference between the sums of ranks is large enough to reject the null hypothesis that the populations are the same.

Hypotheses

As mentioned above, the most general set of hypotheses we can test using the Mann-Whitney U-test is

$$H_0 : \text{population 1 is identical to population 2}$$
$$H_a : \text{populations 1 and 2 are not identical.}$$

If we wanted to test the null hypothesis $H_0 : \mu_1 = \mu_2$ against $H_a : \mu_1 \neq \mu_2$, this would be possible, provided we assumed that the two populations were identical *except for their means*.

Test statistic

The test statistic for this test takes a bit of effort but is very simple. We need to rank the combined samples, and then perform some additional calculations. The first step is to merge and rank the two individual samples, keeping track of which observations belong to which sample. If there are any ties, i.e. some identical observations, just like in the Spearman correlation calculation, the observations get the average rank. This is illustrated in the following example.

Once we have ranked the observations, we sum the ranks for the two samples, to form the statistics S_1 and S_2. If we have n_1 observations in the first sample, and n_2 in the second, the final rank we allocate should be $n_1 + n_2$ (or an average based on it and other ranks). If H_0 is true, and you have equal-sized samples, S_1 will be approximately equal to S_2. However, samples are not always the same size, so we scale these statistics.

To measure the degree that the samples are interspersed, we calculate the scaled S_i

$$U_1 = S_1 - \frac{n_1(n_1 + 1)}{2}$$

for the first sample, and

$$U_2 = S_2 - \frac{n_2(n_2 + 1)}{2}$$

for the second. These corrections deduct the minimum possible sums for each sample. We can check U_1 and U_2, since $U_1 + U_2 = n_1 \times n_2$.

Like the sign test, the test statistic depends on the alternative hypothesis that we are testing.

> For a two-sided test, the test statistic is
> $$\mathcal{U} = \min(U_1, U_2)$$
> i.e. we take the smaller of U_1 and U_2. For a one-sided test, we identify the lower population from H_a (e.g. if $H_a : \mu_1 < \mu_2$, the lower population is population 1) and the test statistic is
> $$\mathcal{U} = \text{the } U \text{ from the lower population.}$$

Rejection region

The test statistic of the Mann-Whitney U-test does not have a known probability distribution, but tables for the rejection region have been developed. Due to our choice of the test statistic (always taking the smaller U or the lower population), the rejection region is always of form $\mathcal{U} \leqslant U_{crit}$, and so we will reject H_0 if the test statistic is *small*. The critical value U_{crit} is obtained from Table D.11 on page 336, for a two-tail test at the 5% or 1% level.

Conclusion

As with previous tests, we draw the conclusion by comparing the test statistic with the rejection region. If the test statistic does lie in the rejection region, we reject H_0 in favour of H_a at the 5% level. Otherwise, we cannot reject H_0 in favour of H_a at the 5% level.

Example 10.6

Q: In an investigation into the distribution of particular native trees two areas of the South Island were targeted. Six quadrats at a specified altitude were selected randomly in each of Nelson Lakes and Mount Aspiring national parks and the number of ribbonwoods (including saplings and seedlings) were counted within each quadrat. The data were as follows:

| | Quadrat | | | | | |
	A	B	C	D	E	F
Nelson Lakes	8	11	13	16	17	24
Mt Aspiring	14	17	18	18	20	25

Perform a Mann-Whitney U-test on whether the median numbers of ribbonwoods at these locations is the same, using $\alpha = 0.05$.

A: The null hypothesis is $H_0 : \mathcal{M}_1 = \mathcal{M}_2$ against the alternative $H_0 : \mathcal{M}_1 \neq \mathcal{M}_2$. Combining the sample, allocating ranks, and sorting out tied observations, we have

Nelson Lakes	8	11	13		16	17					24	
Mt Aspiring				14		17	18	18	20			25
Counter	1	2	3	4	5	6	7	8	9	10	11	12
Ranks: NL	1	2	3		5	6.5					11	
Ranks: Mt A				4			6.5	8.5	8.5	10		12

The two values 17 get the average rank $\frac{6+7}{2} = 6.5$ and the two observations 18 get the average rank $\frac{8+9}{2} = 8.5$. Summing the ranks:

$$S_1 = 1 + 2 + 3 + 5 + 6.5 + 11 = 28.5 \quad \text{and} \quad S_2 = 4 + 6.5 + 8.5 + 8.5 + 10 + 12 = 49.5.$$

The corrected sums are

$$U_1 = S_1 - \frac{n_1(n_1+1)}{2} = 28.5 - \frac{6 \times 7}{2} = 7.5 \quad \text{and} \quad U_2 = S_2 - \frac{n_2(n_2+1)}{2} = 49.5 - \frac{6 \times 7}{2} = 28.5$$

and these add to $7.5 + 28.5 = 36 = n_1 \times n_2$ as required. The test statistic is the smaller of U_1 and U_2, which is 7.5, so $\mathcal{U} = 7.5$. From the U-table, the critical value for the two-sided test with $n_1 = n_2 = 6$ and $\alpha = 5\%$ is $U_{crit} = 5$. The test statistic is not less than this value, so we cannot reject the null hypothesis in favour of the alternative at the 5% level. There is no significant difference in the median numbers of ribbonwoods in the two areas.

Technical aside: *If we add the counting numbers from 1 to n, the sum is*

$$1 + 2 + \cdots + n = \frac{n(n+1)}{2}$$

and a consequence of this is that

$$S_1 + S_2 = \frac{(n_1 + n_2)(n_1 + n_2 + 1)}{2}.$$

This can be a useful check that we haven't made any silly mistakes in our calculation of S_1 and S_2. We also note that if we add n_1 ranks, these must be at least equal to

$$1 + 2 + \cdots + n_1 = \frac{n_1(n_1+1)}{2}$$

which would be the case if all the observations from the first sample were below all the observations from the second sample.

10.4.2 The normal approximation to the U-test

One of the difficulties with the Mann-Whitney U-test is that the tables we must use to obtain the rejection region are fairly obscure. In the event that the sample sizes are not too small, the CLT applies (remember that the test statistic was based on the *sum* of ranks), and the test statistic has an approximate normal distribution.

> If $n_1, n_2 \geqslant 25$, then
> $$Z = \frac{\mathcal{U} - \frac{n_1 n_2}{2}}{\sqrt{\frac{n_1 n_2 (n_1 + n_2 + 1)}{12}}}$$
> has an approximate standard normal distribution.

Here, $\frac{1}{2}n_1 n_2$ is the mean of $\mathcal{U}$, and the denominator of Z in the box above is the standard error of $\mathcal{U}$. The $\mathcal{U}$ statistic is standardised in the usual way to form an approximately normal random variable.

The standardised test statistic can be used to find an approximate p-value for the test, or alternatively, it can be compared with the appropriate rejection region. If the test is two-sided, we compare to the $100\frac{\alpha}{2}$th percentile of the standard normal, and otherwise, to the 100αth percentile. Because of the way we have selected the $\mathcal{U}$ statistic, the test is always a lower tailed test.

Example 10.7

Q: A Mann-Whitney U test statistic is calculated from two highly non-normal samples with $n_1 = 25$ and $n_2 = 27$. The test statistic was found to be $\mathcal{U} = 197$. Conduct a two-sided test at the 5% level.

A: Since both sample sizes are at least 25, we can use the normal approximation for the distribution of the test statistic. The hypotheses are $H_0 : \mathcal{M}_1 = \mathcal{M}_2$ against the two-sided alternative $H_a : \mathcal{M}_1 \neq \mathcal{M}_2$. The test statistic is $\mathcal{U} = 197$ and this has mean $E(\mathcal{U}) = n_1 n_2 / 2 = 25 \times 27 / 2 = 337.5$, and standard error

$$se(\mathcal{U}) = \sqrt{\frac{n_1 n_2 (n_1 + n_2 + 1)}{12}} = \sqrt{\frac{25 \times 27 \times (25 + 27 + 1)}{12}} = \sqrt{2981.25} = 54.60.$$

The z-score for the test is

$$Z = \frac{\mathcal{U} - E(\mathcal{U})}{se(\mathcal{U})} = \frac{197 - 337.5}{54.60} = -2.57.$$

The critical value for this test is $X = -1.96$, since this is the lower value for the two-sided test at the 5% level. Since $-2.57 < -1.96$, the test statistic lies in the rejection region, and we reject the null hypothesis in favour of the alternative at the 5% level.

10.5 Paired comparisons

In many experimental studies, we collect data on the same individuals or the same sites under two different circumstances. We may collect information "before" and "after" an event, administer two different treatments, measure their reactions to two similar products, etc. When we have two samples collected from the *same subjects*, conducting a two-sample *t*-test is unnecessarily conservative. What we are really interested in is the change for each individual, e.g. the reaction time before and after chugging a bottle of beer, and we do not need the results to take into account the differences within the sample. Instead, we perform a *matched pairs* analysis. This is also known as a *paired difference* test.

In actual fact, we can often relax the restriction that the individuals be the same. In the case of people, pairs may consist of twins, or siblings, or two people with the same gender, age and ethnicity. In the case of other "populations", we may match on the basis of some characteristic, e.g. location or size, or combinations of characteristics.

Since we want to look at the change that has occurred between the two samples, we take the difference in each individual's scores.

> The differences are
> $$d_i = x_i - y_i$$
> where x_i is individual i's score in the first sample, and y_i, their score in the second sample.

We get one difference for each individual, so in total, we obtain n differences. Typically, we are interested in testing $H_0 : \mu_1 - \mu_2 = 0$, which translates into a test of $H_0 : \mu_d = 0$, where $\mu_d = \mu_1 - \mu_2$ is the population mean difference. If n is large, or the differences are normally distributed, we use one of the normal-based tests (a *t*-test if the sample size is small), or if the sample is small and non-normal, we use a sign test.

> If the differences d_i are normal, the test statistic is
> $$T = \frac{\bar{X} - \mu_d}{\frac{s}{\sqrt{n}}}$$
> and this has a *t*-distribution with $n - 1$ degrees of freedom. Here $\bar{X}$ and s are the sample mean and standard deviation of the *differences* d_i.

Example 10.8

Q: In order to investigate whether tougher dog control laws and recent media publicity of severe dog attacks had lessened the number of attacks, the *Sunday Star-Times* of 18/4/2004 published data from 18 local authorities on dog attacks, fines, call-outs, etc.

for 2002 and 2003. A random sample of six of the local authorities gave the following data on dog attacks (obviously only those brought to the attention of the authorities):

	North Shore	Auckland	Manukau	Wellington	Christchurch	Dunedin
2002	68	997	691	124	505	86
2003	45	555	676	123	280	102

Use these data to test whether the mean frequency of dog attacks was significantly lower in 2003 than 2002 using a 10% level of significance.

A: Since the data we are given are for *the same authorities* at two points in time, a matched pairs test is appropriate. The null hypothesis is $H_0 : \mu_d = 0$ against $H_a : \mu_d > 0$, where $\mu_d = \mu_{2002} - \mu_{2003}$. The differences are 23, 442, 15, 1, 225, -16, and these have a sample mean of $\bar{x} = 115$ and a standard deviation of $s = 183.11$. The test statistic is

$$T = \frac{\bar{x} - \mu_d}{s/\sqrt{n}} = \frac{115 - 0}{183.11/\sqrt{6}} = 1.54$$

and this has a t-distribution with $n - 1 = 5$ degrees of freedom. The one-sided 10% critical value is 1.476. Since $1.54 > 1.476$, we reject the null hypothesis in favour of the alternative at the 10% level, indicating a decrease in the mean number of dog attacks from 2002 to 2003.

Example 10.9

Q: Looking at the differences in Example 10.8, there is some doubt as to the normality of the differenced data. Repeat the analysis using a sign test.

A: The hypotheses are $H_0 : \mathcal{M}_d = 0$ against $H_a : \mathcal{M}_d > 0$, where $\mathcal{M}_d$ is the population median of the differences. Allocating signs to the differences, we have $+, +, +, +, +, -$, and the test statistic is the number above $m_0 = 0$, i.e. $X = 5$ and this is binomial, with $n = 6$ and $p = 0.5$. The p-value is

$$P(Bin(6, 0.5) \geq 5) = P(Bin(6, 0.5) = 5) + P(Bin(6, 0.5) = 6) = 0.0938 + 0.0156 = 0.1094.$$

Since $0.1094 > 0.10$, we do not reject H_0 in favour of H_a at the 10% level.

10.6 Tests for proportions

Often we want to compare population proportions using two independent samples. The extension from the single sample situation for proportions is identical to how we addressed means from two independent samples: we take the difference between the sample proportions, and sum the individual variances.

Recall that the standardised score we used in the case of means was

$$\frac{\bar{X}_1 - \bar{X}_2 - (\mu_1 - \mu_2)}{\sqrt{\frac{\sigma_1^2}{n_1} + \frac{\sigma_2^2}{n_2}}} \quad \text{derived from} \quad \frac{\bar{X} - \mu}{\frac{\sigma}{\sqrt{n}}}.$$

In the case of a single population, the standardised sample proportion was

$$\frac{\hat{p} - p}{\sqrt{\frac{p(1-p)}{n}}}$$

which we extend to the case of a comparison. Applying the CLT to the sample proportions, and using the same reasoning as with the difference in two sample means, we find

$$Z = \frac{\hat{p}_1 - \hat{p}_2 - (p_1 - p_2)}{\sqrt{\frac{p_1(1-p_1)}{n_1} + \frac{p_2(1-p_2)}{n_2}}} \sim N(0, 1)$$

approximately, *provided n_1 and n_2 are both large*.

10.6.1 Confidence intervals

The form of the confidence interval follows by unravelling the standardised score in the same way as we have done before. Comparing with the single sample situation, we have

$$\frac{\hat{p}_1 - \hat{p}_2 - (p_1 - p_2)}{\sqrt{\frac{p_1(1-p_1)}{n_1} + \frac{p_2(1-p_2)}{n_2}}} \qquad \text{instead of} \qquad \frac{\hat{p} - p}{\sqrt{\frac{p(1-p)}{n}}}$$

and so the confidence interval for $p_1 - p_2$ is of the form

$$\hat{p}_1 - \hat{p}_2 \pm Z\sqrt{\frac{p_1(1-p_1)}{n_1} + \frac{p_2(1-p_2)}{n_2}} \qquad \text{instead of} \qquad \hat{p} \pm Z\sqrt{\frac{p(1-p)}{n}}.$$

As with the single sample case, we must replace the unknown p in the standard error with $\hat{p}$. Here, we replace p_1 by $\hat{p}_1$ and p_2 by $\hat{p}_2$.

Example 10.10

Q: Concerns have been expressed that there are differences in the rates at which certain hospitals offer epidural anæsthesia to women during childbirth. A random sample of 240 women who gave birth in Wellington in 2003 revealed that 96 had epidural anæsthesia, while a similar study in Christchurch showed 68 women in a sample of 176 having epidural anæsthesia. Form a 95% confidence interval for the difference in proportions at these two hospitals.

A: The point estimate for the proportion of women giving birth with epidural anæsthesia in Wellington is $\hat{p}_1 = 96/240 = 0.4$, and for Christchurch is $\hat{p}_2 = 68/176 = 0.386$. The 95% confidence interval for the difference in population proportions is

$$\hat{p}_1 - \hat{p}_2 \pm 1.96\sqrt{\frac{\hat{p}_1(1-\hat{p}_1)}{n_1} + \frac{\hat{p}_2(1-\hat{p}_2)}{n_2}} = 0.4 - 0.386 \pm 1.96\sqrt{\frac{0.4 \times 0.6}{240} + \frac{0.386 \times 0.614}{176}}$$

giving $0.014 \pm 0.095 = (-0.081, 0.109)$. The likely range for $p_1 - p_2$ contains zero, so it is likely that $p_1 = p_2$.

10.6.2 Hypothesis tests

Like the single population case, the standard error in the test statistic is evaluated under the assumption that H_0 is true. When we estimated $p_1 - p_2$ using a confidence interval, we did not have a null hypothesis, so we simply used $\hat{p}_1$ to estimate p_1 and $\hat{p}_2$ to estimate p_2 in the standard error term. In the hypothesis test, we use any values suggested by H_0 when we calculate the standard error.

Hypotheses

The typical set of hypotheses we will test will be $H_0 : p_1 = p_2$ against one of $H_a : p_1 \neq p_2$, $H_a : p_1 < p_2$ or $H_a : p_1 > p_2$. Which alternative hypothesis we use will be determined by background knowledge of the situation, rather than by our samples.

Test statistic

In the testing context, we form the test statistic under the assumption that H_0 is true. Under H_0, $p_1 - p_2$ will typically be zero; however, unlike the single population case, we don't actually have a proposed value for p_1 or p_2. All we claim (using H_0) is that the two proportions are the same.

Since the two proportions are assumed to be the same, we can take advantage of this when we calculate the standard error. We know that $\hat{p}_1 = X_1/n_1$ where X_1 is the number of "successes" in the n_1 "trials" of the first sample, and that X_1 is binomial with parameters n_1 and p_1 (i.e. $X_1 \sim Bin(n_1, p_1)$). Similarly, $\hat{p}_2 = X_2/n_2$ where X_2 is the number of "successes" in the n_2 "trials" of the second sample, and that $X_2 \sim Bin(n_2, p_2)$. If $p_1 = p_2$, then $X_1 + X_2$ is just the number of "successes" in $n_1 + n_2$ trials, and this is $Bin(n_1 + n_2, p)$ where $p = p_1 = p_2$ is the common proportion.

> The *pooled sample proportion* is
> $$p^* = \frac{X_1 + X_2}{n_1 + n_2} = \frac{n_1 \hat{p}_1 + n_2 \hat{p}_2}{n_1 + n_2}$$
> and we use this to estimate the common population proportion $p = p_1 = p_2$.

Using the pooled proportion p^* in the standard error, we calculate the test statistic.

> If $H_0 : p_1 = p_2$, then the test statistic is
>
> $$Z = \frac{\hat{p}_1 - \hat{p}_2 - (p_1 - p_2)}{\sqrt{p^*(1-p^*)\left(\frac{1}{n_1} + \frac{1}{n_2}\right)}} \sim N(0,1)$$
>
> where p^* is the pooled sample proportion.

The denominator of the test statistic results from replacing both p_1 and p_2 by p^*, and taking out the common term:

$$\frac{p^*(1-p^*)}{n_1} + \frac{p^*(1-p^*)}{n_2} = p^*(1-p^*)\left(\frac{1}{n_1} + \frac{1}{n_2}\right).$$

If H_0 is not $p_1 = p_2$, then we use $\hat{p}_1$ in place of p_1 and $\hat{p}_2$ in place of p_2, as in the confidence intervals.

Rejection region and conclusion

Since the test statistic has a standard normal distribution, we obtain the rejection region from the inverse normal table, or from the t-distribution table with $\nu = \infty$ degrees of freedom. Alternatively, we can calculate a p-value for the test using the normal table.

As before, we compare the test statistic to the rejection region, and reject H_0 in favour of H_a if the test statistic lies in the rejection region. Alternatively, we compare the p-value to the level of the test α, and reject H_0 if the p-value is smaller than α.

Example 10.11

Q: Conduct a formal hypothesis test for a difference in proportions using the epidural data in Example 10.10.
A: The hypotheses are $H_0 : p_1 - p_2 = 0$ against $H_a : p_1 - p_2 \neq 0$. Since p_1 and p_2 are assumed to be the same, we estimate them using the pooled sample proportion

$$p^* = \frac{X_1 + X_2}{n_1 + n_2} = \frac{96 + 68}{240 + 176} = \frac{164}{416} = 0.394.$$

The test statistic is

$$Z = \frac{\hat{p}_1 - \hat{p}_2 - (p_1 - p_2)}{\sqrt{p^*(1-p^*)\left(\frac{1}{n_1} + \frac{1}{n_2}\right)}} = \frac{0.4 - 0.386 - 0}{\sqrt{0.394(1-0.394)\left(\frac{1}{240} + \frac{1}{176}\right)}} = \frac{0.014}{0.0485} = 0.29.$$

Since $0.29 < 1.96$ we cannot reject the null hypothesis in favour of the alternative. The sample does not provide significant evidence of a difference. This is consistent with the confidence interval for $p_1 - p_2$.

10.7 Finite populations

As in the single population case, if a population is finite, the variability of the sample mean and sample proportion is smaller than it would be if the population were infinite, due to the fact that the sample may be a sizable portion of the population. We again apply the finite population correction factor (FPCF) to the standard error of the estimates. The FPCF is

$$\sqrt{\frac{N-n}{N-1}}$$

and this applies to the *standard errors* of the individual sample standard errors. Since these are squared (we add the variances) the FPCF appears in the standard error in its squared form.

For tests or estimation using means, the standard error is

$$se(\bar{X}_1 - \bar{X}_2) = \sqrt{\frac{\sigma_1}{n_1}\frac{N_1-n_1}{N_1-1} + \frac{\sigma_2}{n_2}\frac{N_2-n_2}{N_2-1}}$$

where N_1 is the size of the first population and N_2 is the size of the second population. For proportions the standard error is

$$se(\hat{p}_1 - \hat{p}_2) = \sqrt{\frac{p_1(1-p_1)}{n_1}\frac{N_1-n_1}{N_1-1} + \frac{p_2(1-p_2)}{n_2}\frac{N_2-n_2}{N_2-1}}$$

and we use these corrections whenever the population sizes are available.

Exercises

Comparing means, n large

We suggest you use $\alpha = 5\%$ for these tests.

10.1 In a nationwide study on *Perceptions of Personal Safety & Security Amongst Taxi Users*, respondents were asked how often they used a taxi per month. The following table gives sample statistics for the responses:

Group	$\bar{x}$	s	n
Males	6.8	6.6	291
Females	2.4	4.5	519

(a) Give a 95% confidence interval for the difference in average taxi trips per month between males and females.

(b) Conduct a hypothesis test to see whether the data support the claim that males use a taxi more often than females.

(c) Give a p-value for the test in (b), and explain how this leads to the same conclusion as the rejection region approach.

(*Data from* nzta.govt.nz)

10.2 Following Exercise 3.23, *hwa-byung* (HB) is a culturally-bound psychological disorder specific to Korea. A random sample of Korean psychiatric patients was screened for HB and a variety of other psychiatric conditions. Summary statistics for the number of conditions (in addition to HB for the HB group) are:

	$\bar{x}$	s	n
With HB	1.05	0.55	97
Without HB	0.90	0.63	183

(a) Form the 95% confidence interval for the difference in mean number of conditions (in addition to HB, if suffered) for the two groups.
(b) Is the mean number of conditions suffered (apart from HB) higher for the HB sufferers on average? Conduct a formal hypothesis test.
(c) Give a p-value for your test.
(From Min & Suh, The anger syndrome hwa-byung and its comorbidity, Journal of Affective Disorders, 2010)

10.3 A study into public transport usage in Wellington asked participants how many motor vehicles were owned by their households. Summary statistics by mode of travel are:

Travel mode	$\bar{x}$	s	n
Car	1.93	0.74	305
Train	1.77	0.91	319

(a) Calculate a 95% confidence interval for the difference in average number of household vehicles for the car and train users.
(b) Do the data support the contention that train users have less access to a car? Conduct a formal hypothesis test.
(c) Are your interval and test consistent? Explain.
See also Exercise 11.15.
(Data from Thomas, The social environment of public transport, VUW thesis, 2009)

10.4 Using the data from Exercise 2.16, test whether or not the mean number of ants trapped at sites invaded by wasps is significantly lower than the mean at uninvaded sites.
(Thanks to Catherine Duthie, VUW, for the data)

10.5 Two bands that feature heavily on John's MP3 player are Metallica and Pink Floyd. Summary data for the lengths of the songs are in the following table (in minutes):

	$\bar{x}$	s	n
Metallica	6.18	1.58	95
Pink Floyd	5.53	3.08	94

Assume that these collections are random sample from the bands' playlists, and that the population sizes are very large.
(a) Is it also important to assume that the song lengths are normally distributed? Why or why not?
(b) Do the two bands have songs of different length, on average?
See also Exercises 10.28 and 10.59.

10.6 Following Exercise 4.8, rumination (contemplation or reflection) may become persistent and recurrent worrying. A study into rumination in adolescents yields the following summary statistics of rumination scores, by age and gender (male, M, or female, F).

	11 years		13 years	
	M	F	M	F
$\bar{x}$	21.10	20.47	20.17	23.43
s	7.74	6.35	6.62	7.26
n	51	91	170	172

(a) Test whether there is a significant difference in average rumination score for the 11-year-old males and females.
(b) Repeat (a) for the 13 year olds.
(c) Summarise the results of your tests.
(Many thanks to Paul Jose for access to raw data, from Jose & Brown, When does the gender difference in rumination begin? Gender and age differences in the use of rumination by adolescents, Journal of Youth and Adolescence, 2008)

10.7 Following from Exercise 3.36, the typical number of drinks per session of the respondents who had consumed alcohol in the previous four weeks are given in the table:

Drinks	Males	Females
1-5	161	385
6-10	167	267
11-15	129	25
16-20	25	6
20+	11	4

(a) Assuming the appropriate representative value for the "20+" class is 23 drinks, calculate the sample means and standard deviations for the two groups.

(b) Do the males drink significantly many more drinks on average than the females?

See also Exercise 12.27.

(Data from Kypri, Langley, McGee, Saunders & Williams, High prevalence, persistent hazardous drinking among New Zealand tertiary students, Alcohol & Alcoholism, 2002)

Two sample *t*-tests and intervals

We suggest you use $\alpha = 5\%$ for these tests.

10.8 Using the data in Exercise 2.14, compare the average percentage of seats held by women in the highly developed countries with those in the less developed countries.

10.9 How parents of children with Autism Spectrum Disorder (ASD) engage their children when reminiscing about the past was investigated using 12 children with ASD, and a control sample of another 12 children. The frequency of child memory contributions in conversations with their parents was recorded for the two groups with the following statistics:

Group	$\bar{x}$	s	n
With ASD	9.81	7.76	12
Control	9.48	4.30	12

Test whether or not the children with ASD are contributing differently to the conversations than the control children.

See also Exercise 10.20.

(Data from Faust, Parent-child reminiscing: Relationships between parent elaborations, emotion talk and memory contributions of children with Autism Spectrum Disorder, VUW thesis, 2009)

10.10 Sociocultural adaptation is how migrants successfully adapt to life in a new country and culture. Several types of migrant, including international students, expatriates, immigrants, and refugees, were asked to rate their level of competence (1 = not at all competent; 5 = extremely competent) in various behaviours they employed in navigating their new cultural environments. These behavioural skills were categorised into four broad domains: ecological adaptation, communication and social interaction, community involvement, and contextual (work and academic) performance. The table below gives summary statistics for the ratings on communication and social interaction for two subgroups:

	$\bar{x}$	s	n
Expatriates	3.56	0.75	40
International students	3.79	0.67	82

(a) Why is it important that the sample sizes are large in this case?

(b) Form a 95% confidence interval for the difference between the mean ratings for these two groups of migrant.

(c) Is the average for the international students significantly larger than that for the expats?
See also Exercise 10.21.
(Thanks to Jessie Wilson for the data)

10.11 Using the All Blacks' height data in Exercise 3.22, test whether or not the forwards are taller on average than the backs. What assumptions are necessary for your test to be valid?
See also Exercise 10.22.

10.12 Reforestation at Wellington's Otari/ Wilton's Bush since 2002 has been supported by an active group of volunteers. The efforts included data collection and analysis to see which native species best suited planting sites within the area. The following table presents statistics on growth rates (in % per month) of one species, lemonwood (*Pittosporum eugenioides*) at two sites.

Site	$\bar{x}$	s	n
3	6.50	0.59	20
11b (north)	6.31	0.75	14

(a) Calculate a pooled variance estimate using these data. What assumptions are necessary for this to be valid?
(b) Do the data indicate a significant difference in growth rates at the two sites?
See also Exercise 10.24.
(Many thanks to Jonathan Kennett and the Otari volunteers for the data)

10.13 Following from Exercise 10.10, immigrants to New Zealand were asked to rate their level of competence (1 = not at all competent; 5 = extremely competent) in ecological adaptation, communication and social interaction – behaviours they employed in navigating their new cultural environments. The table below gives summary statistics for the ratings:

	$\bar{x}$	s	n
Ecological adaptation	3.92	0.61	45
Community involvement	2.96	0.77	44

Does the sample provide evidence that the immigrants find community involvement more challenging than ecological adaptation?
(Thanks to Jessie Wilson for the data)

10.14 The following table gives the prices of a random sample of cartons of half a dozen eggs. The first group is certified free range or organic. The second group is not, and also excludes certified barn-laid eggs.

Free range or organic			
5.59	5.99	4.69	5.70

Other				
4.29	4.49	3.69	4.69	4.49
2.99	3.20	3.09	3.49	

(a) Assuming the price variance is equal for the two groups, give the estimate of the common value.
(b) Calculate a 95% confidence interval for the difference in population mean prices for certified free-range or organic eggs and others.
(c) Is it plausible that the two means are actually the same? Conduct a formal test.
(d) What assumptions are necessary for your interval and test to be valid? Do they appear to be met?

10.15 The following table summarises total sleep time (in minutes) for two samples: the first of 35 healthy people, and the second of 20 depressed or anxious people.

Group	$\bar{x}$	s
Healthy	346.74	34.49
Depressed/anxious	333.58	40.12

(a) Calculate a 95% confidence interval for the difference between population mean sleep length for healthy and depressed or anxious people.

(b) Do healthy people sleep more than depressed or anxious people? Conduct a formal test.

(c) What assumptions are necessary for your interval and test to be valid?

See also Exercise 10.25.

(*From McNamara et al., Impact of REM sleep on distortions of self-concept, mood and memory in depressed/anxious participants, Journal of Affective Disorders, 2010*)

10.16 In a study of the effects of MDMA (Ecstasy) use on rat brain chemistry, one group of rats received a pre-treatment of saline (the control condition) and another group received a dose of ecstasy. Two weeks later, both groups of rats received a dose of ecstasy. The rats were then placed in a hide box, and the time taken for each rat to emerge from the box was measured. The times are as follows:

Saline/MDMA				
19.80	0.52	7.37	18.22	30.00
29.10	11.43	1.92	13.07	19.45
12.37	30.00	30.00		

MDMA/MDMA				
0.01	0.02	0.08	0.10	0.04
0.65	3.68	1.11	1.97	3.21
0.17	0.03	0.26		

A side-effect of ecstasy is anxiety, and a rat unused to the drug is expected to take longer to emerge from the hide box.

(a) Do the data support the theoretical prediction?

(b) Construct a 95% confidence interval for the difference in mean emergence times for the two treatments. Comment on this interval with reference to the outcome of the test in (a).

See also Exercises 10.18 and 10.26.

(*Thanks to Susan Schenk for access to raw data. From Jones, Brennan, Colussi-Mas, & Schenk, Tolerance to 3,4-methylenedioxymethamphetamine is associated with impaired serotonin release. Addiction Biology, 2010*)

10.17 The following table contains weighted average cost of capital estimates (WACC, in %) for a random sample of New Zealand firms in two sectors: consumer and property.

Consumer							
11.0	7.0	8.1	9.4	10.8	7.0	11.3	7.6
9.1	6.8	7.4	8.3	12.7	8.7	7.3	

Property							
6.8	8.7	7.0	7.1	6.9	7.2	7.2	6.8

Assuming the WACCs are normally distributed with different variances, conduct a test of equality of means.

See also Exercises 10.23 and 10.32.

(*From* pwc.com/nz/en/cost-of-capital)

10.18 Following Exercise 10.16, a further experiment examined the effect of pre-exposure to ecstasy (MDMA) on the influence of a second drug that promotes production of serotonin (mCPP). Again, rats were placed in a hide box, and their emergence times recorded.

Saline/mCPP			
12.54	23.08	2.07	14.00
3.90	14.22	12.00	30.00

MDMA/mCPP			
11.08	9.45	11.36	16.24
6.04	30.00	2.07	20.23

Are the emergence times for rats pre-exposed and not pre-exposed to MDMA significantly different in this experiment? What effect (if any) does the ecstasy appear to have?
(*Thanks to Susan Schenk for access to raw data. From Jones, Brennan, Colussi-Mas, & Schenk, Tolerance to 3,4-methylenedioxymethamphetamine is associated with impaired serotonin release. Addiction Biology, 2010*)

10.19 Following Exercise 2.19, the frequency table below gives summary statistics of the growth rates (in millimetres per day) of *Mytilus californianus* at two sites on the coast of California, USA:

Site	$\bar{x}$	s	n
Boathouse	0.0027	0.0060	12
Jalama	0.0085	0.0102	34

(a) What assumptions are necessary in order to compare the mean growth rates of the mussels at the two sites?
(b) Form a pooled variance estimate.
(c) Conduct the hypothesis test to compare the population means.
(d) Summarise your findings.
(*Thanks to Nicole Phillips for the data*)

Tests for variances

For the following, test the population variances in the specified exercise for equality. Comment on implications of the result for the earlier t-test, whether or not you have done it.

We suggest you use $\alpha = 5\%$ for these tests.

10.20 Child memory contributions (10.9)

10.21 Sociocultural adaptation (10.10)

10.22 All Blacks' heights (10.11)

10.23 Costs of capital (10.17)

10.24 Lemonwood growth rates (10.12)

10.25 Sleep times (10.15)

10.26 Rat emergence times (10.16)

10.27 Following Exercise 4.6:
(a) form the forecast errors for the maximum and minimum temperatures, i.e. calculate the differences.
(b) Calculate the sample standard deviations for the maximum forecast errors (differences), and the minimum forecast errors (differences).
(c) Using the standard deviations from (b), conduct a hypothesis test of equality of variance for the maximum and minimum forecast errors.
See also Exercise 13.10.
(*Thanks to Paula Acethorp and the Meteorological Service of New Zealand Limited metservice.com for the data*)

Mann-Whitney U-test

We suggest you use $\alpha = 5\%$ for these tests.

10.28 Following Exercise 10.5, the lengths of 12 songs of each band are given in below (in minutes, to 2dp).

Metallica					
3.57	3.67	5.02	5.20	5.77	6.08
6.60	6.83	7.22	7.88	8.45	8.80
Pink Floyd					
1.55	2.78	3.77	4.00	4.17	4.50
4.65	6.35	6.78	8.62	8.90	16.52

(a) Draw a back-to-back stemplot of the two samples and comment on any features.
(b) Conduct a Mann-Whitney U-test to respond to the question "do the two bands have different length songs, on average?"
See also Exercise 10.59.

10.29 Invasive wasps need protein late in the summer to feed developing larvae. One source of this protein is beetles, and we expect to see fewer beetles late in the summer at sites where wasps have invaded. The following table gives the numbers of beetles trapped at sites which have been invaded by wasps, and not. The numbers early in the summer (E) and late in the summer (L) are given. There are 20 counts per combination.

Invaded				Uninvaded			
E	E	L	L	E	E	L	L
21	6	1	1	7	8	2	5
9	18	2	1	33	7	3	21
12	7	1	1	7	7	2	5
8	4	1	1	1	2	4	8
5	6	1	1	2	5	4	3
8	3	1	1	5	9	1	5
7	3	2	1	3	5	12	7
7	2	1	1	12	6	4	1
9	2	7	1	8	18	3	3
5	5	1	1	5	7	6	3

(a) Explain why a t-test is not appropriate for comparing the early and late summer counts at either of the site types.
(b) For the *invaded sites only*, perform a Mann-Whitney U-test for a decrease in the average number of beetles.
(c) Repeat the Mann-Whitney U-test using counts from the *uninvaded sites*.
(d) Discuss any implications the second test might have for the first.
(*Thanks to Catherine Duthie for the data.*)

10.30 The British National Corpus is a supposedly representative collection of British English texts, and contains about 100 million words of running text. Analysis of this text for the plural versions of a list of nouns ending in "f" gives the following frequencies (in cases where the noun is indistinguishable from the verb, a random sample of passages gives an estimate of the number of noun usages):

Plural	Count	Plural	Count
calves	424	briefs	132
elves	386	chiefs	931
knives	653	fifes	2
lives	7353	oafs	8
loaves	147	proofs	236
sheaves	70	reefs	244
shelves	1159	reliefs	184
thieves	927	safes	37
wives	1867	serfs	175
wolves	648	waifs	21

(a) Draw boxplots of the two samples. Why is a two sample t-test not appropriate?
(b) Perform a Mann-Whitney U-test to test the hypothesis that the more frequently occurring nouns use the "-ves" plural form.
(*Data thanks to Laurie Bauer*)

10.31 A random sample of New Zealanders was asked "As well as a classification, films and games often come with a descriptive note (e.g. contains violence). When choosing a film, video, DVD or game how important is the descriptive note in your decision to watch it?" A second random sample was asked "When choosing a film, video, DVD or game for your child to watch how important is the descriptive note in your decision?" The responses were on an 11-point Likert scale from 0 (not at all important) to 10 (very important). The scores were:

Influence on personal choice							
1	1	1	5	5	5	6	6
8	8	9	10	10	10	10	

Influence on choice for children							
1	5	7	8	8	8	9	10
10	10	10	10	10	10	10	

(a) Why is a two-sample t-test not appropriate for these data?
(b) Why is a matched test not appropriate for these data?

(c) Conduct a Mann-Whitney U-test of the suggestion that adults pay more consideration to the descriptive note when choosing for their child(ren).

(Data from censorship.govt.nz)

10.32 In Exercise 10.17, an assumption of normality for the two populations was made.
(a) Draw stemplots of the two samples.
(b) Is the normality assumption justified in either case?
(c) Conduct a Mann-Whitney U-test to compare the two groups.
(d) Comment on the suitability of this test relative to the t-test in Exercise 10.17.

Paired comparisons

We suggest you use $\alpha = 5\%$ for these tests.

10.33 The effects of music therapy on the level of anxiety in elderly with psychiatric disorders were studied. Subjects' anxiety levels were measured before (pre) and after (post) the therapy using the "State Trait Anxiety Inventory". The scores for the experimental group and a control group are:

Experimental			Control		
Subj	Pre	Post	Subj	Pre	Post
1	25	27	1	21	27
2	41	25	2	40	39
3	74	60	3	67	71
4	38	25	4	34	35
5	50	31	5	50	48
6	40	29	6	42	40
7	32	31	7	20	23
8	36	24	8	36	36
9	20	23	9	22	24

(a) Why is a matched pairs analysis a valid way of comparing the scores for the control group, and likewise for the experimental group?

(b) Test the change in scores for the *control* group for significance using a t-test.
(c) Test the change in scores for the *experimental* group for significance using a t-test. Specifically, test for an improvement.
(d) Both t-tests require the differences to be normal. Check this assumption for the experimental group.
(e) Redo the test for the experimental group using a sign test.
(f) Summarise the results.

(Data from Castelino, The effect of single sessions of music therapy on the level of anxiety in older persons with psychiatric disorders, NZ School of Music thesis, 2009)

10.34 Using the population growth rate data from Exercise 2.10:
(a) Draw a stemplot of the difference in rates for the two years. Comment on the shape of this distribution.
(b) Is the average growth rate different for the two years? Use a t-test.
(c) Repeat the analysis using a sign test.
(d) Summarise your findings.

10.35 A study of training benefits for acutely fed versus overnight-fasted endurance cyclists indicated a significant improvement in peak oxygen utilisation ($VO_{2,peak}$, in litres per minute) in the fasted group. In addition, the following statistics are reported for the $n = 7$ fasted-trained subjects:

	$\bar{x}$	s
Pre-training	3.52	0.76
Post-training	3.86	0.86

(a) Using the statistics provided, test for an increase in peak oxygen utilisation for fasted athletes.

(b) If the original data were available, why would a matched pairs analysis be suitable in this instance. What would the benefits of this test be?

(Data from Stannard, Buckley, Edge & Thompson, Adaptations to skeletal muscle and endurance exercise training in the acutely fed versus overnight-fasted state, Journal of Science and Medicine in Sport, 2010)

10.36 Mauritius and Tunisia are two of the most successful African nations with massive growth and development in the last few decades. A researcher investigating corruption expected Mauritius to be *less* corrupt than Tunisia, with higher scores on the Corruption Perceptions Index (transparency.org). In 2009, New Zealand was ranked least corrupt with a score of 9.4, while Somalia ranked last, with a score of 1.1. Figures for both countries are in the following table.

	Mauritius	Tunisia
1998	5.0	5.0
1999	4.9	5.0
2000	4.7	5.2
2001	4.5	5.3
2002	4.5	4.8
2003	4.4	4.9
2004	4.1	5.0
2005	4.2	4.9
2006	5.1	4.6
2007	4.7	4.2
2008	5.5	4.4
2009	5.4	4.2

(a) On what basis is a matched comparison appropriate for these data?
(b) Calculate differences and plot them using a stemplot. Comment on their features.
(c) Compare the mean rating for the two countries using a t-test.
(d) Now conduct a sign test of the researcher's hypothesis.
(e) Summarise the evidence presented by these data.

(Thanks to Simon Carey, VUW)

10.37 Many situations require organisms to make choices involving the detection or identification of sometimes subtle stimuli, e.g. an animal must decide whether or not a plant is toxic or safe to eat. In an experiment, human subjects were presented with a choice which was sometimes punished. The disparity level (low or high) reflects how easily detectable the option which leads to punishment is. The scores in the table are for 12 subjects and measure their sensitivity to the "punisher ratio". Positive scores indicate systematic choices away from the more frequently punished alternative.

Subject	Low	High
1	0.27	0.39
2	0.13	−0.06
3	0.49	0.79
4	−0.04	−0.15
5	0.13	0.41
6	0.36	0.14
7	0.30	0.70
8	0.46	0.07
9	0.42	0.40
10	0.04	0.07
11	0.23	0.42
12	0.34	0.82

(a) Form the differences and use a stemplot to show that these may be reasonably assumed normal.
(b) Perform a matched t-test to see whether subjects are indeed more sensitive with the high disparity level.

(Data from Lie & Alsop, Stimulus disparity and punisher control of human signal-detection performance, Journal of the Experimental Analysis of Behavior, 2010)

10.38 The smoking rates for males and females in countries in the Asia-Pacific region were given in Exercise 4.15.
  (a) Draw a stemplot of the difference in percentages for males and females. Comment on the shape of this distribution.
  (b) Use a t-test to comment on the claim that males smoke more than females.
  (c) Repeat using a sign test.
  (d) Summarise your findings.

10.39 The following retail website brand awareness figures are the percentage of survey respondents who recognised the website brand:

Website	% 2009	% 2010
farmers.co.nz	0.7	3.1
thewarehouse.co.nz	4.2	12.7
woolworths.co.nz	2.0	5.1
harveynorman.co.nz	1.0	0.0
dse.co.nz	3.9	6.0
noelleeming.co.nz	3.4	4.8
1-day.co.nz	5.2	7.3
ezibuy.co.nz	4.2	5.4
mightyape.co.nz	1.7	1.9
foodtown.co.nz	2.2	2.4
smilecity.co.nz	3.9	4.1
apple.com	2.0	2.1

  (a) Why is a paired comparison justified for these two samples?
  (b) Why is a t-test inappropriate for these two samples?
  (c) Test the hypothesis that website brand recognition is increasing for retail websites generally.
(Data from Marketing, 2010)

10.40 Studies have found that consumers react to credit card logos when estimating the price of a good in a pamphlet. The following table gives average estimated prices for a range of products in a catalogue where no credit card logo is visible (CCA, in \$). Other subjects were asked to price the same objects but they were able to see a credit card logo "left over from another experiment" (CCP, in \$).

Product	CCA	CCP
Sweatshirt	24.05	22.00
Bag	32.05	27.50
Camera	268.50	251.50
Toaster	121.05	101.25
DVD player	206.95	161.45
Watch	82.75	55.75
Lamp	32.10	27.75
Dress	25.25	21.35
Walkman	109.75	74.75
Stereo	282.70	229.00
Monopoly	41.50	30.75
Painting	338.20	301.25

  (a) Calculate differences, and draw a stemplot of them.
  (b) Use your stemplot to justify a matched sign test over a matched t-test.
  (c) Perform a sign test to examine whether or not the credit card logo significantly influences the New Zealand students' average prices.
  (d) Do the results support or conflict with the previous literature on this issue?
(Data from Lie, Hunt, Peters, Veliu & Harper, The "negative" credit card effect: credit cards as spending-limiting stimuli in New Zealand, The Psychological Record, 2010)

10.41 The Reserve Bank of New Zealand regularly publishes forecasts of key economic variables. In the following table, a measure of its forecast accuracy for annual consumers' price index (CPI) inflation in one year's time is compared with its forecasts of annual CPI inflation in two years' time. These forecasts are also produced by external (private sector) forecasters, and these are in the table (forecasts A through H), along with the performance of the average forecast (Ave).

Forecaster	1 year	2 year
A	0.55	0.76
B	0.72	0.89
C	0.61	0.90
D	0.66	0.83
E	0.87	1.00
F	0.78	0.96
G	0.64	0.85
H	0.90	0.97
Ave	0.61	0.84
RBNZ	0.66	0.64

(a) Are the 2-year forecasts worse than the 1-year forecasts?, i.e. test whether $\mu_d = \mu_2 - \mu_1 > 0$ using a t-test.

(b) Are the assumptions of your t-test met?

(c) Repeat the test using a sign test.

(*Data from Labbé and Pepper, Assessing recent external forecasts, RBNZ, 2009*)

10.42 The table below gives the proportion of the named ingredient in a random sample of products. On the left, the content (in %) of a premium brand, and on the right, the equivalent content (in %) of a "home brand" alternative.

Product	Prem	Home
Apple sauce	99.8	92
Baked beans	50	44
Beetroot	67	51
Chopped tomatoes	60	60
Coconut cream	80	51
Corned beef	98	44
Creamed corn	80	35
Creamed rice	42	10
Fish fingers	55	59
Guacamole (avocado)	1	4
Hazelnut spread	13	13
Hummus (chick pea)	69	48
Olive oil spread	19	26
Peach slices	66	56
Raspberry jam	50	40
Sandwich tuna	73	76
Tinned asparagus	65	65
Weetbix	97	92

(a) Why is a matched pairs analysis valid in this instance?

(b) Why is a t-test not valid in this instance?

(c) Conduct a sign test to check the validity of a claim that "home brand" products contain less of the primary ingredient than a premium alternative.

Comparing proportions

We suggest you use $\alpha = 5\%$ for these tests.

10.43 Following Exercise 9.42, 5634 of the sample were male, and 7358 were female. Among the sample, 1.6% of the males report suffering from dysthymia, a form of depression. The proportion among the females was 2.6%.

(a) Are the sample proportions too close to zero for the normal approximation to be used?

(b) Calculate the pooled proportion, and show that this is closer to the females' proportion than to the males'. Explain why.

(c) Is the incidence of dysthymia significantly higher in females?

(*Data from* moh.govt.nz)

10.44 The Australasian Survey of Student Engagement sampled 4800 students from New Zealand universities, and 25,800 Australian students. The survey found 19% of Kiwi students study more than 21 hours per week, while only 15% of Australian students do. Are these proportions significantly different?

10.45 *Te Rau Hinengaro: The New Zealand Mental Health Survey* assessed the mental health of 12,992 New Zealanders using the Composite International Diagnostic Interview (CIDI 3.0). The survey found 39.5% of the sample had met criteria for a DSM-IV mental disorder at some time in their life before interview.

(a) Give a 95% confidence interval for the population proportion of New Zealanders meeting the criteria for a DSM-IV mental disorder at some time in their life.

(b) How large a sample would be needed to estimate the proportion to within ±0.1% (approximately 4000 New Zealanders)?

(*Data from* moh.govt.nz)

10.46 A subject in an appraisal of a reading treatment programme exhibited "pure" phonological dyslexia, with deficient "nonword" reading but normal irregular-word reading. Pre-training, the individual correctly read 27 of 50 nonwords, while post-training, the individual correctly read 41 of 50 nonwords. Is this a significant improvement?

(*From Rowse & Wilshire, Comparison of phonological and whole-word treatments for two contrasting cases of developmental dyslexia, Cognitive Neuropsychology, 2007*)

10.47 In the 2007/08 Active NZ survey, 1857 men and 2586 women were interviewed. 24.9% of those men, and 25.8% of those women had acted as volunteers over the sample period. The most common roles were as a volunteer coach, trainer, teacher or instructor. Does this sample suggest the population proportion of New Zealand men who volunteer was smaller than the proportion of women? (*Data from* sparc.org.nz)

10.48 Researchers compared people's memories of an event after they discussed the event either with their romantic partner or a stranger. Pairs of subjects watched slightly different versions of a movie, and they discussed some details from the movie, but not others. Because the movie versions were different, conversations often involved misinformation about the other's movie. Members of couples disputed what their partner told them 39 times out of 76, while members of stranger pairs disputed what their partner told them 36 times out of 78.

(a) Calculate the two sample proportions and a pooled sample proportion.

(b) Using the sample evidence, are strangers less likely to dispute what their partner tells them?

(*Data from French, Garry & Mori, You say tomato? Collaborative remembering leads to more false memories for intimate couples than for strangers, Memory, 2008*)

10.49 The donation box at City Gallery Wellington was manipulated for several months to learn about visitor response to different visual cues. In one part of the experiment, each donation was immediately removed from the box so the next visitors would see the box empty. When this was in effect, 104 out of 5585 visitors made a donation. Is this proportion significantly different from that of another sample, where 182 out of 5394 visitors made a donation into the box with about $100 in it – mostly coins?

(*From Martin & Randal, How is donation behaviour affected by the donations of others? Journal of Economic Behavior and Organization, 2008*)

10.50 A 2009 report on *Cultural Indicators for New Zealand* identified 11,922 cultural activities in the main centres, and 3976 outside the main centres. Of these, the proportions involving theatre were 12.1% and 6.5% respectively.

(a) Give a 95% confidence interval for the difference in population proportions for main centre and nonmain centre cultural activities involving theatre.

(b) Are the population proportions likely to be different?

(*Data from* mch.govt.nz)

10.51 *Kai Tiaki Nursing New Zealand* (April 2010) reports "the number of registered nurses identifying as Māori steadily declined from 2820 (7.13 per cent of the total number of RNs) to 2754 (7.13%)." Given sample sizes of registered nurses of 38,683 for 2005, and 38,626 for 2007, test this decline for significance.

10.52 The Dunedin Multidisciplinary Health and Development Study is a long-running cohort study of 1037 children born in Dunedin in 1972-1973. It is a very important source of information on the health, development and behaviour of young people. In a 2006 investigation of the general validity of this study across New Zealand, results from the Dunedin Study are compared with a sample of similarly aged subjects in the New Zealand Health Study.

Of 481 females in the Dunedin Study 64 were ex-smokers at age 26; in the NZ Health Survey 31 out of 194 similarly aged females were ex-smokers. Are these sample proportions significantly different? Give and interpret the p-value for this test.

(*Figures published in the New Zealand Medical Journal, 2006*)

10.53 Typically, the lowest denomination bank notes show the poorest quality. Following from Exercise 9.32, 25% of the $5 notes were "poor" quality, while 5% of $10 are poor quality. Do the sample data support the expectation?

(*From Langwasser, Recent trends and developments in currency – 2009, RBNZ, 2010*)

10.54 In a 2008 survey of arts participation in New Zealanders aged 15 years and over, 22 out of 243 Māori people surveyed had been a participant in the performing arts, with the majority of these in a concert or music performance. Of 176 Pacific people in the same survey, 37 had participated in the performing arts. Are the proportions significantly different?

(*Data from* creativenz.govt.nz)

10.55 Tests for chlamydia performed on 6614 pregnant women in New Zealand between 1999 and 2003 yielded the following results:

Age group	Positive	n
< 25 years	12.2%	985
⩾ 25 years	2.3%	5629

Are these proportions significantly different? (*Data from* moh.govt.nz)

10.56 Following from Exercise 10.52, the proportion of the 496 males in the Dunedin Study who exercised more than 5 hours per week was 30.8%; the equivalent proportion of the 98 similarly aged males in the NZ Health Survey was 18.2%.

(a) Form a 95% confidence interval for the difference in the proportions of Dunedin 25-26 year old males who exercise at least 5 hours per week, and similar males nationwide.

(b) Test whether or not the proportion of Dunedin 25-26 year old males who exercise at least 5 hours per week is greater than the proportion of similar males nationwide.

(*Figures published in the New Zealand Medical Journal, 2006*)

10.57 *Seen and heard: children's media use, exposure and response* summarises the results of a survey of 604 children aged 6 to 13 years and their primary caregivers. The children were asked whether or not they had watched television on the previous day, and if so, when, and had they done any other activity at the same time (e.g. homework, played)? Of the 365 children who had watched TV before school, 20% had done nothing else while watching. Of the 289 who watched TV after dinner, 22% had done nothing else while watching. Are these proportions significantly different?

(*Data from* bsa.govt.nz)

10.58 Following Exercise 10.57, the children aged 9-13 were asked if they had seen any TV content that disturbed them (e.g. violence: killing, blood and guts, etc; sexual content: rude things, kissing etc; bad language and others). The following table summarises a collection of responses from 249 9-11 year olds and 137 12-13 year olds:

Content	9-11	12-13
Violence	28%	29%
Sexual content	22%	19%
Scary/spooky things	24%	14%
Bad language	15%	10%
Misc (e.g. suffering)	16%	24%

(a) Form 95% confidence intervals for the difference in population proportions for the various categories.
(b) What might you conclude from those confidence intervals about the age-group differences?

(*Data from* bsa.govt.nz)

Finite populations

We suggest you use $\alpha = 5\%$ for these tests.

10.59 Redo the test you performed in Exercise 10.5 acknowledging that Metallica have produced 123 songs, and Pink Floyd 162. How do these population sizes affect your test?

(encycmet.com *and* wikipedia.org)

10.60 Health researchers wish to estimate the average global proportion of smokers, by sex. Assuming the Western Pacific nations selected in Exercise 3.11 are a sufficiently random sample:
(a) Calculate 95% confidence intervals for the mean proportion of male smokers and of female smokers.
(b) Assuming the total number of countries is 193, redo your confidence intervals to take this into account.
(c) Display your confidence intervals on a number line, and comment.
(d) What effect does the population size have on your intervals?

(*Data from* wikipedia.org)

Chapter 11

Many populations

The natural continuation of Chapters 9 and 10 is to consider the case where we wish to compare more than two populations at once. We could conduct pairwise comparisons using tests from Chapter 10; however, when we do this we cannot ensure the overall level of the tests is equal to α. As a result, we use specially designed tests that allow us to compare all the samples at once.

In the case where the sample means are normally distributed (either due to the CLT, or to underlying normality), we use a test called an *analysis of variance*, usually abbreviated to *ANOVA*. We will see why a test for *means* has such a name in the next section. This ANOVA test applies when we have large samples, or when we have small samples which are normally distributed. The second case will be much more common, since it is typically too expensive to collect many large samples.

When we cannot assume normality, the ANOVA test can lead us to draw the wrong conclusion, and we must use a non-parametric test called the *Kruskal-Wallis* test, which, like the Mann-Whitney U-test, is based on ranks.

Both tests will test the null hypothesis that k population means are identical, where k is a number greater than 2. In the case of ANOVA, we can conduct the test when $k = 2$ but the p-value we obtain is identical to the much simpler t-test of the same hypotheses.

11.1 Small and normal sample test for means

As mentioned above, it will be an uncommon occurrence that we have $k > 2$ samples with at least 30 observations each. As a consequence, we'll rarely be able to use the CLT to provide normality for our sample means, and we will have to assume that we have normal data. In practice, this means having samples which are symmetric, unimodal, and have no outliers, features which can be checked using histograms or stemplots, and boxplots.

11.1.1 The ANOVA test

The test we use is called *analysis of variance*, which at first glance is a strange name to give to a test of k population means. The reason for this name is that we actually compare two estimates of variance in order to discover whether all the means are the same.

The first estimate of variance is found by estimating variability based on a single estimate of the mean, i.e. one that applies to all samples. The second variance is analogous to the *pooled variance* on page 218, in which we estimate individual sample means. If these two variances are similar, this indicates estimating individual means doesn't make much difference and would tend to support the hypothesis that the population means are equal. If the variances are quite different, this indicates estimating individual sample means *is* necessary and lends support to the suggestion that the means are not equal. This is demonstrated in Figures 11.1 and 11.2 where we see that *the pooled population has a much greater spread when the means are unequal*. The difference in the variances is examined formally using the F-distribution.

Figure 11.1   Pooled sample when the population means are equal

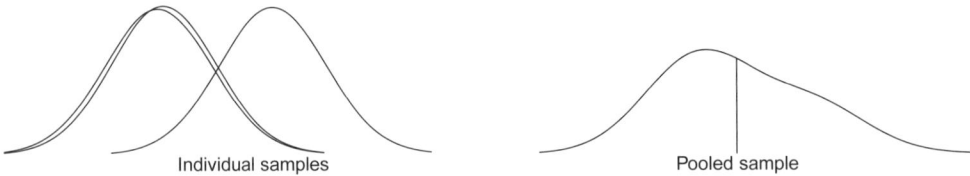

Figure 11.2   Pooled sample when the population means are not equal

For the purposes of the ANOVA test, we must assume:

- that the underlying populations are normal
- that the populations have equal variances.

Hypotheses

The null hypothesis we wish to test is $H_0 : \mu_1 = \mu_2 = \cdots = \mu_k$. With two samples, we could replace the '=' by '$\neq$' to form the alternative hypothesis. Doing that in this case would lead to $\mu_1 \neq \mu_2 \neq \cdots \neq \mu_k$; however, this is a much stronger statement than what we want. We do not require that *all* the means are different; rather, we will reject H_0 if *any one* of the means is different from the others. This is difficult to write using symbols, so we resort to words. The hypotheses we test are

H_0 : all k population means are equal
H_a : at least one population mean is different.

An alternative expression for H_a is that "not all the means are the same".

Test statistic

As with the Mann-Whitney U-test, quite a bit of work is involved in calculating the test statistic. We must first collect independent random samples from each of the k populations. From the first population, we will have n_1 observations

$$X_{11}, X_{12}, X_{13}, \ldots, X_{1n_1}.$$

These are written in the form X_{ij} where the i tells us in this case that the observations are from the first population, and the j counts the observation in this ith group, e.g. X_{13} is the third observation in the first group. The second sample would be

$$X_{21}, X_{22}, X_{23}, \ldots, X_{2n_2}$$

and so on, through to the kth sample, which would be

$$X_{k1}, X_{k2}, X_{k3}, \ldots, X_{kn_k}.$$

Each of these samples has an associated size, mean, and standard deviation, which are denoted n_i, $\bar{X}_i$ and s_i respectively. It is also useful to record the sum of the observations in each group. We call these the totals, and write the total for group i as T_i. The data can be conveniently laid out as follows.

			Group				
	1	2	3	$\cdots$	k		
	X_{11}	X_{21}			X_{k1}		
	X_{12}	X_{22}			X_{k2}		
	$\vdots$	$\vdots$			$\vdots$		
		X_{2n_2}			X_{kn_k}		
	X_{1n_1}						
Sample size	n_1	n_2	n_3	$\cdots$	n_k	$n = \sum n_i$	
Total	T_1	T_2	T_3	$\cdots$	T_k	$T = \sum T_i$ = grand total	
Mean	$\bar{X}_1$	$\bar{X}_2$	$\bar{X}_3$	$\cdots$	$\bar{X}_k$	$\bar{\bar{X}}$ = grand mean	
SD	s_1	s_2	s_3	$\cdots$	s_k	$\mathcal{S}$ = overall standard deviation	

The total number of observations is $n = n_1 + \cdots + n_k$, and $\bar{\bar{X}}$ and $\mathcal{S}$ are the mean and standard deviation of the entire data set, i.e. treated as if it is a single sample of n observations.

In the ANOVA test, we measure variance in two ways:

- The mean square for the treatments, MST, is a measure of the inherent variability in the data *plus* any variability due to the different groups (historically, the groups were receiving different *treatments*).
- The mean square for the errors, MSE, measures the inherent variability in the data alone. [These differences can be thought of as "errors" from the respective group means and the mean square is equivalent to the pooled variance when comparing two groups.]

If the alternative hypothesis H_a is true, then

$$\text{MST} = \text{MSE} + \text{group effects}$$

and the MST will exceed MSE due to the group effects. If H_0 is true, there will be no group effects, and MST and MSE will be roughly equal. We test H_0 by comparing the two estimates of variance, i.e. we test $H_0 : \sigma_T^2 = \sigma_E^2$ against $H_a : \sigma_T^2 > \sigma_E^2$ (where T denotes treatment or group, and E denotes error). We can use the F-test for this; however, we need to calculate the mean squares (variances) first.

> The variance estimators (mean squares) are both based on a sum of squares divided by its degrees of freedom, i.e.
>
> $$\text{MST} = \frac{\text{SST}}{k-1} \quad \text{and} \quad \text{MSE} = \frac{\text{SSE}}{n-k}$$
>
> where SST is the sum of squares for the treatments, and SSE is the sum of squared errors. To avoid confusion, we do not use σ^2 for these.

As with the sum of squares in the sample variance, these sums of squares have two mathematical forms, one that is easy to calculate, and another that can be used to explain what the sum of squares is measuring.

> The following are useful computational formulae:
>
> $$\text{SST} = \sum_{i=1}^{k} \frac{T_i^2}{n_i} - \frac{T^2}{n}$$
>
> $$\text{SSE} = \sum X_{ij}^2 - \sum_{i=1}^{k} \frac{T_i^2}{n_i}$$

The mean squared error is an extension of the pooled variance used in the two-sample t-test, generalised to $k > 2$ groups. The MSE can be called the *pooled variance*, and vice versa. Following the development in Chapter 10, we can write a further form for SSE, namely

$$\text{SSE} = \sum_{i=1}^{k} (n_i - 1) s_i^2.$$

This states that the sum of squared errors can be found by working out $(n_i - 1)s_i^2$ for each group, and adding. If the sample variances are more readily available than the group totals, we may choose to use this formula over the recommended computational formula above.

Finally, we calculate the test statistic F, which is a ratio of the two variances (mean-squares).

> The test statistic for the ANOVA test is
> $$F = \frac{\text{MST}}{\text{MSE}} = \frac{\text{SST}/(k-1)}{\text{SSE}/(n-k)}$$
> and this has an F-distribution with $k-1$ and $n-k$ degrees of freedom.

The F-distribution compares variances, and a variance is simply a sum of squares divided by its degrees of freedom. The F-distribution has two degrees of freedom parameters, one corresponding to the numerator ($k-1$ in the ANOVA case) and the second corresponding to the denominator ($n-k$).

To make the job of calculating the ANOVA test statistic easier, it is common to use the ANOVA table layout in Table 11.1.

Source of variability	Degrees of freedom	Sum of squares	Mean square	Test statistic
Between groups	$k-1$	SST	$\text{MST} = \dfrac{\text{SST}}{k-1}$	$F = \dfrac{\text{MST}}{\text{MSE}}$
Within groups	$n-k$	SSE	$\text{MSE} = \dfrac{\text{SSE}}{n-k}$	
Total	$n-1$	Total SS		

Table 11.1   The ANOVA table

The ANOVA table is arranged with three rows: the first contains details for the variability *between* the groups, and the second the variability *within* the groups. The final row is not needed for the test, but is a useful check that the degrees of freedom and sums of squares have been calculated correctly.

The total number of degrees of freedom, shown in the final row of the "Degrees of freedom" column, is $n-1$. This is based on our n observations, less one for estimation of the common population mean suggested by H_0. The second row of the degrees of freedom column is based on individual estimation of the k group means. Again, we start with n degrees of freedom from our n observations, and we lose one for each group mean that we estimate (k in total), leaving us with $n-k$ degrees of freedom. The first row of the table is best thought of as the residual degrees of freedom, i.e. $n-1$ less $n-k$. Intuitively, we can think of the remaining $k-1$ degrees of freedom as the k degrees of freedom we have when we specify k individual group means, less one degree of freedom because we constrain these to have (weighted) mean $\overline{\overline{X}}$.

The "Sum of squares" column in the ANOVA table contains SST and SSE. In the final row, we include Total SS, which is $n - 1$ times the total variance S^2, i.e.

$$\text{Total SS} = (n-1)S^2 = \sum (X_{ij} - \overline{\overline{X}})^2$$

where S is the standard deviation of the combined data. The two different ways of calculating Total SS are mathematically equivalent; however, the first is much easier to evaluate than the second (which is easier to interpret). Total SS measures the variability of the combined samples assuming H_0 is true. Since SST + SSE = Total SS, we can use this fact to check SST and SSE. Alternatively, you may prefer to calculate Total SS and one of SSE and SST, then find the final sum of squares by subtraction.

Rejection region

The rejection region will be obtained from the $F(k-1, n-k)$ distribution. It follows from discussion on page 250 that we will want to reject H_0 when the F statistic is large. In this case, the test is a *one-sided* test, and we will reject when $F > F_\alpha$ as indicated in Figure 11.3.

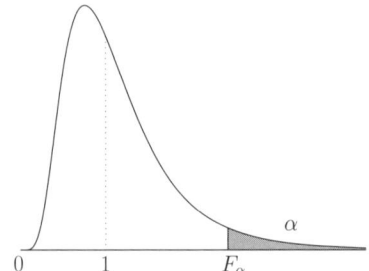

Figure 11.3 The ANOVA rejection region

To obtain F_α, we must specify a value of α (usually 5%) and use the corresponding table (Table D.8 on page 333 for 5%) to look up the critical value. It will always exceed the value $F = 1$.

Conclusion

As with previous tests, the conclusion is drawn by comparing the test statistic to the rejection region.

- If the test statistic lies in the rejection region, we reject the null hypothesis in favour of the alternative at the $100\alpha\%$ level.

If we reject H_0, this implies that the population means are not identical, and we would estimate the individual means using the respective sample means.

- If the test statistic does not lie in the rejection region, we cannot reject the null hypothesis.

When we cannot reject H_0, this implies that the data are drawn from the same population. We have assumed that the populations are normal, and that variances of the populations from which the samples are drawn are equal. Further, the test has shown that the means are also likely to be equal, and since the normal distribution has only two parameters: the mean and the variance, the distributions are all identical.

Example 11.1

To compare fungus growth in four different media A, B, C and D (where A is thought to be superior at promoting growth), four samples were grown in each medium, although two later became contaminated and had to be discarded. The area of fungus in each sample after ten days was measured by counting the number of squares covered on a grid. The results are:

Medium	Data				T_i	n_i	s_i^2
A	8	14	9	12	43	4	7.583
B	7	9	5		21	3	4
C	6	4	7	8	25	4	2.917
D	4	7	5		16	3	2.333

We can confirm that $T = 43 + 21 + 25 + 16 = 105$, with $n = 14$ and $k = 4$.

We test $H_0 : \mu_A = \mu_B = \mu_C = \mu_D$ against the alternative that at least one population mean is different.

We calculate
$$\text{SSE} = \sum (n_i - 1)s_i^2 = 3(7.583) + 2(4) + 3(2.917) + 2(2.333) = 44.17$$
and $\text{MSE} = \frac{\text{SSE}}{n-k} = 4.417$. Also,
$$\sum \frac{T_i^2}{n_i} = \frac{43^2}{4} + \frac{21^2}{3} + \frac{25^2}{4} + \frac{16^2}{3} = 850.8333,$$
so
$$\text{SST} = \sum \frac{T_i^2}{n_i} - \frac{T^2}{n} = 850.8333 - \frac{105^2}{14} = 63.33 \quad \text{and} \quad \text{MST} = \frac{\text{SST}}{k-1} = 21.111.$$

The test statistic is $F = \text{MST}/\text{MSE} = 21.111/4.417 = 4.779$, and the completed ANOVA table looks like this:

	df	SS	MS	F
Between groups	$4 - 1 = 3$	63.33	21.11	4.779
Within groups	$14 - 4 = 10$	44.17	4.417	
Total	$14 - 1 = 13$			

The test is a one-sided one, and using the $F(3, 10)$ table, the critical value is 3.7083 at the 5% level. We will reject H_0 at the 5% level of significance if the test statistic exceeds this value.

It does, and so we reject the null in favour of the alternative. Looking back at the data, we notice that sample A's areas are typically larger than those for the other media.

Technical aside: *The ANOVA computational formulae do not show us what the sums of squares are measuring. The ANOVA sums of squares are also given by*

$$SST = \sum_{i=1}^{k}(\bar{X}_i - \bar{\bar{X}})^2$$

$$SSE = \sum(X_{ij} - \bar{X}_i)^2.$$

The first formula helps us to see that MST = SST/(k − 1) measures the variability of the group means about the grand mean. If this is large, we would expect to reject the null hypothesis that the population means are equal.

11.1.2 The model-based approach

The F-test used to test $H_0 : \mu_1 = \cdots = \mu_k$ can be motivated using a more general, model-based approach. The model we use is similar in form to the regression model of Chapter 4 and allows us to make more advanced comparisons based on different beliefs about the populations.

For the ANOVA test, we have two competing models for the data. The first of these is called the *reduced model*, which is the model that holds if H_0 is true. In this case, we denote the common mean by μ.

The reduced model for the ANOVA test is

$$X_{ij} = \mu + \epsilon_{ij}$$

where the ϵ_{ij} are independent normal random variables with zero mean and variance σ^2.

Note that all the observations X_{ij} have the same distribution, since the mean and variance is the same for all i and j.

The other model we specify is the *complete model*. This model reflects the situation where H_a is true, and it is necessary to specify different means for each of the populations.

> The complete model for the ANOVA test is
>
> $$X_{ij} = \mu_i + \epsilon_{ij}$$
>
> where the ϵ_{ij} are independent normal random variables with zero mean and variance σ^2.

Notice that the difference between the complete and reduced models is very slight: the reduced model has mean μ (a single number for all groups), whereas the complete has μ_i (one for each group). This allows the observations from population i to have a mean μ_i, specific to that distribution, and generally, for the population means to differ.

Estimates of μ and the μ_i are found using the same ideas as in least squares regression. In particular, we minimise the residual sum of squares $\sum e_{ij}^2$ where e_{ij} is the estimate of ϵ_{ij}. This amounts to using the sample means for the estimates. So, we estimate μ by the grand mean $\overline{\overline{X}}$ and the individual means μ_i by the respective sample means $\bar{X}_i$, i.e.

$$\hat{\mu} = \overline{\overline{X}} \quad \text{and} \quad \hat{\mu}_i = \bar{X}_i$$

for the reduced and complete models respectively. The resulting residual sums of squares are

$$RSS_{reduced} = \sum(X_{ij} - \hat{\mu})^2 = \sum(X_{ij} - \overline{\overline{X}})^2 = \text{Total SS}$$

for the reduced model, and

$$RSS_{complete} = \sum(X_{ij} - \hat{\mu}_i)^2 = \sum(X_{ij} - \bar{X}_i)^2 = \text{SSE}$$

for the complete model, where Total SS and SSE have the same definitions as in Table 11.1. In the reduced model, we started with n observations and estimated a single parameter μ, leaving us with $n - 1$ degrees of freedom. For the complete model, we estimated k means, giving us a total of $n - k$ degrees of freedom.

To conduct the test, we calculate the test statistic based on the reduced and complete sums of squares $RSS_{reduced}$ and $RSS_{complete}$.

> The test statistic for the comparison between the reduced and complete models is
>
> $$F = \frac{(RSS_{reduced} - RSS_{complete})/(df_{reduced} - df_{complete})}{RSS_{complete}/df_{complete}}$$
>
> and this has an F-distribution with $df_{reduced} - df_{complete}$ and $df_{complete}$ degrees of freedom.

It follows from the general form of the test statistic, that the model based ANOVA test statistic is identical to the previous form, since

$$\frac{RSS_{reduced} - RSS_{complete}}{df_{reduced} - df_{complete}} = \frac{\text{Total SS} - \text{SSE}}{(n-1) - (n-k)} = \frac{\text{SST}}{k-1}$$

and $RSS_{complete}/df_{complete} = \text{SSE}/(n-k)$ giving exactly the same formula as before. If the test statistic is large, we favour the complete model over the reduced model, and since the complete model is consistent with H_a, we reject the null H_0 for large values of the test statistic, exactly as before.

11.1.3 Interpreting the ANOVA result

When we reject the null hypothesis, it is natural to ask "why did we reject the null?" In the case of three samples, we may have rejected H_0 because one population is lower than the others, or because one is higher than the others, or alternatively, because all three populations have different means. When we have more than three samples, the possibilities multiply!

- If we do not reject the null hypothesis, the appropriate estimate for every population mean is the global sample mean $\overline{\overline{X}}$.

- If we do reject H_0, the estimate for the population mean of the ith group is $\bar{X}_i$, the sample mean of the ith group.

If we rejected H_0, we can form individual confidence intervals for the group means.

A 95% confidence interval for μ_i is given by

$$\bar{X}_i \pm t \frac{s_p}{\sqrt{n_i}}$$

where t is the appropriate value of the t-distribution with $n-k$ degrees of freedom, and where $s_p^2 = \text{MSE}$ is the pooled variance.

Note that we use the pooled variance as the estimate of σ^2 for every population, and use $n-k$ degrees of freedom for the t-value.

Example 11.2

Q: Following from Example 11.1, calculate 95% confidence intervals for the population mean area of fungus growing in media A and D.

A: The form of the 95% confidence interval for μ_A is $\bar{x}_A \pm t s_p/\sqrt{n_A}$ where t is from the t-distribution with $n-k = 10$ degrees of freedom, and $s_p = \sqrt{MSE} = \sqrt{4.417}$. From the t-table, we find $t = 2.228$, and so the 95% confidence interval is

$$\bar{x}_A \pm 2.228 \frac{s_p}{\sqrt{n_A}} = 10.75 \pm 2.228 \frac{\sqrt{4.417}}{\sqrt{4}} = 10.75 \pm 2.34 = 8.41, 13.09.$$

11.1 Small and normal sample test for means

Similarly, the 95% confidence interval for μ_D is

$$\bar{x}_D \pm 2.228 \frac{s_p}{\sqrt{n_D}} = 5.333 \pm 2.228 \frac{\sqrt{4.417}}{\sqrt{3}} = 5.333 \pm 2.703 = 2.63, 8.04.$$

These intervals do not overlap, indicating a difference between these two population means.

We can investigate *pairwise* differences using confidence intervals.

> A 95% confidence interval for $\mu_i - \mu_j$ (with $i \neq j$) is given by
>
> $$\bar{X}_i - \bar{X}_j \pm ts_p \sqrt{\frac{1}{n_i} + \frac{1}{n_j}}$$
>
> where t is the appropriate value of the t-distribution with $n - k$ degrees of freedom, and where $s_p^2 = $ MSE is the pooled variance.

If we calculate the confidence intervals for all pairs, any that do not contain zero suggest that those particular population means differ.

Example 11.3

Q: Following from Example 11.1, calculate a 95% confidence interval for the difference in population mean area of fungus growing in media A and D.

A: Again, using the t-distribution with $n - k = 10$ degrees of freedom, the 95% confidence interval for $\mu_A - \mu_D$ is

$$\bar{x}_A - \bar{x}_D \pm ts_p \sqrt{\frac{1}{n_A} + \frac{1}{n_D}} = 10.75 - 5.333 \pm 2.228\sqrt{4.417}\sqrt{\frac{1}{4} + \frac{1}{3}} = 5.417 \pm 3.576$$

giving an interval of $(1.84, 8.99)$. This interval does not contain zero, so the sample data indicates the two population means differ, consistent with earlier results.

An alternative to calculating all of the pairwise confidence intervals is a graphical one, in which we draw *half confidence intervals* for each population mean. These half intervals replace the t-value by $t = 1$, which is roughly half the 95% t-value obtained from the t-table, with $n - k$ degrees of freedom.

> The 95% half confidence interval for μ_i is given by
>
> $$\bar{X}_i \pm \frac{s_p}{\sqrt{n_i}}.$$

To aid comparison, we plot the half confidence intervals on a number line, and any that do not overlap suggest a difference in the population means for those groups.

> **Example 11.4**
>
> Q: Continuing from Example 11.1, calculate half confidence intervals for the four population means, and display these on a number line.
> A: The half confidence intervals have the form $\bar{x}_i \pm s_p/\sqrt{n_i}$ and are:
>
> For A: $\qquad 10.75 \pm \dfrac{\sqrt{4.417}}{\sqrt{4}} = 10.75 \pm 1.051 = 9.70, 11.80.$
>
> For B: $\qquad 7 \pm \dfrac{\sqrt{4.417}}{\sqrt{3}} = 7 \pm 1.213 = 5.79, 8.21.$
>
> For C: $\qquad 6.25 \pm \dfrac{\sqrt{4.417}}{\sqrt{4}} = 6.25 \pm 1.05 = 5.20, 7.30.$
>
> For D: $\qquad 5.333 \pm \dfrac{\sqrt{4.417}}{\sqrt{4}} = 5.333 \pm 1.213 = 4.12, 6.55.$
>
> These can be displayed on a number line as follows:
>
>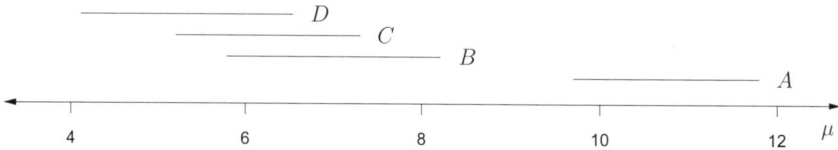
>
> where we see clear separation between (B, C, D), and A. This indicates A has a significantly higher mean than the other three groups, which do not differ strongly.

11.1.4 Residuals and assumptions

Like all statistical tests, for the results of an analysis of variance to be reliable, we require that certain assumptions hold. In particular, for an analysis of variance to yield reliable results we require that the underlying populations from which the data are taken are *normally* distributed. We can check to see if this requirement is at least roughly true by calculating what are called the *residuals*. They are similar to the residuals of Chapter 4.

> The residual for X_{ij}, the jth piece of data in the ith group, is
> $$e_{ij} = X_{ij} - \bar{X}_i.$$

So, the residuals are the data less the mean of the group they are in.

Each piece of data has a residual and all the residuals taken together should have the shape of a normal distribution with mean zero. The method we use to check this is simple:

- Calculate all the residuals.
- Plot the residuals using a stem plot, boxplot or histogram.
- If the data are asymmetric, or contains outliers, we have evidence against normality.

> More sophisticated techniques, such as drawing a *qq*-plot, become available as you progress into more advanced statistical methods.

Example 11.5

Using the data from Example 11.1, we calculate the residuals for each observation. The mean for group A is $\bar{X}_1 = 10.75$, so the residual for the first piece of data is $e_{11} = X_{11} - \bar{X}_1 = 8 - 10.75 = -2.75$. This indicates that the observation was 2.75 *below* the average for that group.

The remaining residuals are:

A	B	C	D
−2.75	0	−0.25	−1.33
3.25	2	−2.25	1.67
−1.75	−2	0.75	−0.33
1.25		1.75	

Note that residuals can be positive or negative, and also that the total of the residuals for each group is approximately zero (it will be approximate if you have rounded any residuals).

Ordering these residuals from least to greatest we get:

$$-2.75, -2.25, -2, -1.75, -1.33, -0.33, -0.25, 0, 0.75, 1.25, 1.67, 1.75, 2, 3.25.$$

We have n = 14 of them and the median is worked out by calculating $14 \times 0.5 = 7$, a whole number, so the median is halfway between the seventh and eighth observations, i.e. halfway between −0.25 and 0 and is −0.125. The lower quartile is obtained by $14 \times 0.25 = 3.5$, which is not a whole number and so we choose the fourth observation, i.e. $LQ = -1.75$. Similarly, the upper quartile is fourth from the top end of the data and is 1.67. So a boxplot of the residuals looks like this:

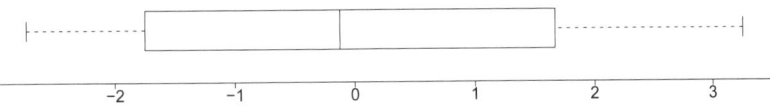

This is highly symmetric so we have no reason to doubt the underlying normality of our data and can report the results of the ANOVA with some confidence.

11.2 Small and non-normal sample test for medians

In the case of small, non-normal samples, we can neither assume normality, nor rely on the CLT to provide it. In Chapter 9 we developed the sign test, which was based on whether each observation was above or below the hypothesised population median. In Chapter 10, we used the Mann-Whitney U-test, which is based on the ranks of the observations, and how these ranks are distributed between the samples. In the case of three or more samples, we use a test called the *Kruskal-Wallis* test, which is also based on the observed distribution of the ranks.

Hypotheses

As with the Mann-Whitney U-test, the most general null hypothesis that we can test is that all k populations (with $k \geq 3$) have identical distributions, against the alternative that at least one is different. If we can assume that the populations are identical apart from their locations, we can test the null hypothesis that the k medians are equal, against the alternative that they are not all equal. Using the notation $\mathcal{M}_i$ to denote the population median of the ith group, we can write the hypotheses as:

$$H_0 : \mathcal{M}_1 = \mathcal{M}_2 = \cdots = \mathcal{M}_k$$
$$H_a : \text{at least one of the } \mathcal{M}_i \text{ is different.}$$

These hypotheses are identical to those for the ANOVA test, except that they are written in terms of the population *medians* rather than the population means. If the populations are symmetric, then the means and medians coincide, and the hypotheses will be equivalent.

Test statistic

Like the ANOVA test, our data consist of k samples, with sizes $n_1, n_2, \ldots, n_k$. These sample sizes are not necessarily equal. In order to calculate the test statistic, we must first:

- pool and rank the data
- allocate ranks, and deal with any ties
- sum the ranks for each group, giving $R_1, R_2, \ldots, R_k$.

11.2 Small and non-normal sample test for medians

As with the calculation of Spearman's correlation in Section 4.2.2 and the Mann-Whitney U-statistic in Section 10.4.1, any tied observations get the average rank.

As with the sums S_1 and S_2 in the Mann-Whitney U-test, the size of the sums R_i will depend on the number of observations in the sample. In this test we base the test statistic on the *average rank*, rather than deducting the minimum sum (as was the case in the Mann-Whitney test).

The Kruskal-Wallis test statistic is

$$K = \frac{12}{n(n+1)} \left(\sum_{i=1}^{k} \frac{R_i^2}{n_i} \right) - 3(n+1)$$

and this has an approximate chi-squared distribution* with $k-1$ degrees of freedom.
*"chi" is pronounced "kai" as in the Māori word for "food".

The factor $12/(n(n+1))$ in the test statistic has complicated origins, but it is used to ensure the chi-squared distribution is appropriate. The chi-squared distribution is similar in nature to the F-distribution: it is always positive, and asymmetric. Here, each sample provides a degree of freedom, but we lose one because the R_i must add to $n(n+1)/2$, where $n = n_1 + \cdots + n_k$.

Rejection region

Since the test statistic has an approximate chi-squared distribution with $k-1$ degrees of freedom, denoted χ^2_{k-1}, we use this distribution to obtain the rejection region for the test. When H_0 is true, we expect small values of the test statistic, since the average ranks will be close to the overall average. Since large values of the test statistic will lend evidence toward H_a, we reject the null in the *upper tail* of the chi-squared distribution only, i.e. when $K > \chi^2_{crit}$. We obtain the critical value χ^2_{crit} from Table D.7 on page 332 by specifying the degrees of freedom (for the row) and the upper tail probability (for the column).

Conclusion

Once again, we draw the conclusion for the test by comparing the test statistic to the rejection region. If the test statistic lies in the rejection region, we reject the null hypothesis in favour of the alternative at the $100\alpha\%$ level. Otherwise, we cannot reject the null.

Example 11.6

Q: Fourteen rugby league players testing two high-protein diet supplements are allocated at random to three groups. Groups A and B are each put on one of the two supplements. Group C is a control group. All the subjects followed the same diet differing only in whether or not a dietary supplement was taken. Weight gains (in grams) over the first week were:

A	B	C
152	40	−34
80	92	61
124	−28	−100
139	51	80
82	52	

Using a Kruskal-Wallis test, decide whether there is sufficient evidence for differing median weight gains across the three groups.

A: The null hypothesis is $H_0 : \mathcal{M}_A = \mathcal{M}_B = \mathcal{M}_C$ against the alternative that at least one median differs from the others. The pooled and sorted data, with ranks, are in the following table:

Data:	−100	−34	−28	40	51	52	61	80	80	82	92	124	139	152
Group:	C	C	B	B	B	B	C	C	A	A	B	A	A	A
Rank:	1	2	3	4	5	6	7	8.5	8.5	10	11	12	13	14

where the common scores of 80 get the average rank $\frac{8+9}{2} = 8.5$. Summing these ranks for each group, we get:

$$R_A = 8.5+10+12+13+14 = 57.5,\ R_B = 3+4+5+6+11 = 29,\ R_C = 1+2+7+8.5 = 18.5$$

which add to $\frac{14 \times 15}{2} = 105$ as required. We calculate

$$\sum \frac{R_i^2}{n_i} = \frac{57.5^2}{5} + \frac{29^2}{5} + \frac{18.5^2}{4} = 915.0125$$

so the test statistic is

$$K = \frac{12}{n(n+1)} \left(\sum \frac{R_i^2}{n_i} \right) - 3(n+1) = \frac{12}{14(14+1)} \times 915.0125 - 3 \times (14+1) = 7.286$$

and this has a chi-square distribution with $k - 1 = 2$ degrees of freedom. The 5% upper-tail value from this distribution is 5.991, and since $7.286 > 5.991$ we reject the null hypothesis in favour of the alternative at the 5% level. Looking back at the data, we see the greatest scores were for Group A, indicating higher weight gain in this group.

Exercises

The ANOVA test

We suggest you use $\alpha = 5\%$ for these tests.

11.1 A random sample of 50 New Zealanders, 10 each from Auckland (AKL), Hamilton (HLZ), Wellington (WLG), Christchurch (CHC) and Dunedin

(DUD) completed Tobacyk's 1988 *Revised Paranormal Belief Scale*. The following are their mean scores of responses to 26 paranormal phenomena. Higher scores mean greater belief in the paranormal.

	AKL	HLZ	WLG	CHC	DUD
	1.81	1.42	1.19	3.23	1.15
	1.19	1.38	1.88	1.35	2.69
	1.15	1.23	1.69	1.69	1.85
	3.81	3.77	3.54	1.35	4.04
	2.69	3.35	1.38	1.62	1.96
	0.96	1.38	2.15	1.38	3.46
	2.00	1.50	1.31	1.12	3.31
	1.23	1.85	1.38	1.46	1.85
	3.27	3.69	1.65	1.15	1.15
	2.58	1.23	1.31	4.27	2.62
T_i	20.69	20.80	17.48	18.62	24.08
s_i	0.99	1.07	0.70	1.04	0.98

(a) Complete the ANOVA table for these data.
(b) Conduct the ANOVA test, stating your hypotheses and conclusion clearly.
(c) Calculate half confidence intervals for these cities and show them on a number line. Discuss your graph.
(d) What is the best point estimate for the average paranormal belief score for Wellingtonians? Why?

(*Special thanks to Marc Wilson for access to these data*)

11.2 A study on the habitat of two species of limpets (*Benhamina obliquata* and *Siphonaria australis*) reports the following ANOVA table:

Source	MS	df	F
Tidal condition	1.095	2	41.466
Error	0.026	56	

A *p*-value of < 0.001 is also reported.
(a) Explain how the F statistic in the table is found.
(b) Find the rejection region for the test, and verify the conclusion implied by the *p*-value.

(*Data from Russell & Phillips, Synergistic effects of ultraviolet radiation and conditions at low tide on egg masses of limpets* (Benhamina obliquata *and* Siphonaria australis) *in New Zealand, Marine Biology, 2009*)

11.3 A laboratory study examined the survival time of a colony of Argentine ants when introduced into a colony of native ants (*Monomorium antarcticum*) consisting of queens and 200 workers. The size of the invading colony, called propagule size, was manipulated, to provide a sample of survival times for each of three groups consisting of two queens and 10, 100 or 200 workers. Summary statistics are:

Source	df	SS
Propagule size	2	3925.18
Residual	24	14323.25

(a) Complete the ANOVA table.
(b) Using the ANOVA table, perform a test for equality of mean survival times.
(c) What assumptions are needed for the ANOVA test performed here to be reliable?

(*Data from Sagata & Lester, Behavioural plasticity associated with propagule size, resources, and the invasion success of the Argentine ant* Linepithema humile, *Journal of Applied Ecology, 2009*)

11.4 The following data are the perceived social support scores of 66 thirteen year olds grouped according to how many other people are in their immediate family. Families consisting of two other people were classified as "small", those with four other people as "medium" and those with six others as "large".

				Small family				
19	20	16	21	23	18	23	26	26	18
12	31	12	12	25	17	28	27	14

				Medium family				
11	11	12	15	13	10	11	15	22	22
12	27	21	15	20	14	21	13	18	26
17	19	14	19	21	21	16	17	35	22
16	23	21	11	19	19	17	21	18	22

			Large family			
11	21	11	16	8	18	17

(a) Draw stemplots of the three samples, and comment on the suitability of the ANOVA test for these data.
(b) Complete the ANOVA table.
(c) Conduct a formal hypothesis test for equality of mean perceived scores.

(Many thanks to Paul Jose for access to raw data, from Jose & Brown, When does the gender difference in rumination begin? Gender and age differences in the use of rumination by adolescents, Journal of Youth and Adolescence, 2008)

11.5 A study into hospital admission rates in bipolar patients investigated seasonal variation in these rates. They reported "$F(3, 128) = 4.515, p < 0.005$".
(a) How can you tell the data were grouped into four seasons?
(b) Can you deduce the total number of hospital admissions?
(c) The test statistic was 4.515. What was the critical value for the ANOVA test, and what was the conclusion?
(d) Is your conclusion consistent with the stated p-value? Explain.

(Results published in Bipolar Disorders, 2004)

11.6 A study of bacteria in Antarctic sea ice looked at production by cells collected from melted ice cores. The response variable is the time point estimate of radioactive leucine (in pico-moles per litre per hour) incorporated after 48 hours by bottom-ice bacteria incubated at three saline concentrations (salinity, in parts per thousand ‰):

Salinity	$\bar{x}$	s	n
10	84.7	14.4	3
33	739.9	156.3	3
55	488.3	87.4	3

(a) Using the data provided, calculate the sample totals T_i, the grand total T, and the treatment sum of squares.
(b) Complete the ANOVA table for these data.
(c) Test for equality of mean times for these three salinity levels.
(d) Comment on the suitability of the assumptions of your test.

(Thanks to Andrew Martin and Ken Ryan for the data; from Martin, Hall & Ryan, Low salinity and high-level UV-B radiation reduce single-cell activity in Antarctic sea ice bacteria, Applied and Environmental Microbiology, 2009)

11.7 A Malaghan Institute of Medical Research study examined the response of the immune system in mice to a vaccine with priming and without. In the primed mice, the vaccine was added to mouse immune cells in culture, and these were injected into the mice. One week later, this process is repeated, with the vaccine at four different concentrations: zero (the control), 5 nM, 50 nM and 5 μM. The primed mice receive the vaccination, as do a similar group of non-primed mice who received no vaccine a week earlier. These injected cells now become "target cells" for the immune system. The following table gives the ratio of control cells to target cells for mice with each concentration vaccine, normalised using a control mouse:

5 nM	50 nM	5 μM
3.38	8.75	50.00
3.91	11.30	16.00
0.78	11.50	10.00
5.20	26.00	44.00
2.00	10.00	33.00

(a) Draw up an ANOVA table for these data.
(b) Test whether the observed differences in the ratios are significant at the 5% level of significance.
(c) Find 95% half confidence intervals for the mean ratios for each of the concentrations and display on a number line. Comment on what this indicates.
(d) Find a 95% confidence interval for the difference between the mean ratio for the mice with concentrations 5 nM and 5 μM. Comment on the claim that there is no difference in the mean ratios for these two groups.

(*Thanks to Peter Ferguson from the Malaghan Institute for the data*)

11.8 A study on the effects of playing violent video games on the anger levels of adolescents categorised subjects according to their anger levels before and after the game: increase, decrease, or no change. The following table summarises the pre-game anger levels for the three groups:

	Increase	Decrease	No change
n	22	8	77
$\bar{x}$	11.64	22.50	12.05
s	2.79	5.71	3.51

(a) Construct the ANOVA table to compare the pre-game anger levels for these three groups.
(b) Are the anger levels significantly different?
(c) State the assumptions of your test, and explain how they might be checked.

(*Data from Unsworth, Devilly & Ward, The effect of playing violent video games on adolescents: should parents be quaking in their boots?, Psychology, Crime and Law, 2007*)

11.9 Following from Exercise 4.42, the table below lists the double glazing cost savings (in $ per annum) for ten identical houses, assuming location in Dunedin, Christchurch, Wellington and Auckland. Negative savings are costs.

Dun	Chch	Welly	Auck
336	246	188	−3
118	56	16	−120
228	147	95	−78
70	9	−30	−161
137	83	49	−66
295	278	149	−41
198	114	61	−120
190	118	71	−79
103	61	34	−57
192	123	79	−70

(a) Complete the ANOVA table for these data.
(b) Are the average savings significantly different?
(c) Calculate half confidence intervals for the average saving in each city, and draw them on a number line. Comment on your diagram with reference to the outcome of your test.

(*Data from Smith, A cost benefit analysis of secondary glazing as a retrofit alternative for New Zealand homes, VUW thesis, 2009*)

11.10 Following from Exercise 11.9, the houses are the *same* houses, so the samples are matched.
(a) Form differences for Dunedin, Christchurch and Wellington relative to Auckland, e.g. subtract Auckland's savings from Dunedin's. For the first house this will be $336 - (-3) = 339$.

(b) Perform an ANOVA test for the three samples.
(c) What conclusions can you draw?

11.11 Little is known about the the ecotoxic effects of prescription drugs and self-care products, many of which end up in coastal waters through human sewerage. In particular, psychotropic medications have the potential to significantly alter the behaviour of a wide range of animals. Investigating this phenomenon, reseachers exposed shrimps (*Echinogammarus marinus*) to five concentrations of the anti-depressant fluoxetine: none for the control, fluoxetine concentrations of 0.01, 0.1, 1 and 10 micrograms per litre. Specifically, the phototaxis scores in the table below describe the direction of swimming of six shrimps in each treatment over a period of time.

		Concentration		
0	0.01	0.1	1	10
0.4	2.9	3.7	4.3	3.6
0.8	3.2	1.7	4.0	2.9
0.3	3.0	5.9	4.4	2.8
1.5	1.9	4.4	4.2	1.5
0.9	4.1	4.5	3.3	2.5
0.3	2.7	5.7	0.8	3.4

(a) Complete the ANOVA table.
(b) Using the ANOVA table, perform a test for equality of mean geotaxis for the three concentrations.
(c) Calculate half confidence intervals for the three means, and display on a number line. Does your diagram confirm the outcome of the ANOVA?

(*Based on Guler & Ford, Anti-depressants make amphipods see the light, Aquatic Toxicology, 2010*)

11.12 To investigate the concern that breakfast cereal marketed towards young children (e.g. using cartoon characters on the packaging) has higher sugar content than adults' cereals, random samples of both types of cereal are collected. In addition to New Zealand data for these cereals, products available in Canada and marketed towards children are also sampled. The total amount of sugar (listed as *Carbohydrates – sugars*, in grams per 100 grams) listed on the packaging is given below:

Kids' cereals				
8.1	16.5	17.8	33.4	35.4
38.0	44.0			

Adults' cereals				
2.3	2.8	7.9	13.6	22.7
24.0	29.5	31.2	32.0	

North American kids' cereals				
32.3	33.3	33.3	38.7	39.3
41.4	43.3	53.3		

(a) Complete the ANOVA table for these data.
(b) Is there sufficient evidence to suspect different population means across the three groups?
(c) Calculate half confidence intervals and show these on a number line.
(d) Summarise your findings.

11.13 Following Exercise 9.44, tail shedding is a defence mechanism against predation used by lizards when other methods do not work. How quickly this occurs is very important for the lizards' survival. Adults of three species of native New Zealand skink (*Oligosoma* spp.) were studied, and latency of autotomy, i.e. how long it took for them to lose their tails in a threatening situation, recorded (in seconds). The table below summarises the results:

	$\bar{x}$	s	n
O. aeneum	69.3	72.1	13
O. polychroma	68.2	76.9	24
O. zelandicum	35.2	48.2	17

(a) Complete the ANOVA table.
(b) Using the ANOVA table, perform a test for equality of mean autotomy times for the three species.
(c) Calculate half confidence intervals for the three means, and display on a number line. Does your diagram confirm the outcome of the ANOVA?

(*Thanks to Kelly Hare for the data. From Hare & Miller, Frequency of tail loss does not reflect innate predisposition in temperate New Zealand lizards, Naturwissenschaften, 2010*)

11.14 In response to complaints that "movies are far too long these days", the following sample of movie lengths was obtained. The movies were selected from imdb.com's Top 250 Films as voted by users, and grouped according to their release date (pre-90s, 1990s, post-90s). The times (in minutes) are below:

Pre-90s	1990s	Post-90s
175	142	103
200	154	152
161	195	201
96	146	130
133	139	178
121	106	113
207	118	179
102	136	
112	127	
109	110	
175		
130		
$\bar{x}$ = 137.3	143.4	150.9
s = 25.6	38.7	37.0

(a) Comment on the sampling method for these data. What are the likely populations?

(b) Complete the ANOVA table for the data.
(c) Test for a difference in means.
(d) Form half confidence intervals for the group means and draw these on a number line. Comment.

(*Data from* imdb.com)

11.15 Following Exercise 10.3, a study into public transport usage in Wellington used a social comfort scale to assess participants' levels of discomfort in social settings (low scores correspond to a higher level of discomfort). Summary statistics by mode of travel are:

	$\bar{x}$	s	n
Car	2.23	0.52	305
Train	2.40	0.61	319
Bus	2.40	0.52	216

(a) Complete the ANOVA table.
(b) Using the ANOVA table, perform a test for equality of mean discomfort levels for the three groups.
(c) Calculate half confidence intervals for the three means, and display on a number line. Does your diagram confirm the outcome of the ANOVA?

(*Data from Thomas, The social environment of public transport, VUW thesis, 2009*)

11.16 Bipolar Disorder has been associated with irregularities in the endocrine system. A study used Magnetic Resonance Imaging (MRI) to measure the pituitary gland volume (PGV, in mm^3) of patients with Bipolar Disorder, their immediate relatives and age-matched controls. Summary statistics for the PGV are presented below.

	$\bar{x}$	s	n
Healthy controls	682	158	52
Bipolar patients	793	122	29
Patients' relatives	678	170	49

(a) Complete the ANOVA table.
(b) Using the ANOVA table, perform a test for equality of mean pituitary gland volume for the three groups.
(c) Calculate half confidence intervals for the three means, and display on a number line. Does your diagram confirm the outcome of the ANOVA?

(*Data from Takahashi et al., Pituitary volume in patients with bipolar disorder and their first-degree relatives. Journal of Affective Disorders, 2010*)

11.17 The following multiple EBIT (earnings before interest and taxes) figures are from a report on firms listed on the New Zealand Stock Exchange in three industry sectors:

Energy	Ports	Property
22.2	22.5	−4.7
11.5	17.7	28.0
43.2	−143.7	78.2
6.4	17.5	−36.1
15.5	12.4	−11.1
10.8		−19.7
		−10.6
		−15.4

(a) Complete the ANOVA table.
(b) Using the ANOVA table, perform a test for equality of mean multiple EBIT for the three groups.
(c) Calculate half confidence intervals for the three means, and display on a number line. Does your diagram confirm the outcome of the ANOVA?

(*From* pwc.com/nz/en/cost-of-capital)

Kruskall-Wallis

We suggest you use $\alpha = 5\%$ for these tests.

11.18 The following table contains a random sample of individual donations paid into a donation box which had been manipulated by researchers. The "notes" box was pre-loaded with $100 containing a large number of $5 bills, the "coins" box was pre-loaded with $100 containing a large number of silver coins, and the "empty" donation box was literally emptied after every donation.

Notes	Coins	Empty
5.00	1.00	2.00
10.00	1.50	2.00
5.00	0.70	1.00
2.00	1.10	1.50
1.50	1.90	5.00
2.00	2.00	2.00
20.00	1.00	2.00
5.00	1.50	1.00
1.00	5.00	2.00

(a) Why is the ANOVA test an inappropriate one with which to compare the population mean donation per donor into these three "donation boxes"?
(b) Perform a Kruskall-Wallis test using these samples.
(c) Summarise your findings.

(*Data from Martin & Randal, How is donation behaviour affected by the donations of others? Journal of Economic Behavior and Organization, 2008*)

For the following exercises, conduct a test for equal medians for the stated date.

11.19 Paranormal belief scores (11.1)
11.20 Support scores (11.4)
11.21 Control to target cell ratios (11.7)
11.22 Double glazing cost savings (11.9)
11.23 Shrimp phototaxis (11.11)
11.24 Sugar content of cereals (11.12)
11.25 Movie lengths (11.14)
11.26 Multiple EBIT (11.17)

Chapter 12

Tests for categorical data

Categorical data has been displayed graphically using bar charts in Section 2.4, and used in Chapter 7 when calculating proportions, i.e. counting the number of sample members with a particular characteristic in a category and dividing by the total sample size.

In this chapter, we investigate two ways to analyse categorical data. The first is a goodness of fit test, where we see if the distribution of observed frequencies in categories is consistent with some prior idea about the data.

The second situation we examine is when we have bivariate categorical data. In Chapter 4 we estimated a linear relationship between two *measurements*, and quantified the strength of that relationship. Regression does not apply to categorical variables, so in this chapter we look at a test for a link between two categorical variables.

12.1 One-way chi-square

The goodness of fit test is a special case of a *one-way chi-square* test. [Note that "chi" is pronounced "kai" as in the Māori word for "food".] This test can be used to analyse the shape of a barplot, i.e. we can test whether the set of proportions over the categories fits a particular belief about the situation.

12.1.1 The general test

Imagine that we have a situation where there are several possible outcomes, e.g. the six major religious groups in New Zealand. If we consider a sample of 100 New Zealanders, all of whom belong to one of these six groups, our sample size n is 100, and we have allocated them among $k = 6$ categories.

In general, we assume that we have n observations distributed over k categories, as in the above example. These k categories must include all possible outcomes. Each category has an associated population proportion $p_i = P(\text{individual belonging to category } i)$. Because the categories cover all possibilities, the sum of their probabilities is 1, i.e. $\sum p_i = 1$.

Hypotheses

The most general null hypothesis we can test is $H_0 : \{p_1 = p_1^*, p_2 = p_2^*, \ldots, p_k = p_k^*\}$ where $\sum p_i^* = 1$, i.e. the probability of falling into the first category is p_1^*, into the second is p_2^*, etc. The alternative hypothesis is H_a : at least one proportion is not as specified in H_0. [Since the proportions must add to 1, one of the k proportions is fixed when we specify the other $k - 1$ proportions, so if H_a is true, we must actually have two or more proportions differing from their H_0 values.]

A special case of the null hypothesis is $H_0 : p_1 = p_2 = \cdots = p_k$ where all the categories have the same probability. An alternative form of this particular hypothesis, due to the requirement that $\sum p_i = 1$, is $H_0 : p_i = \frac{1}{k}$ for all i, and so we can test whether a set of sample proportions is significantly different.

Test statistic

We now collect the sample data, and this will consist of the n observations split up into the k categories.

> We define the number of observations in category i to be o_i and call this the *observed* frequency.

Since we have n observations in the sample, the sum of the observed frequencies is $\sum o_i = n$.

In addition to the observed frequencies provided by the sample, we can obtain *expected* frequencies from H_0. We do this by focusing on one category at a time, and view falling into this category as a success and not falling into it as a failure. Since the probability of falling into the first category is p_1, the number of individuals in the first category is binomial with n trials, and probability of success p_1, with a mean number of $n \times p_1$ (see Chapter 7).

> The expected frequency in category i is $e_i = n \times p_i$, where p_i is the population proportion for category i, obtained from H_0.

Since the proportions p_i sum to 1, the expected frequencies sum to n, as do the observed frequencies. If H_0 is true, we have the probabilities about right, and we would expect the observed frequencies to be close to the expected frequencies. So it makes sense to calculate the differences $o_i - e_i$. As with the sample variance, we do not care whether e_i exceeds o_i or vice versa, so we square these differences giving $(o_i - e_i)^2$. The test statistic is formed by scaling these contributions appropriately (in this case, dividing by the expected frequency e_i), and then summing the ratios up.

> The test statistic is
> $$\chi^2 = \sum_{i=1}^{k} \frac{(o_i - e_i)^2}{e_i}$$
> and this has an approximate chi-square distribution, with $k - 1$ degrees of freedom.

The test statistic is always positive, unless every expected frequency equals its observed frequency, in which case the test statistic is zero.

The degrees of freedom are due to the number of proportions we are free to specify. In this case, we have k categories, but we also have a constraint that $\sum p_i = 1$. As a consequence, we can only freely choose $k-1$ proportions, i.e. once we have set $p_1^*, p_2^*, \ldots, p_{k-1}^*$ in H_0, the final proportion must be obtained using $p_k^* = 1 - p_1^* - \cdots - p_{k-1}^*$.

Rejection region

The chi-square test statistic has a chi-square distribution with $\nu = k - 1$ degrees of freedom. We obtain the rejection region for this test from the chi-square table on page 332. As usual, we must specify α, and since the table is arranged by the right-hand probability $P(X > x)$, we use the column headed α for the critical value. We choose the row according to the number of degrees of freedom we have, and the value we obtain is the critical value for the test.

Conclusion

We compare the test statistic to the rejection region. If the test statistic lies in the rejection region, we reject the null hypothesis in favour of the alternative at the $100\alpha\%$ level. If the test statistic does not lie in the rejection region, we cannot reject the null hypothesis, indicating that the relative frequencies of the sample data do not differ significantly from the proportions specified in H_0.

Example 12.1

Q: A hopeless gambler is worried that her opponent is playing with an unfair die. She "borrows" the die and rolls it 120 times, with the following results:

Number	1	2	3	4	5	6
Frequency	23	18	24	19	25	11

Test whether or not the die is fair, at the 5% level.

A: The null hypothesis is that $p_i = 1/6$ for each i, i.e. each side of the die is equally likely to come up, with the alternative hypothesis that the p_i are not all equal.

The expected frequencies are all $n \times p_i = 120 \times 1/6 = 20$, so the test statistic is

$$\chi^2 = \sum \frac{(o_i - e_i)^2}{e_i} = \frac{(23 - 20)^2}{20} + \frac{(18 - 20)^2}{20} + \cdots + \frac{(11 - 20)^2}{20}$$

$$= \frac{3^2}{20} + \frac{(-2)^2}{20} + \frac{4^2}{20} + \frac{(-1)^2}{20} + \frac{5^2}{20} + \frac{(-9)^2}{20} = 6.8$$

and this has a chi-square distribution with $k - 1 = 6 - 1 = 5$ degrees of freedom. The critical value with $\alpha = 0.05$ is 11.070 and so we cannot reject the null hypothesis.

Although we did not reject H_0, notice that there were many more "odds" than "evens" (we could do an hypothesis test to check the proportions of "odds" for example), so the gambler could be correct in her concerns about the die.

12.1.2 Assumptions

In order for the chi-square distribution to be a good approximation to the distribution of the test statistic, we need a few conditions to be met. As in all our statistical tests, it is critical that we have a random sample, and in this particular case, we require that each observation falls into one and only one category. In addition to this requirement, at least 80% of the *expected* frequencies that we calculate should be greater than 5.

Once we have a random sample, and the data organised into categories, we can usually fix any problems with small expected frequencies using a process called *amalgamation*. This is where we combine categories so that the expected frequencies are large enough. The categories we combine should be "near" to one another, in the context of what we are testing for, e.g. we might combine the categories corresponding to the number of individuals living in a household so that we have three categories: small, medium and large. This process is further illustrated in Example 12.2 in the next section.

12.1.3 The goodness of fit test

In addition to testing $H_0 : \{p_1 = p_1^*, p_2 = p_2^*, \ldots, p_k = p_k^*\}$ where the p_i^* add to 1 and are specified in advance, we may use the one-way chi-square test to conduct a *goodness of fit* test. A goodness of fit test can answer a question like: "are the sample data drawn from a binomial distribution with proportion $p = 0.5$?" or "is the population normally distributed?" The difference between these two questions is that, in the first, the parameters of the binomial distribution are specified, whereas in the second, we merely specify the shape of the distribution. The goodness of fit test can handle both.

Specification of the hypotheses in terms of the distribution of the data automatically provides the p_i^* as in the previous section. This is demonstrated in the following example. In this case, the remainder of the test is identical to the one-way test in Section 12.1.

Example 12.2

A short multichoice test, with seven questions, has had an average score of 60% when used in the past. Ninety-two students sat the test, with the following distribution:

Score	0	1	2	3	4	5	6	7
Frequency	0	1	4	11	31	19	18	8

If the average score is still 60%, and the questions are independent, with the same probability of success, the individual scores will be binomial. Use these data to test the hypothesis that the total score on the test is binomial, with $n = 7$ and $p = 0.6$.

Using the binomial table on page 325, we obtain the probabilities $p_i = P(X = x_i)$ for the $B(7, 0.6)$ distribution. These can be multiplied by $n = 92$ to find the expected frequencies e_i, as shown in the following table.

Score = x_i	0	1	2	3	4	5	6	7
p_i	0.0016	0.0172	0.0774	0.1935	0.2903	0.2613	0.1306	0.0280
$e_i = np_i$	0.15	1.58	7.12	17.81	26.71	24.04	12.02	2.58

We note that three out of eight categories (37.5%) have expected frequencies less than 5. This is greater than the permissible 20% so we must amalgamate. Combining the categories 0, 1 and 2, the probability for this category becomes $p = 0.0016 + 0.0172 + 0.0774 = 0.0962$ giving an expected frequency of $e = 92 \times 0.0962 = 8.85$. (Alternatively, we could have added $0.15 + 1.58 + 7.12$.) This leaves one category from the new total of six with an expected frequency less than 5, so we can continue with the test.

Note that the corresponding observed frequency for the amalgamated category is $0 + 1 + 4 = 5$.

Score	0,1,2	3	4	5	6	7
Observed o_i	5	11	31	19	18	8
Expected e_i	8.85	17.81	26.71	24.04	12.02	2.58

The null hypothesis is H_0 : the scores have a binomial distribution with $n = 7$ and $p = 0.6$. The alternative hypothesis is that the scores do not have the stated binomial distribution. If we reject the null hypothesis, the test will not tell us why this was the case. It is possibly due to a poor choice of p, or alternatively, that the binomial distribution is inappropriate (perhaps the questions are not independent, or perhaps the probability of success is different for each question).

An alternative expression for the null hypothesis is to list the probabilities $p_i = P(X = x_i)$ shown in the table above. This is much more cumbersome, and it is less obvious that they are obtained from the binomial distribution. Nonetheless, it reminds us that the test is no different from the general one-way chi-square test.

The test statistic is

$$\chi^2 = \frac{(5-8.85)^2}{8.85} + \frac{(11-17.81)^2}{17.81} + \cdots + \frac{(8-2.58)^2}{2.58} = 20.39$$

and this has an approximate chi-square distribution with $k - 1 = 6 - 1 = 5$ degrees of freedom. The critical value at the 5% level is 11.070 and so we reject the null hypothesis in favour of the alternative at the 5% level.

12.1.4 Optional advanced material

We must make a change to the one-way chi-square test if we conduct a goodness of fit test in which we estimate population parameters from the data. In the case of estimating parameters, we must deduct one degree of freedom for each parameter we estimate (in addition to the degree of freedom we already deduct for the constraint that $\sum p_i = 1$). For example, if we are testing for a normal distribution using k categories, and we estimate both μ and σ, we use $k - 1 - 2 = k - 3$ degrees of freedom.

The population mean and standard deviation for the normal distribution are estimated using the sample mean and variance respectively. If we do not have the original data but only the categorical data, we must use the approximation of Section 3.2.3 based on the midpoints of the categories.

Estimating the binomial proportion is best done by noting that the mean of the binomial distribution is k times p, where k is the number of categories, and is equal to the number of trials in the binomial experiment. The sample mean estimates kp and so we can estimate p by dividing the sample mean by k. Note that the number of trials in the binomial experiment is $k =$ the number of categories, not the sample size n.

Example 12.3

We repeat the test in Example 12.2 for the case where the proportion is unspecified.

The sample mean score is

$$\bar{X} = \frac{1}{n}\sum f_i x_i = \frac{1}{92}(0 \times 0 + 1 \times 1 + 4 \times 2 + \cdots + 8 \times 7) = 4.62.$$

This gives the estimate of the population proportion to be

$$p = \frac{\bar{X}}{k} = \frac{4.62}{7} = 0.660.$$

We note that this is higher than the 0.6 we tested in Example 12.2 consistent with our rejecting the null hypothesis previously.

Unfortunately, we do not have the binomial table for $p = 0.66$. A computer gives expected frequencies:

0.05 0.66 3.82 12.37 24.01 27.97 18.10 5.02

which yield a chi-square statistic of $\chi^2 = 6.88$ (with amalgamation of 0, 1, and 2 as before, leaving six categories). The critical value from the chi-square distribution with $k - 1 - 1 = 6 - 1 - 1 = 4$ degrees of freedom is 9.488 (where we deducted one degree of freedom because we estimated p), and so we do not reject the null hypothesis in this instance. It appears that the scores do have a binomial distribution, but with a higher probability of success than predicted by the instructor, i.e. $p = 0.6$ is too low for this class.

12.2 Contingency tables

The strength of the relationship between two variables can be measured in different ways depending on whether data is quantitative or categorical. In Chapter 4, we looked at the strength of a linear relationship between two quantitative variables using Pearson's correlation, and also at the strength of the relationship between two ordinal variables using Spearman's correlation. *Contingency tables* are used in problems in which data is classified by two or more categorical variables and we wish to know if these classifications are independent or if there is some relationship between them.

As with the earlier chi-square tests, we need not restrict ourselves to categorical variables. We can also use contingency tables if one or both of the variables are continuous, by dividing the possible values of the continuous variable(s) into non-overlapping intervals that cover the possible range of the data. This is identical to constructing the classes of a frequency table for a histogram and our categories become the classes that we have.

Example 12.4

Sixty-four prison inmates who have all been convicted of crimes of violence are given the opportunity to undergo either a drug treatment or a counselling treatment in order to manage their anger. Some prisoners refused either option. After a six-month period on the treatment of their choice the inmates were clinically assessed as to whether there had been any improvement in their anger control. The results were as follows:

	Treatment type		
	Drug treatment	Counselling	No treatment
Improved	10	8	6
Not improved	6	16	18

The two variables: treatment and improvement, are categorical variables, and they are nominal, i.e. not ordered.

It is usual to add row and column totals to this table along with the overall total so the working table will look like this:

	Drug treatment	Counselling	No treatment	Total
Improved	10	8	6	24
Not improved	6	16	18	40
Total	16	24	24	64

We are interested in whether or not the treatment type has any bearing on the result of the treatment, and can conduct a hypothesis test based on the chi-square distribution.

We could represent this data in a stacked bar chart but it is not very easy to get a sense of what is going on.

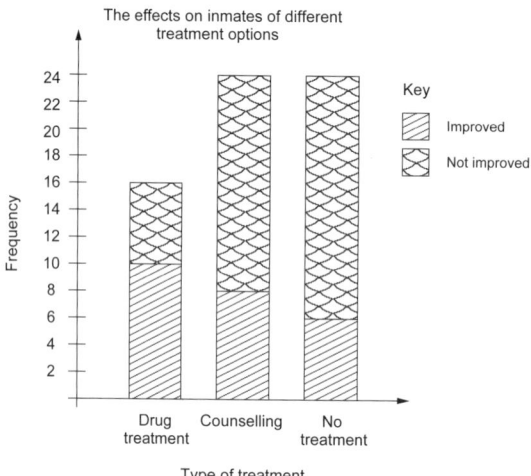

If treatment and improvement were independent, then we would expect the proportion of inmates who improved to be the same regardless of treatment. Overall 24 offenders improved, i.e. $24/64 = 0.375 = 37.5\%$ of the total. Since 16 had the drug treatment, if there were no relationship we would expect 0.375×16 to improve, i.e. 6 out of 16. Similarly, since, overall, 40 showed no improvement, we expect the proportion of the total $40/64 = 0.625 = 62.5\%$ not to improve. Of those who took part in the drug treatment, we would expect $0.625 \times 16 = 10$ individuals not to improve.

12.2.1 The general $r \times c$ case

The contingency table in Example 12.4 is in the general form reproduced in Figure 12.1. The individual data are classified according to the category they fall in for variable A (treatment type in the example) and for variable B (response in the example), and frequencies are reported for the combinations of these categories, e.g. drug treatment and improvement, counselling and improvement, through to no treatment and no improvement.

12.2 Contingency tables

	Variable A				
Variable B	A_1	A_2	$\cdots$	A_c	Total
B_1	o_{11}	o_{12}	$\cdots$	o_{1c}	r_1
B_2	o_{21}	o_{22}	$\cdots$	o_{2c}	r_2
$\vdots$	$\vdots$	$\vdots$	$\ddots$	$\vdots$	$\vdots$
B_r	o_{r1}	o_{r2}	$\cdots$	o_{rc}	r_r
Total	c_1	c_2	$\cdots$	c_c	n

Figure 12.1 The contingency table

The observed frequencies are denoted o_{ij}, where the i denotes the row (the category of variable B) and the j denotes the column (the category in variable A). Each row has a total, and for row i, this total is r_i. Likewise, each column has a total, and for column j, this is c_j.

In summary:

o_{ij} = the observed number in the ijth cell, i.e. the cell in the ith row and jth column

r_i = the observed number of individuals in the ith row, i.e. the total of row i

c_j = the observed number in the jth column, i.e. the total of column j.

The row and column totals each add to n, the total sample size, which is also equal to the sum of the observed frequencies, i.e.

$$n = \sum r_i = \sum c_j = \sum o_{ij}.$$

Hypotheses

The null hypothesis for the contingency table test is that variable A and variable B are independent. That is, an individual being classified in a category in variable A has no bearing on the category they are in for B, and vice versa. In Example 12.4, this corresponds to an improvement being unrelated to the type of treatment received. The alternative hypothesis is that the variables are not independent, so knowledge of which category an individual is in for variable A gives us some information about their likely classification for variable B; e.g. knowing someone is receiving drug treatment may lead us to suspect they are more likely to show an improvement than not.

Test statistic

The null hypothesis specifies that the two categorical variables are independent, and this has the implication that the probability that an observation lies in categories A_j

and B_i is equal to the probability that the observation lies in category A_j times the probability that the observation lies in category B_i, i.e.

$$P(\text{observation in } A_j \text{ and } B_i) = P(\text{observation in } A_j) \times P(\text{observation in } B_i).$$

Multiplication of the probabilities of independent events was probability Rule 6 in Section 6.2.

From the contingency table, we observe c_j observations in category A_j, and so we estimate $P(\text{observation in } A_j)$ by the sample proportion c_j/n. Similarly, we observe r_i observations in category B_i, and so we estimate $P(\text{observation in } B_i)$ using the sample proportion r_i/n. Multiplying these sample proportions we have the estimate of $P(\text{observation in } A_j \text{ and } B_i)$ given by

$$\frac{c_j}{n} \times \frac{r_i}{n} = \frac{r_i \times c_j}{n^2}.$$

From the one-way chi-square test in Section 12.1, we know that the *expected* frequency in a category is the probability for that category times the sample size, n. This gives

$$n \times \frac{r_i \times c_j}{n^2} = \frac{r_i \times c_j}{n}.$$

If H_0 is true, the expected frequency for the contingency table is given by

$$e_{ij} = \frac{r_i \times c_j}{n}$$

where r_i is the number in row i, c_j the number in column j, and n is the total number of observations.

Example 12.5

The expected frequencies for the data in Example 12.4 are calculated as follows:

$$e_{11} = \frac{r_1 \times c_1}{n} = \frac{24 \times 16}{64} = 6 \quad \text{and} \quad e_{21} = \frac{r_2 \times c_1}{n} = \frac{40 \times 16}{64} = 10, \text{ etc.}$$

Calculating the remaining *expected frequencies*, and displaying these in a table:

	Drug treatment	Counselling	No treatment	Total
Improved	6	9	9	24
Not improved	10	15	15	40
Total	16	24	24	64

Unlike the previous example, the expected values are not usually whole numbers and you should not round them to be so. Usually, one decimal place will be enough, although in rare cases, it can be impossible to reconcile the row and column totals with this degree of rounding. Increasing the accuracy of the expected frequencies usually solves this problem.

The test statistic is based on comparison of the observed and expected frequencies, and if the difference between them is large, this is evidence against H_0. As in the one-way test, we add the scaled, squared differences in these frequencies, using exactly the same formula as before.

> The test statistic for the contingency table test is
> $$\chi^2 = \sum \frac{(o_{ij} - e_{ij})^2}{e_{ij}}$$
> and this has an approximate chi-squared distribution with $(r-1) \times (c-1)$ degrees of freedom.

The degrees of freedom are given by $(r-1) \times (c-1)$. Where does this come from? It is the number of expected frequencies that we can freely specify. Since each row of expected frequencies must add to the appropriate row total, and there are c expected frequencies per row, we can specify $(c-1)$ in each row. The last one is obtained by subtracting the others from the row total. Since there are r rows, once we have specified the e_i for $(r-1)$ of them, the final figure in each column can be found by subtracting from the column total. In total, we have $(r-1) \times (c-1)$ expected frequencies to specify (we use the formula $r_i \times c_j / n$), and the others we find by subtraction. This is illustrated in Figure 12.2. The best way to calculate the degrees of freedom is to remove a row and column from the contingency table, and count the remaining cells.

			Variable A		
Variable B	A_1	A_2	$\cdots$	A_{c-1}	A_c
B_1					e_{1c}
B_2					e_{2c}
$\vdots$		$(r-1) \times (c-1)$ cells			$\vdots$
B_{r-1}					$e_{r-1,c}$
B_r	e_{r1}	e_{r2}	$\cdots$	$e_{r,c-1}$	e_{rc}

Figure 12.2 The degrees of freedom calculation

Rejection region

If there is perfect agreement between the observed and expected frequencies, then we would expect the test statistic to equal zero. The more the observed and expected frequencies disagree, the *greater* the value of the test statistic, and the more of a relationship there is between the two variables. Consequently, this is a one-sided upper-tail test, and we obtain the critical value from the chi-squared distribution with $(r-1) \times (c-1)$ degrees of freedom.

Since the test statistic is based on a sum of positive terms, it can sometimes be useful to determine the rejection region before the test statistic, particularly when there are some really large differences present in the table. In doing so, we may find that we can greatly simplify our calculations. Calculating the remaining terms in the sum will only push our test statistic further into the rejection region and will not alter our decision.

Conclusion

If the test statistic lies in the rejection region, we reject the null hypothesis in favour of the alternative at the $100\alpha\%$ level, and conclude that the two categorical variables are not independent. If the test statistic is not in the rejection region, we cannot reject the null hypothesis, and we conclude that the variables are independent. This means that the category an individual falls into for one variable is completely unrelated to the category it falls into for the other.

Example 12.6

We now complete the test for the data in Example 12.4. The null hypothesis is H_0 : treatment type and improvement are independent, and the alternative is that they are not independent. It is common to combine observed and expected frequencies into a single display, with

$$o_{ij} \big/ e_{ij}$$

in each cell of the contingency table as follows:

	Drug treatment	Counselling	No treatment	Total
Improved	10 / 6	8 / 9	6 / 9	24
Not improved	6 / 10	16 / 15	18 / 15	40
Total	16	24	24	64

The test statistic is

$$\chi^2 = \frac{(10-6)^2}{6} + \frac{(8-9)^2}{9} + \frac{(6-9)^2}{9} + \frac{(6-10)^2}{10} + \frac{(16-15)^2}{15} + \frac{(18-15)^2}{15}$$
$$= \frac{4^2}{6} + \frac{(-1)^2}{9} + \frac{(-3)^2}{9} + \frac{(-4)^2}{10} + \frac{1^2}{15} + \frac{3^2}{15}$$
$$= 2.6667 + 0.1111 + 1 + 1.6 + 0.0667 + 0.6 = 6.044$$

and this has an approximate χ^2 distribution with $(r-1) \times (c-1) = 1 \times 2 = 2$ degrees of freedom. Using $\alpha = 0.05$, the critical value is $\chi^2 = 5.991$, and we will reject H_0 if the test statistic exceeds this value.

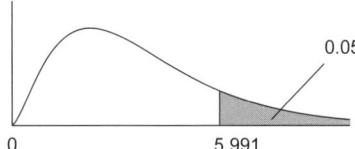

Since 6.044 is greater than 5.991, the test statistic lies in the rejection region, and we must reject H_0 in favour of H_a. The data suggests that improvement *is* related to the type of treatment administered. Looking back at the contingency table, we see that indeed, more of the subjects improve on the drug treatment than over the whole sample, while fewer improve with no treatment.

This example is completed using a spreadsheet in Appendix C.3.1.

In order for the contingency table test to be accurate, we again require that *at least 80% of the expected frequencies exceed* 5. If this is not the case, amalgamation must be performed by combining multiple rows or columns, and recalculating the e_i and the degrees of freedom accordingly.

12.2.2 Testing in the 2×2 case

In the case of a 2×2 contingency table, i.e. one with two rows and two columns, the test statistic needs a small change in order for the chi-square distribution to be a good one for the test statistic. The change is called *Yates' continuity correction* and it is in the same spirit as the continuity correction used in Section 8.4 for the normal approximation to the binomial distribution.

> For a 2×2 contingency table, Yates' continuity correction is used, and the test statistic is
>
> $$\chi^2 = \sum \frac{(|o_{ij} - e_{ij}| - 0.5)^2}{e_{ij}}$$
>
> and this has an approximate chi-squared distribution with 1 degree of freedom.

This test statistic is an easy one to implement, since for a 2×2 table, the absolute difference $|o_i - e_i|$ is the same for all four cells. We must subtract 0.5 from this number, before squaring, and dividing by the individual e_i. The hypotheses, distribution of the test statistic, rejection region, and conclusion follow the same format as the general test.

Example 12.7

Q: Survey data are collected from owners of small and medium businesses on whether the economy will improve, stay stable, or deteriorate over the next six months. The data collected were:

	Improve	Stable	Deteriorate	Total
Small-business owners	6	18	8	28
Medium-business owners	3	12	13	32
Total	9	30	21	60

Use the data to test whether the business size and owner's confidence are related.

A: The null hypothesis is H_0 : firm size and owner's opinion are *independent* against the alternative that they are related. The expected values are calculated by

$$e_{ij} = \frac{r_i \times c_j}{n}, \quad \text{e.g.} \quad e_{11} = \frac{28 \times 9}{60} = 4.2.$$

Completing the remaining entries, the contingency table is:

	Improve	Stable	Deteriorate	Total
Small-business owners	6 / 4.8	18 / 16	8 / 11.2	32
Medium-business owners	3 / 4.2	12 / 14	13 / 9.8	28
Total	9	30	21	60

We note that two of the e_{ij} are below 5, and this is a proportion of $33\frac{1}{3}\%$. This means we must amalgamate. Combining the "Improve" and "Stable" categories, we have the new contingency table

	Improve or Stable	Deteriorate	Total
Small-business owners	24 / 20.8	8 / 11.2	32
Medium-business owners	15 / 18.2	13 / 9.8	28
Total	39	21	60

where now $6 + 18 = 24$ small-business owners predict the economy either to improve or worsen, with an expected frequency of $4.8 + 16 = 20.8 = \frac{32 \times 39}{60}$. This table is now 2×2 so we must use Yates' continuity correction in the test statistic. For any cell, $|o_{ij} - e_{ij}| = 3.2$, so $|o_{ij} - e_{ij}| - 0.5 = 2.7$ for each cell. The test statistic is

$$\chi^2 = \sum \frac{(|o_{ij} - e_{ij}| - 0.5)^2}{e_{ij}} = \frac{2.7^2}{20.8} + \frac{2.7^2}{11.2} + \frac{2.7^2}{18.2} + \frac{2.7^2}{9.8} = 2.146.$$

The critical value from the chi-square distribution with 1 degree of freedom is 3.841, and since $2.146 < 3.841$, we cannot reject H_0 in favour of H_a at the 5% level. We have not found a significant relationship between business size and its owner's opinion.

Exercises

One-way chi-square

We suggest you use $\alpha = 5\%$ for these tests.

12.1 A fair die is rolled giving the following observed frequencies:

Outcome	1	2	3	4	5	6
Frequency	43	58	59	50	45	45

(a) Calculate expected frequencies for each of the outcomes.
(b) Perform the chi-square test to see whether the observed and expected frequencies differ significantly.

12.2 Select any 100 digits from the random number table in Appendix D (Table D.1), and tabulate the observed frequencies.
(a) If each digit (0 to 9) is equally likely, what is the expected frequency of each digit?
(b) Conduct a formal test for equal proportions of each digit.
(c) If 100 students perform the same test with different samples from Table D.1, how many would you expect to reject the null hypothesis?

12.3 The New Zealand short-tailed bat is a very ancient species and the only remaining member of the family *Mystacinidae* known to still survive. As part of a captive breeding programme, two short-tailed bats were bred. The female bat was homozygous dominant for dark fur colour (DD) and homozygous recessive for long ear length ($\ell\ell$). The male bat was homozygous recessive for light fur colour (dd) and homozygous dominant for short ear length (LL). These bats produced four offspring. These offspring were then crossed to produce a second generation. The following table shows the observed phenotypes (appearance) of the 48 second-generation bats and the number of each phenotype we would expect according to Mendelian inheritance laws.

Appearance	Obs.	Exp.
Dark fur, short ears	25	27
Dark fur, long ears	7	9
Light fur, short ears	12	9
Light fur, long ears	5	3

Do the observed data support the expected distribution of independently inherited appearance traits?

12.4 Following from Exercise 2.21, the table below gives observed and expected frequencies for the average reign length (in years) of the 161 Chinese emperors. The rate parameter of the exponential distribution has been estimated from the data.

Reign	Observed	Expected
0-(5)	58	50.7
5-(10)	24	34.7
10-(15)	24	23.8
15-(20)	18	16.3
20-(25)	15	11.2
25-(30)	6	7.6
30-(35)	5	5.2
35-(40)	3	3.6
40-(45)	2	2.5
45-(50)	3	1.7
50-(55)	1	1.2
55-(60)	1	0.8
60-(65)	1	0.5
65+	0	1.2

(a) Why is amalgamation necessary before doing the χ^2 test?
(b) Give (amalgamated) observed and expected frequencies for a 40+ category.

(c) Perform the goodness of fit test, noting that one parameter has been estimated from the data.
(*Thanks to Estate Khmaladze, John Haywood & Ray Brownrigg for the data*)

12.5 An instructor is unimpressed with his class's performance in a multichoice test. The test consisted of 20 questions, and of the four answers provided for each question, the instructor had included two which were "clearly wrong" and he expected no one to select, and two which he expected the students to choose correctly about 60% of the time. He believed the questions were independent.

Score	Observed	Expected
≤ 5	0	} 2.5
6	3	
7	22	5.6
8	17	13.7
9	49	27.3
10	55	45.1
11	49	61.5
12	62	69.2
13	63	63.9
14	33	47.9
15	20	28.7
16	12	} 19.6
≥ 17	0	

(a) Under the instructor's "null hypothesis" explain why the scores will have a binomial distribution with $n = 20$ and $p = 0.6$.
(b) Assuming (a), find expected frequencies for all possible scores (0 to 20), and show how the expected frequencies in the table above are obtained.
(c) Is further amalgamation necessary before the chi-square test can be performed?
(d) Conduct the chi-square test of the instructor's belief.
(e) Do you observe any patterns in the scores to explain your findings?

12.6 American author Malcolm Gladwell (*Outliers*, 2008) proposes a theory that Canadian hockey players are more likely to be born early in the year than later due to vagaries of the selection system. The 2009 All Blacks squad had birth month frequencies (f) according to the following table:

Month	f	Month	f	Month	f
Jan	3	May	2	Sep	1
Feb	4	Jun	0	Oct	5
Mar	4	Jul	2	Nov	0
Apr	4	Aug	2	Dec	4

(a) Calculate expected frequencies assuming the months all have equal length.
(b) Is amalgamation necessary? If so, combine into six groups of two months.
(c) Do the All Blacks' data exhibit any deviation from what would be expected if there was no selection bias? (*Data from* allblacks.com)

Contingency tables

We suggest you use $\alpha = 5\%$ for these tests.

12.7 In a nationwide study on *Perceptions of Personal Safety & Security Amongst Taxi Users*, respondents were asked "where do you normally sit in a taxi?". The following table gives a subset of the responses:

	Males	Females
Front seat	163	151
Right rear seat	15	36
Left rear seat	81	265

Test gender and seating position for independence.
(*Data from* nzta.govt.nz)

12.8 The number of New Zealand heritage sites destroyed, removed or relocated, by region and 2-year period, are given in the following table:

	05/06	07/08
Northland	2	0
Auckland	3	6
Waikato	3	1
Bay of Plenty	0	0
Hawke's Bay	5	3
Takanaki/Manawatu	2	4
Wellington	7	2
Nelson/Marlborough	1	0
Canterbury	7	3
West Coast	1	2
Otago	9	0
Southland	0	0

(a) Why is amalgamation of the rows in this table necessary before a test of independence is performed?
(b) Amalgamate the table using three groups of four regions, e.g. combine Northland, Auckland, Waikato and Bay of Plenty into one group, etc.
(c) Using your amalgamated table, test for independence between region and year.
(d) Comment on methodology, and your findings.
(*Data from* mch.govt.nz)

12.9 A study to track trends in drug use in the New Zealand population between 1998 and 2006 asks participating cannabis users to describe their cannabis consumption relative to the previous year (i.e. 2003 use compared with use in 2002). The following contingency table is obtained from a random sample of 100 respondents from the last two surveys:

	Stopped	Less	Same	More
2003	28	31	30	11
2006	15	43	30	12

Does this sample provide evidence of changing cannabis use patterns?
(*Data from Wilkins & Sweetsur, Trends in population drug use in New Zealand: findings from national household surveying of drug use in 1998, 2001, 2003, and 2006,* The New Zealand Medical Journal, *2008*)

12.10 Following from Exercise 10.52, exercise patterns (in minutes per week) of males in the Dunedin Study are compared with those of similarly aged males in the NZ Health Survey. The data are as follows:

Time spent exercising	Dunedin Study	NZ Health Survey
0 mins	138	42
< 150 mins	95	14
150-300 mins	110	19
> 300 mins	153	23

Test exercise amount and survey for independence. What conclusions can you draw?
(*Figures published in The New Zealand Medical Journal, 2006*)

12.11 The following table gives the number of firms delisted from the stock exchanges of the countries listed in 2002 and 2007 respectively:

Country	2002	2007
Australia	66	93
Ireland	6	4
New Zealand	12	7
Norway	12	18
Singapore	18	10

Are country and year independent?
(*Data from* msd.govt.nz)

12.12 The following table gives a summary of a sample of adults' participation in the arts in 2008, by level of formal education:

Education	None	≤ 12	> 12
School	576	202	232
Post-school	305	150	168
Post-graduate	126	71	77
Other	77	42	56

Test education level and participation in the arts for independence.

(Data from creativenz.govt.nz)

12.13 A study into a txt-based aid for people trying to stop smoking recorded previously used aids. These are summarised in the following table for Māori and non-Māori subjects.

	Māori	non-Māori
Nicotine replacement	37	185
Medication	4	35
Quitline	20	66
Other	2	31

(a) Is amalgamation necessary before performing a test of independence? If so, combine the (non-nicotine) medication and "other" categories.
(b) Conduct the test of independence.
(c) Summarise any valid patterns in the table.

(Data from Bramley et. al., Smoking cessation using mobile phone text messaging is as effective in Māori as non-Māori, New Zealand Medical Journal, 2005)

12.14 A survey of 2124 New Zealanders recorded their attendance of the arts and their ethnicity. The results are summarised in the following table:

Ethnicity	None	Low	Med	High
NZ European	269	299	448	478
Māori	29	44	56	117
Pacific Island	25	35	48	69
Asian	49	49	66	43

Test ethnicity and arts attendance for independence. Comment on any patterns evident in the data.

(Data from creativenz.govt.nz)

12.15 A sample of New Zealand school children was asked what superpower they would most like. The table below summarises their responses.

	Male	Female
Invisibility	1115	2155
Strength	1223	460
Telepathy	844	3411
Flight	1868	3545
Time travel	3056	2267
Other	85	112

Test gender and preferred superpower for independence.

(Data from censusatschool.org.nz)

12.16 A sample of New Zealand school children was asked which is their dominant hand. The table below summarises their responses.

	Right	Left	Both
Female	10118	1205	576
Male	6651	920	58

(a) Using the complete data, test gender and hand dominance for independence.
(b) Repeat your test without the ambidextrous (Both) category.

(Data from censusatschool.org.nz)

12.17 Following from Exercise 6.5, there are observed patterns in food-storage behaviour among pairs of New Zealand robins *Petroica australis*. The following table describes the ultimate fate of food items stored by the male and female birds within a pair.

	Cached by male	Cached by female
Not retrieved	376	122
Retrieved by male	41	18
Retrieved by female	104	13

Test for independence between the gender of the bird which originally cached the food, and its retrieval.
(*Data from Burns & van Horik, Sexual differences in food re-caching by New Zealand robins, Journal of Avian Biology, 2007*)

12.18 A random sample of New Zealanders was asked their favoured political party, and whether it is likely that "Princess Diana was killed by British secret service in order to prevent continued Royal scandal". The responses are as follows:

	Unlikely	Don't know	Likely
Labour	766	213	253
National	962	243	305
Greens	324	91	88
NZ First	28	8	21
Act	148	10	17
United Future	15	1	1
Maori Party	21	8	7

(a) Assuming independence between political party and opinion, calculate expected frequencies for this table.
(b) Is amalgamation necessary in this instance? If so, combine United Future and National voters.
(c) Conduct the test for independence, and summarise your finding.
(*Special thanks to Marc Wilson for the data*)

12.19 Deaths and serious injuries to trespassers on New Zealand's rail corridor over the period 1994-2003 are tabulated below:

Time	M	Tu	W	Th	F	Sa	Su
0600-1759	11	11	19	22	13	7	8
2200-0559	5	9	8	17	19	6	2

Test day of the week and time of accident for independence.
(*Data from ltsa.govt.nz*)

12.20 The following table summarises the number of patents registered in 1895 by type: agricultural (A), pastoral (Ps), dairy, fishing and forestry (DFF), and preservation (Pr), and region.

	A	Ps	DFF	Pr
Auckland	2	5	6	1
Hawke's Bay	1	1	0	1
Taranaki	0	2	1	0
Wellington	9	17	4	4
Nelson	0	0	0	0
Marlborough	0	0	2	0
West Coast	0	0	1	0
Canterbury	14	6	8	0
Otago	6	24	8	2

(a) Why is amalgamation necessary before conducting a test of independence?
(b) Combine the rows to make North and South Island categories. Are the assumptions now met? Amalgamate further if necessary.
(c) Using your amalgamated table conduct a test of independence between region and patent type.
(d) Repeat your analysis for the Wellington and Canterbury registered patents only.
(*Thanks to Rebecca Craigie & Les Oxley for the data*)

12.21 The donation box at City Gallery Wellington was manipulated for several months to learn about visitor response to different visual cues. The researchers had a collection of "seed money" which would be used to give the appearance of previous donations. Each had a total of $100: one with a $50 bill in it, another with 13 $5 bills, and the third almost entirely coins. The three treatments are in the table below, along with the composition of donations obtained while each seed was in force.

Seed	Bills	Gold	Silver
$50	22	93	139
$5	19	100	168
Coins	7	126	316

Test seed contents and composition of donations for independence. If you reject independence, what patterns can you discern in the data?
(*From Martin & Randal, How is donation behaviour affected by the donations of others? Journal of Economic Behavior and Organization, 2008*)

12.22 The New Zealand Arrestee Drug Abuse Monitoring programme investigated drug and alcohol use in a sample of 2206 detainees in 2005. The following table contains information about those detainees who were using drugs at the time of their arrest. They were asked to what extent they believed the use of these drugs contributed to their involvement in the illegal activities for which they were arrested.

	All/ a lot	Some	A little/ not at all
Cannabis	29	21	145
Alcohol	202	43	114

Does this sample provide evidence for dependence between the drug type and its role in crime involvement?
(*Data from* police.govt.nz)

12.23 Due to environmental concerns, rising oil price, and rising population, using a bicycle for short trips is becoming increasingly beneficial. About 750,000 New Zealanders cycle once a month, with about 20% of those riding nearly every day. However, cycle use is not without its hazards, particularly on road networks which are designed primarily for cars. The following statistics on deaths and injuries were reported in mid-2010:

	Death	Serious	Minor
2005	12	124	633
2006	9	152	687
2007	12	182	704
2008	10	188	713

Do the data indicate any change in death and injury patterns over time, i.e. are year and severity of reported accident independent?
(*Data from* stuff.co.nz)

12.24 Following Exercise 10.57, 597 of the children's parents were asked if they knew of a time after which it is not suitable for their children to watch TV. The responses are in the following table:

Time	6-8	9-11	12-13
Before 8:30 pm	44	27	10
8:30 pm	27	67	36
8:31-9:00 pm	8	15	11
9:01-9:30 pm	8	32	21
9:31-10:00 pm	4	10	14
10:01-10:30 pm	6	15	11
After 10:30 pm	11	20	15
No/don't know	103	63	19

(a) Is amalgamation necessary before performing the test for independence? If so, combine time categories until the assumption is met.
(b) Test time and age group for independence. Is there evidence that the two are linked?
(*Data from* bsa.govt.nz)

12.25 Following Exercise 9.27, infection rates of the "swine-flu" virus in New Zealand were investigated. The following gives the number of positive tests by age group and ethnicity:

Age	Māori	Pacific	Other
1-4	8	14	33
5-19	22	21	59
20-39	10	18	33
40-59	13	11	32
60+	9	9	55

Test age group and ethnicity for independence.
(Data from Bandaranayake, Bissielo, Huang & Wood, Seroprevalence of the 2009 influenza A (H1N1) pandemic in New Zealand, 2010)

12.26 A random sample of New Zealanders was asked if they think that "NASA faked the first moon landings for publicity". The responses, by political preference, are as follows:

	No	Yes
Labour	804	279
National	995	322
Greens	337	109
NZ First	34	14
Act	149	8
United Future	16	1
Māori Party	17	9

(a) Assuming independence between political party and opinion, calculate expected frequencies for this table.
(b) Is amalgamation necessary in this instance? If so, combine United Future and National voters.
(c) Conduct the test for independence, and summarise your finding.
(Special thanks to Marc Wilson for the data)

12.27 Following Exercises 3.36 and 10.7, the alcohol-related negative consequences reported by the survey participants are as follows:

Consequence	Males	Females
Fights	84	41
Emotional outbursts	84	229
Blackouts	190	236
Difficulty concentrating	90	88
Drink-driving	82	37

Test gender and negative consequence for independence.
(Data from Kypri, Langley, McGee, Saunders & Williams, High prevalence, persistent hazardous drinking among New Zealand tertiary students, Alcohol & Alcoholism, 2002)

12.28 A report for the Ministry for the Environment assessed lake quality around New Zealand. The water quality is assessed on a scale from Impacted (1 in the table below) to Pristine (6 below).

Region	1	2	3	4	5	6
Auckland	1	1	4	1	0	0
Bay of Plenty	0	1	3	4	4	0
Canterbury	2	0	2	1	12	9
Greater Wellington	5	0	26	9	2	0
Hawke's Bay	0	1	0	0	0	0
Manawatu-Wanganui	0	0	0	1	0	0
Northland	0	1	1	0	0	0
Otago	0	1	3	1	2	3
Southland	0	0	1	0	3	0
Taranaki	0	0	0	1	0	0
Tasman District	0	0	0	0	0	1
Waikato	8	2	4	0	1	0
West Coast	0	0	0	1	1	0

(a) Why is amalgamation necessary in this case?
(b) Combine the rows to make North Island and South Island (Canterbury, Otago, Southland, Tasman, West Coast) categories.
(c) Combine columns 1 and 2, 3 and 4, and 5 and 6.
(d) Using your 2×3 table, test region and lake quality for independence.
(e) If the purpose of this analysis was to examine the effects of agricul-

ture and industry on lake water quality, what might be a better way of amalgamating the rows? Propose new category names.
(*Data from* mfe.govt.nz)

2 × 2 tables

We suggest you use $\alpha = 5\%$ for these tests.

12.29 A random sample of individuals was asked whether they think they are better than average, or worse than average, at their job. The responses were:

	Female	Male
Worse than average	691	286
Better than average	761	746

Test gender and response for independence. Comment on your result.
(*Special thanks to Marc Wilson for the data*)

12.30 Following Exercise 12.11, test for independence using only the Australian and New Zealand data.

12.31 A random sample of individuals was asked whether they think they are more or less smart than the average person. The responses were:

	Female	Male
Not as smart as average	563	195
Smarter than average	956	949

Test gender and response for independence. Comment on your result.
(*Special thanks to Marc Wilson for the data*)

12.32 A random sample of males and another of females was asked how often they tell "white lies" (lies without malice), and black lies (lies designed to get the teller something). The results are in the table below:

	White		Black	
	F	M	F	M
< monthly	406	296	1464	866
≥ monthly	1527	1045	469	473

(a) Using the entire table, test for independence between gender and frequency of lying.
(b) For the white lies only, test for independence between gender and frequency of lying.
(c) For the black lies only, test for independence between gender and frequency of lying.
(d) These results are obtained through survey data. Comment on the awkwardness of this sampling method in this particular context.

(*Special thanks to Marc Wilson for the data*)

12.33 The littering behaviour of people sitting on Victoria University's quad steps was monitored for a period of time. Subjects were classified as litterers or non-litterers. Of those witnessed littering, behaviour was classified as either active (subjects walked across the quad and dropped the rubbish on the ground) or passive (subjects left rubbish on the ground and failed to pick it up when leaving). In addition, any litter was classified as either cigarette or non-cigarette. A total of 452 people was observed.

(a) The following patterns were observed for the cigarette litter:

	Active	Passive
Littered	44	78
Non-littered	51	8

Test the type of littering (active or passive) and the littering outcome for independence.

(b) The non-cigarette litter was "disposed of" in the following manner:

	Active	Passive
Littered	10	47
Non-littered	120	94

Test the type of littering (active or passive) and the littering outcome for independence.

(c) Draw an overall conclusion based on your tests.

(*Data from Sibley & Liu, Differentiating active and passive littering: a two-stage process model of littering behavior in public spaces, Environment and Behavior, 2003*)

12.34 Observation of visitors to an art gallery identified that they are more likely to place money in a donation box on a Sunday than on any other day of the week. The following table contains data for a random sample of visitors observed on one weekend.

	Donors	Visitors
Saturday	10	453
Sunday	10	259

(a) Since each donor is also included in the visitor tally, explain why the assumptions of the contingency table analysis are violated by the table above.

(b) Form the 2 × 2 table with "Donors" and "Non-donors" as the columns.

(c) Test the 2 × 2 table in (b) for independence between day and donation.

(d) Is the claim above supported by the result of your test?

(*Martin & Randal, How Sunday, price, and social norms influence donation behaviour. Journal of Socio-Economics, 2009*)

12.35 New Zealand Sign Language is one of many languages around the world that exhibits variation in its numbering system. A sample of 109 Deaf people from around NZ is used to examine patterns in sign choice as influenced by age, region and gender. Among the sample, two major variants were observed for the number 11:

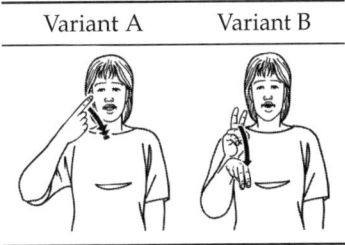

Variant A	Variant B

(nine instances of other variants have been discarded). The following table summarises the data:

Age group	A	B
15-29	18	16
30-44	6	23
45+	11	26
Gender	A	B
Female	20	23
Male	15	42
Region	A	B
North	27	16
South	8	49

Test for independence between variant and:
(a) age group;
(b) gender;
(c) region.

(*Data from McKee, McKee & Major, Sociolinguistic Variation in NZSL Numerals, 2008. Line drawings thanks to A Dictionary of New Zealand Sign Language, 1997, Graeme Kennedy ed.*)

Chapter 13

Inference and the regression line

In Chapter 4 we estimated the linear relationship between two variables Y and X. The underlying model for these data was assumed to be

$$Y_i = \alpha + \beta X_i + \epsilon_i$$

where Y_i is the dependent variable, α and β are unknown (population) values for the intercept and slope of the line, X_i is the independent variable, and the ϵ_i are independent and identically distributed errors, with zero mean, and constant variance σ^2.

Using the method of least squares, we can estimate the parameters α and β to form the regression line

$$\hat{Y}_i = a + bX_i$$

where the estimated slope and intercept are

$$b = \frac{S_{XY}}{S_{XX}} = \frac{\sum XY - \frac{1}{n}\sum X \sum Y}{\sum X^2 - \frac{1}{n}\left(\sum X\right)^2} \quad \text{and} \quad a = \bar{Y} - b\bar{X}$$

respectively. As in previous chapters, we have sample estimates of population parameters, and we can use these sample estimates to conduct hypothesis tests on those population parameters.

13.1 Testing the slope

The most common hypothesis test conducted on the regression line is that $\beta = 0$. In other words, the true slope of the regression line is zero, i.e. there is no linear relationship. In this case, it means that the "true" relationship is actually

$$Y_i = \alpha + \epsilon_i = \mu + \epsilon_i$$

so that the dependent variable does not depend on X_i, and is simply a random sample with mean $\mu = \alpha$ and constant variance σ^2. If we test whether this is likely to be true, this is equivalent to testing whether Y depends on X in this linear way.

Figure 13.1 shows two instances where this behaviour might be expected. The first indicates that Y does not vary with X, and regardless of the size of X, Y is the same. In the second plot, Y varies randomly with X, which was the situation with a random cloud of points, and $r = 0$ (refer to Figure 4.5). The third plot shows a clear relationship between X and Y, indicating some sort of dependence.

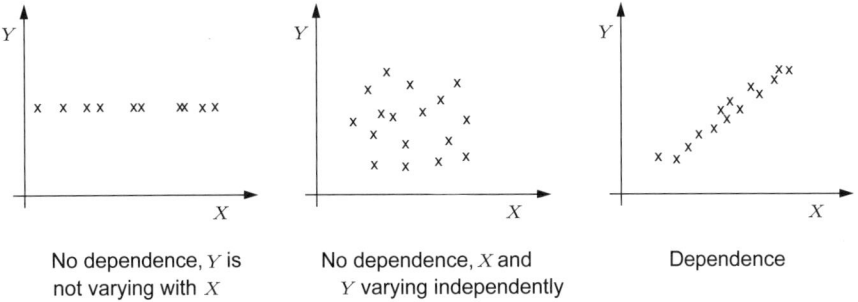

Figure 13.1 Dependence of Y on X

We are not restricted to testing $H_0 : \beta = 0$. We may also have a different value for the slope, or alternatively for the intercept, or both.

Example 13.1

Suppose we have two weighing devices. The first is very accurate, and produces the weight X_i. The other has recently been developed, and this produces the weight Y_i. Although the second device is not as accurate as the first, it still should be correct on average. Here, we expect $Y_i = X_i + \epsilon_i$, i.e. the second weight is equal to the first one, plus some random error. In this case, we are interested in testing whether or not $\alpha = 0$ and $\beta = 1$.

Since testing the slope is most common, we outline the details of this test first.

Hypotheses

As mentioned above, testing $H_0 : \beta = 0$ against the alternative $H_a : \beta \neq 0$ is very common. This tells us whether or not Y is related to X in a linear way. If we reject H_0, it indicates X is useful. Otherwise, it implies there is no linear relationship between the two variables. In general, we can test any value for the population slope, so the null hypothesis is $H_0 : \beta = \beta_0$, where β_0 is a number we specify. We choose one of the following alternative hypotheses: $H_a : \beta \neq \beta_0$, $H_a : \beta > \beta_0$ and $H_a : \beta < \beta_0$, depending on what our prior view is about the likely relationship between Y and X.

Test statistic

We now collect the paired data (X_i, Y_i) for $i = 1, \ldots, n$. These should be plotted on a scatterplot, and if a linear relationship is appropriate, the regression line should be estimated. In order to continue, we need to assume that the errors ϵ_i are normally distributed. This assumption can be examined by checking that the *residuals*

$$e_i = Y_i - \hat{Y}_i$$

are approximately normally distributed.

This assumption is in addition to the requirement that the errors have constant variance. We can check the validity of these assumptions using a plot of e_i against $\hat{Y}_i$, which should exhibit no pattern, and a boxplot of the residuals e_i will allow us to check for symmetry, and absence of outliers.

If the ϵ_i are normal, then it follows that the slope coefficient b is also normally distributed. Under the null hypothesis, it has mean β and standard error given by

$$se(b) = \frac{\sigma}{\sqrt{S_{XX}}}$$

where σ^2 is the common variance of the ϵ_i and S_{XX} is as before.

We will need to estimate σ^2 in order to calculate the standard error, and complete the standardisation of b. The estimates of the errors ϵ_i are the residuals e_i. We estimate σ^2 by the sample variance of the residuals, deducting two degrees of freedom since we have estimated two parameters (α and β). The estimate is

$$\hat{\sigma}^2 = \frac{\sum(e_i - \bar{e})^2}{n - 2} = \frac{\sum e_i^2}{n - 2}$$

since the residuals sum to zero, with $\bar{e} = 0$. Fortunately, we do not have to calculate every residual in order to estimate σ^2, as there is a much better computational formula.

The estimate of the variance of the errors is given by

$$\hat{\sigma}^2 = \frac{S_{YY} - b \times S_{XY}}{n - 2}$$

where

$$S_{YY} = \sum Y^2 - \frac{1}{n}\left(\sum Y\right)^2 \quad \text{and} \quad S_{XY} = \sum XY - \frac{1}{n}\sum X \sum Y.$$

Then we standardise b, and replace σ by $\hat{\sigma}$ in the standard error.

> The test statistic is
> $$T = \frac{b - \beta}{\left(\frac{\hat{\sigma}}{\sqrt{S_{XX}}}\right)}$$
> and this has a t-distribution with $n - 2$ degrees of freedom.

Rejection region

Since the test statistic has a t-distribution with $n - 2$ degrees of freedom, we obtain the rejection region from the t-table on page 331. The exact form of the rejection region also depends on the level of the test (usually 5%), and the alternative hypothesis H_a.

Conclusion

If the test statistic lies in the rejection region, we reject the null hypothesis in favour of the alternative at the $100\alpha\%$ level (in this context α is the level of the test). If the test statistic does not lie in the rejection region, we cannot reject the null hypothesis. If the null hypothesis we are testing has $\beta = 0$, and we do not reject the null, this indicates that the dependent variable Y *does not* depend linearly on X. Note that we cannot rule out a non-linear relationship if we cannot reject H_0.

Example 13.2

Q: Using the oak tree data in Table 4.1, test whether or not the age of the tree is a useful predictor of diameter at breast height.

A: Recall from Example 4.5 that the estimated regression line for the oak tree data is $\hat{y} = 1.0841 + 0.1715x$, with $\bar{x} = 22$, $n = 14$, $S_{XX} = 2028$, $S_{YY} = 67.69429$ and $S_{XY} = 347.8$. We wish to test $H_0 : \beta = 0$ against $H_a : \beta \neq 0$. The estimate of the error variance σ^2 is

$$\hat{\sigma}^2 = \frac{S_{YY} - bS_{XY}}{n - 2} = \frac{67.69429 - 0.1715 \times 347.8}{14 - 2} = \frac{8.047}{12} = 0.671.$$

[Note that the residuals in Example 4.10 give $\sum e_i^2 = 8.0729$ instead of 8.047 without any rounding errors.] The test statistic is

$$T = \frac{b - \beta}{\frac{\hat{\sigma}}{\sqrt{S_{XX}}}} = \frac{0.1715 - 0}{\frac{\sqrt{0.671}}{\sqrt{2028}}} = \frac{0.1715}{0.0182} = 9.43$$

and this has a t-distribution with $n - 2 = 12$ degrees of freedom. The rejection region is $T < -2.179$ and $T > 2.179$ and since $9.43 > 2.179$ we reject H_0 in favour of H_a at the 5% level.

13.2 Testing slope and intercept simultaneously

In Example 13.1 we saw a situation where we might want to test both the intercept and slope simultaneously, i.e. $H_0 : \{\alpha = \alpha_0 \text{ and } \beta = \beta_0\}$ against H_a : at least one of these is wrong. We could do individual hypothesis tests for the slope and intercept, although in this case it is difficult to control the level of the combined tests. The alternative is to use the F-test approach of Section 11.1.2. We must fit two models, the complete model, and the reduced model. From each we obtain the residual sum of squares on which we base the test statistic.

> The residual sum of squares for the complete model is
> $$RSS_{complete} = \sum(Y_i - a - bX_i)^2 = \sum e_i^2 = S_{YY} - b \times S_{XY}.$$

The reduced model uses the values suggested by $H_0 : \{\alpha = \alpha_0 \text{ and } \beta = \beta_0\}$. In this case, we have no parameters to estimate, and so the residuals are $e'_i = Y_i - \alpha_0 - \beta_0 X_i$.

> The residual sum of squares for the reduced model is
> $$RSS_{reduced} = \sum(Y_i - \alpha_0 - \beta_0 X_i)^2.$$

The test statistic has the same form as the one in Section 11.1.2; however, in this case $df_{reduced} - df_{complete}$ will always be two, since the complete model always features two more parameters than the reduced model. As in the previous section, the complete model has $n-2$ degrees of freedom.

> The test statistic is
> $$F = \frac{(RSS_{reduced} - RSS_{complete})/2}{RSS_{complete}/(n-2)}$$
> and this has an F-distribution with 2 and $n-2$ degrees of freedom.

The $RSS_{reduced}$ will always be bigger than $RSS_{complete}$ since the latter is the smallest RSS of any line through the data (it is based on the least squares line). If the difference is small, then the reduced model will be a reasonable description of the data, and we would tend not to reject H_0. If the difference is large, we *will* wish to reject H_0, and this will occur with large values of the test statistic only.

The rejection region is obtained in the usual way, by using the level of the test and the distribution of the test statistic, along with the one-sided nature of this F-test, to obtain the critical value.

The test is completed by comparing the test statistic with the rejection region. If the test statistic lies in the rejection region, we reject H_0 in favour of H_a at the $100\alpha\%$ level. Otherwise, if the test statistic does not lie in the rejection region, we cannot reject the null in favour of the alternative.

13.3 Prediction intervals

In Section 4.3.3, we saw how the regression line can be used to form predictions. The regression line is
$$\hat{Y} = a + bX$$
and this can be interpreted as the value of Y we would *expect* for a given value of X (here, "expect" is used in both the usual and statistical senses). X may be a shop's advertising expenditure for the month, and the regression line might be used to predict the month's sales. The values of $\hat{Y}$ that we obtain by inserting values of X into the equation of the line, are point estimates of the actual values. Using the ideas of Chapter 9 and subsequent chapters, we can develop *prediction intervals*, which are actually confidence intervals for the dependent variable.

> Given X_0 and an estimated regression line $\hat{Y} = a + bX$, the variable $\hat{Y}_0 = a + bX_0$ has mean $\alpha + \beta X_0$, and standard error given by
> $$se(\hat{Y}_0) = \sigma\sqrt{1 + \frac{1}{n} + \frac{(X_0 - \bar{X})^2}{S_{XX}}}.$$

The mean of the prediction makes sense, since $Y_i = \alpha + \beta X_i + \epsilon_i$, and the error ϵ_i has mean zero. The standard error is pretty nasty looking, but we note that "$\sigma \times 1$" corresponds to the standard deviation of the error ϵ_i, and the rest is because we have estimated α and β. Since $(X_0 - \bar{X})^2$ gets bigger the further we move from $\bar{X}$, the standard error gets larger as we move away from $\bar{X}$, resulting in wider prediction intervals for X values a long way from the sample mean.

As with other small sample confidence intervals, we use the familiar form: point estimate $\pm$ (value from the t-distribution) $\times$ (standard error of the point estimate). We use the t-distribution with $n-2$ degrees of freedom, since we replace σ by $\hat{\sigma} = \sqrt{\sum e_i^2/(n-2)}$.

> Prediction intervals for Y_0 are given by
> $$\hat{Y}_0 \pm t\hat{\sigma}\sqrt{1 + \frac{1}{n} + \frac{(X_0 - \bar{X})^2}{S_{XX}}}$$
> where t is the appropriate value from the t-distribution with $n-2$ degrees of freedom, and $\hat{Y}_0 = a + bX_0$.

We can form prediction intervals for the value of X_i in the sample, or alternatively for values of X that are not in the sample data. These are illustrated in Figure 13.2 which show that the interval is at its narrowest at $\bar{X}$.

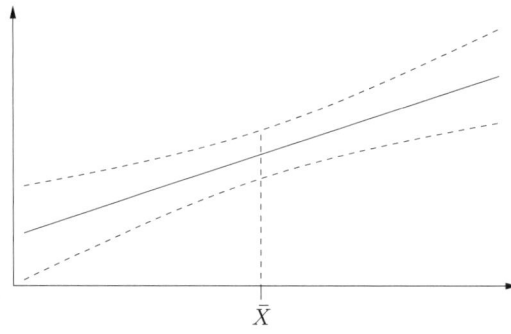

Figure 13.2 Widening prediction intervals

Example 13.3

Q: In Example 4.8, a prediction was made for the diameter at breast height of a 25-year-old oak tree. Provide a 95% confidence interval for this diameter.

A: The prediction was $\hat{y} = a + bx = 1.0842 + 0.1715 \times 25 = 5.3716$. In addition, we have found $\hat{\sigma} = 0.819$, $\bar{x} = 22$, and $S_{XX} = 2028$. The 95% confidence interval for y_0, corresponding to $x_0 = 25$ is

$$\hat{y}_0 \pm 2.179 \hat{\sigma} \sqrt{1 + \frac{1}{n} + \frac{(x_0 - \bar{x})^2}{S_{XX}}} = 5.37 \pm 2.179 \times 0.819 \times \sqrt{1 + \frac{1}{14} + \frac{(25-22)^2}{2028}}$$

$= 5.37 \pm 1.85$, giving the interval $(3.52, 7.22)$.

Exercises

Testing the slope for significance

We suggest you use $\alpha = 5\%$ for these tests.

For the following questions, test the estimated slope coefficient for significance using data from the indicated exercise.

13.1 Female legislators (4.5)

13.2 Depression scores (4.8)

13.3 Oil spill size (restrict to spills less than 200,000 tonnes) (4.16)

13.4 Nitrate concentration (4.25)

13.5 Kea blood-lead concentration (4.27)

13.6 Changes in unemployment rate (4.29)

13.7 Karapoti Classic finishers (5.21)

13.8 Miro heights (5.19)

13.9 Large earthquakes (5.24)

Testing for non-zero slopes

13.10 Following Exercises 4.6 and 10.27:
 (a) Using forecast temperature as the x-variable, estimate the regression

line for the minimum (actual) temperatures.
(b) Is the estimated slope in (c) significantly different from one?
(c) Using forecast temperature as the x-variable, estimate the regression line for the maximum (actual) temperatures.
(d) Is the estimated slope in (a) significantly different from one?
(e) What do the results of your tests say about the performance of the forecasters?

13.11 Following Exercise 4.39 on double packs at the supermarket:
(a) Estimate the regression line for the data, using the single price as the x-variable.
(b) Test the hypothesis that in general there is a cost saving when you buy the double size over two of the single size.
Hint: test $H_0 : \beta = 2$ against a suitable alternative.

13.12 Following Exercise 4.40 on average age of first use of ten drugs:
(a) Estimate the regression line for the data, using the males' average ages as the x-variable.
(b) Show that $\hat{\sigma} = 2.18$.
(c) Test whether or not the slope coefficient is significantly greater than one.
(d) Each of the averages above is based on a different (sub-) sample size. What bearing (if any) might this have on the assumptions of your test?

13.13 Following Exercise 4.42 on the savings from double-glazing, is the estimated slope significantly different from one?

13.14 The intended use of a set of industrial scales is measuring weights of objects with typical range 210-240 kilograms (kg). As part of the calibration process, objects with known *exact* weights are repeatedly weighed, with measurements given below:

Actual	w_1	w_2	w_3
200	212.5	212.1	209.8
210	220.9	222.7	222.9
220	229.3	231.0	228.5
230	240.9	243.6	242.9
240	250.5	253.0	252.5
250	260.8	258.7	260.2

(a) Draw a scatterplot of the data.
(b) Estimate the linear regression line for these data and add it to your plot, along with the line $y = x$.
(c) Interpret your regression line.
(d) Is the estimated slope coefficient significantly different from one?
(e) Suggest how the scales should be reprogrammed to provide weights.

13.15 The following table gives the total average volume of rain per season in the Waikato River catchments (in kilometres cubed).

	Summer	Winter
Taupo	1.18	2.36
Waipa	0.71	1.81
Upper	0.60	1.51
Middle	0.68	1.88
Lower	0.44	1.36

(a) Estimate the regression line for these data, using summer rainfall as the predictor variable.
(b) Is the slope coefficient significantly different from zero?
(c) Is the slope coefficient significantly different from one?
(d) Interpret your results.
(Data from Dravitzki, Precipitation in the Waikato River catchment, VUW Thesis, 2009)

13.16 Following from Exercise 4.26, we examine the relationship between fuel price and public transport usage. In the table below are the increase in average fuel price from the previous year (in cents) and the percentage increase in bus passenger numbers:

Year	Fuel price change (c)	Bus passenger increase (%)
2004	8	3.5
2005	13	5.7
2006	25	3.8
2007	1	−1.1
2008	16	7.6
2009	−11	−4.5

(a) One might expect that as fuel price increases, so too does the number of bus passengers. Is there a significant negative relationship between fuel price change and bus passenger increase?

(b) Sometimes changes in the economy take some time to alter behaviour. Estimate the relationship between bus passenger increase and the previous year's fuel price. Repeat the hypothesis test.

(c) Discuss your findings.

(Thanks to Greater Wellington Regional Council for the data)

13.17 The following table gives Victoria University of Wellington's estimated carbon dioxide (CO_2) emission profile in 2006 and 2007.

	2006	2007
Business Travel	3233	1906
Commuting Students	3830	3830
Commuting Staff	1291	1291
Electricity	4502	4095
Gas	3328	3063
Diesel Generators	2.27	1.20
Waste	292	282
Paper	456	280

(a) Regress the 2007 figures on the 2006 figures. Interpret the coefficients.

(b) Is the slope coefficient significantly different from one?

(Data from Kodikara, An assessment of whether a carbon neutral initiative can successfully be implemented at Victoria University of Wellington, VUW thesis, 2008)

Prediction intervals

For the following exercises, calculate and carefully interpret 95% prediction intervals at the two values specified.

13.18 Rumination scores of 5 and 20 (4.8)

13.19 Changes in GDP of 0.01 and 0.04 (4.29)

13.20 Single packs priced at $2.50 and $13 (4.39)

13.21 Miro height in July 2006, and July 2010 (5.19)

13.22 Weights of 235 kg and 270 kg (13.14)

13.23 Fuel price changes of ±15% (13.16)

Chapter 14

Review of estimation and testing

In this short chapter, we list assumptions, test statistics and their distributions for the most common hypothesis tests covered in this book. In the cases where confidence intervals are also available, we give their form.

The tests are arranged into the following order: large samples, small samples, for normal and non-normal data, and bivariate relationships.

14.1 Large samples

A *large sample* is one for which $n \geqslant 30$.

Test for a mean

We assume $\sigma \approx s$. The test statistic is

$$Z = \frac{\bar{X} - \mu}{\frac{s}{\sqrt{n}}}$$

and this has an approximate $N(0, 1)$ distribution. Use Tables D.4 and D.5 on pages 329 and 330. We are 95% confident that true mean μ lies between

$$\bar{X} \pm z \frac{s}{\sqrt{n}}$$

where $z = 1.96$.

Test for two means

We assume $\sigma \approx s$ for each population. If the samples are *independent*, the test statistic is

$$Z = \frac{\bar{X}_1 - \bar{X}_2 - (\mu_1 - \mu_2)}{\sqrt{\frac{s_1^2}{n_1} + \frac{s_2^2}{n_2}}}$$

and this has an approximate $N(0, 1)$ distribution. Use Tables D.4 and D.5 on pages 329 and 330. We are 95% confident that the difference in the population means $\mu_1 - \mu_2$ lies between

$$\bar{X}_1 - \bar{X}_2 \pm z \sqrt{\frac{s_1^2}{n_1} + \frac{s_2^2}{n_2}}$$

where $z = 1.96$.

Test for a proportion

The sample proportion is $\hat{p} = X/n$ where X is the number of successes out of the n observations. We require

$$0 < \hat{p} \pm 3\sqrt{\hat{p}(1-\hat{p})/n} < 1$$

and

$$0 < p \pm 3\sqrt{p(1-p)/n} < 1$$

for the test to be valid. The test statistic is

$$Z = \frac{\hat{p} - p}{\sqrt{\dfrac{p(1-p)}{n}}}$$

and this has an approximate $N(0,1)$ distribution. Use Tables D.4 and D.5 on pages 329 and 330. We are 95% confident that true proportion p lies between

$$\hat{p} \pm z\sqrt{\frac{\hat{p}(1-\hat{p})}{n}}$$

where $z = 1.96$.

Test for two proportions

We require

$$0 < \hat{p}_i \pm 3\sqrt{\hat{p}_i(1-\hat{p}_i)/n} < 1$$

and

$$0 < p_i \pm 3\sqrt{p_i(1-p_i)/n} < 1$$

for each population. If $H_0 : p_1 = p_2$, the test statistic is

$$Z = \frac{\hat{p}_1 - \hat{p}_2 - (p_1 - p_2)}{\sqrt{p^*(1-p^*)\left(\dfrac{1}{n_1} + \dfrac{1}{n_2}\right)}}$$

and this has an approximate $N(0,1)$ distribution, where p^* is the pooled sample proportion. Use Tables D.4 and D.5 on pages 329 and 330. We are 95% confident that true difference in proportions $p_1 - p_2$ lies between

$$\hat{p}_1 - \hat{p}_2 \pm z\sqrt{\frac{\hat{p}_1(1-\hat{p}_1)}{n_1} + \frac{\hat{p}_2(1-\hat{p}_2)}{n_2}}$$

where $z = 1.96$.

14.2 Small samples

14.2.1 Normal data

We assume at least one of the samples we are using is small, i.e. has $n < 30$. We must also assume that any small sample is normally distributed. Check this assumption by plotting data in a histogram, stemplot or boxplot.

Test for a mean

The test statistic is

$$T = \frac{\bar{X} - \mu}{\dfrac{s}{\sqrt{n}}}$$

and this has a t-distribution with $n - 1$ degrees of freedom. Use Table D.6 on page 331. We are 95% confident that true mean μ lies between

$$\bar{X} \pm t\frac{s}{\sqrt{n}}$$

where t is from the t-distribution with $n - 1$ degrees of freedom.

Test for two means

We assume that $\sigma_1^2 = \sigma_2^2$, and we estimate this common σ^2 using the pooled variance s_p^2.

If the samples are *independent*, the test statistic is

$$T = \frac{\bar{X}_1 - \bar{X}_2 - (\mu_1 - \mu_2)}{\sqrt{\frac{s_p^2}{n_1} + \frac{s_p^2}{n_2}}}$$

and this has a t-distribution with $n_1 + n_2 - 2$ degrees of freedom. Use Table D.6 on page 331. We are 95% confident that the difference in the population means $\mu_1 - \mu_2$ lies between

$$\bar{X}_1 - \bar{X}_2 \pm t\sqrt{\frac{s_p^2}{n_1} + \frac{s_p^2}{n_2}}$$

where t is from the t-distribution with $n_1 + n_2 - 2$ degrees of freedom.

If the samples are *dependent*, difference the pairs consisting of x_i from sample one, and y_i for the corresponding observation in sample two, to form the differences $d_i = x_i - y_i$. The n differences have sample mean $\bar{X}$ and population mean μ_d, sample standard deviation s and population standard deviation σ_d. We assume $\sigma_d \approx s$. The test statistic is

$$T = \frac{\bar{X} - \mu_d}{\frac{s}{\sqrt{n}}}$$

and this has a t-distribution with $n - 1$ degrees of freedom. Use Table D.6 on page 331. We are 95% confident that true mean μ_d lies between

$$\bar{X} \pm t\frac{s}{\sqrt{n}}$$

where t is from the t-distribution with $n - 1$ degrees of freedom.

Test for two variances

The test statistic is

$$F = \frac{s_\ell^2}{s_s^2}, \quad \text{where} \quad s_\ell^2 > s_s^2$$

and this has an F-distribution with $n_\ell - 1$ and $n_s - 1$ degrees of freedom. Use Table D.9 on page 334 for a 5%, two-sided test (i.e. use the $2\frac{1}{2}$% table).

Test for $k > 2$ means

We assume the k groups have common variance, and estimate this using $s_p^2 = $ MSE. The test statistic is

$$F = \frac{\text{MST}}{\text{MSE}}$$

and this has an F-distribution with $k - 1$ and $n - k$ degrees of freedom, where $n = \sum n_i$ is the total number of observations. Use Table D.8 on page 333 for a 5% test.

If *we reject H_0*, a 95% confidence interval for the population group mean μ_i is

$$\bar{X}_i \pm t\frac{s_p}{\sqrt{n_i}}$$

where t is from the t-distribution with $n - k$ degrees of freedom. A 95% confidence interval for the population difference in means $\mu_i - \mu_j$ is

$$\bar{X}_i - \bar{X}_j \pm ts_p\sqrt{\frac{1}{n_i} + \frac{1}{n_j}}$$

where t is from the t-distribution with $n - k$ degrees of freedom.

If *we do not reject H_0*, the point estimate for the common mean μ is the grand mean $\bar{\bar{X}}$.

14.2.2 Non-normal data

Test for a median

The test statistic of the sign test is $X = $ the number of observations above m_0 if the test is one-sided, or the smaller of this number and the number below m_0 if the test is two-sided. The test statistic has a binomial distribution with n trials and probability of success $p = 0.5$. Use the binomial tables on pages 325 and 326.

Test for two medians

The Mann-Whitney U-test test statistic is $\mathcal{U} = U_1$ for a one-sided test with $\mathcal{M}_1 < \mathcal{M}_2$ in the alternative hypothesis, or the smaller of U_1 and U_2 for a two-sided test. Use Table D.11 on page 336.

Test for $k > 2$ medians

The Kruskal-Wallis test statistic is

$$K = \frac{12}{n(n+1)} \left(\sum_{i=1}^{k} \frac{R_i^2}{n_i} \right) - 3(n+1)$$

and this has an approximate chi-squared distribution with $k-1$ degrees of freedom. Use Table D.7 on page 332.

14.3 Bivariate data

Test for independence between two categorical variables

We assume mutually exclusive categories, and that at least 80% of the expected frequencies exceed 5. The test statistic is

$$\chi^2 = \sum \frac{(o_{ij} - e_{ij})^2}{e_{ij}}$$

and this has a chi-squared distribution with $(r-1) \times (c-1)$ degrees of freedom. Use Table D.7 on page 332.

If $r = c = 2$, use Yates' continuity correction, and the test statistic is

$$\chi^2 = \sum \frac{(|o_{ij} - e_{ij}| - 0.5)^2}{e_{ij}}$$

which has a chi-squared distribution with 1 degree of freedom.

Test for a linear relationship between two quantitative variables

Assume the model

$$Y_i = \alpha + \beta X_i + \epsilon_i$$

where the ϵ_i are independent $N(0, \sigma^2)$ random variables. The regression line is

$$\hat{Y} = a + bX$$

with

$$b = \frac{S_{XY}}{S_{XX}} \quad \text{and} \quad a = \bar{Y} - b\bar{X}.$$

The test statistic is

$$T = \frac{b - \beta}{\left(\frac{\hat{\sigma}}{\sqrt{S_{XX}}} \right)}$$

where

$$\hat{\sigma}^2 = \frac{S_{YY} - bS_{XY}}{n - 2}$$

and this has a t-distribution with $n - 2$ degrees of freedom. Use Table D.6 on page 331.

Part IV

Appendices

Appendix A

Self-assessment guide and mathematical basics

Using this guide

If, by the end of the second or third time you have worked through this appendix, you are still uncomfortable with its contents, you should seek remedial mathematical help. The earlier you can get this help the better. Without it, your ability to cope with the contents of this book will be severely impacted – as early as Chapter 4.

Basic operations

The basic operations are: addition, subtraction, multiplication and division. Their symbols are: $+$ $-$ $\times$ $\div$

You should be familiar with the following rules:

$a + b = c$ and $b + a = c$
$a + (b + c) = (a + b) + c$
$a \times b = b \times a$
$a \times (b \times c) = (a \times b) \times c$
$a \times (b + c) = a \times b + a \times c$
$a \div b \neq b \div a$

e.g.
$3 + 5 = 8$ and $5 + 3 = 8$
$4 + (7 + 2) = (4 + 7) + 2 = 13$
$6 \times 9 = 9 \times 6 = 54$
$2 \times (5 \times 4) = (2 \times 5) \times 4 = 40$
$6 \times (2 + 5) = (6 \times 2) + (6 \times 5) = 42$
$1 \div 2 \neq 2 \div 1$

A negative number multiplied or divided by a positive number is negative, and vice versa.

e.g.
$-3 \times 4 = -12$ and $3 \times -4 = -12$
$-3 \div 4 = 0.75$ and $3 \div -4 = -0.75$

A negative number multiplied or divided by another negative number is positive.

e.g.
$-3 \times -4 = 12$
$-3 \div -4 = 0.75$

Inequalities

$x \neq y$ reads x is not equal to y
$x > y$ reads x is greater than y

$x < y$ reads x is less than y
$x \geq y$ reads x is greater than or equal to y
$x \leq y$ reads x is less than or equal to y
$x \approx y$ reads x is approximately equal to y

If $x - y \geq 0$ then $x \geq y$
If $x - y$ is positive then $x > y$

e.g.
$3 - 2 = 1$, so $3 > 2$

Exponents (powers, indices)

x^2 reads x-squared
x^3 reads x-cubed
x^n reads x to the power n

e.g.
$3 \times 3 = 3^2 = 9$
$3 \times 3 \times 3 \times 3 = 3^4 = 81$
$3^1 = 3$
$3^0 = 1$

Basic rules of manipulating indices:

$$x^a \times x^b = x^{a+b} \qquad \frac{x^a}{x^b} = x^{a-b}$$

A common operation is the square root:
$x^{\frac{1}{2}} = \sqrt{x}$ = square root of x, since
$(\sqrt{x})^2 = x^{\frac{1}{2}} \times x^{\frac{1}{2}} = x^{\frac{1}{2} + \frac{1}{2}} = x^1 = x$

Order of operations

In a complicated expression with many operations, we need a rule to make sure the expression is evaluated correctly. The rule is usually called BEDMAS, an acronym for Brackets, Exponents, Division, Multiplication, Addition and Subtraction.

We perform those operations in that order.

<u>1st</u>, do computations in brackets   B
<u>2nd</u>, evaluate any exponents   E
<u>3rd</u>, do any multiplication or division   DM
<u>4th</u>, do any addition or subtraction   AS

e.g.
$$(2 \times 7 + 2 - 1)^2 + 11 \times 3 - 4$$

- First, we must evaluate the expression in brackets using the other rules $2 \times 7 + 2 - 1 = 14 + 2 - 1 = 15$ (first the multiplication, then the rest). This leaves us with $15^2 + 11 \times 3 - 4$.
- The highest priority operation is now the exponent $15^2 = 225$. The problem simplifies to: $225 + 11 \times 3 - 4$.
- Multiplication comes next: $11 \times 3 = 33$ leaving $225 + 33 - 4$.
- Evaluating the addition and subtraction (in any order), the answer is 254.

Confirm these results:

1. $(4 + 3)^2 \times 7 - 5 = 338$
2. $4 + 3^2 \times 7 - 5 = 62$
3. $(4 + 3)^2 \times (7 - 5) = 98$
4. $4 + 3 \times 7 - 5 = 20$

The next two examples are representative of the most difficult BEDMAS problems in the book. Try them yourself before working through the following steps:

e.g.
$$\sqrt{\frac{6.3^2}{11} + \frac{9.1^2}{17}} = 2.03 \text{ (2dp)}$$

- First, we reformat to discover the hidden brackets, namely

$$\left(\frac{6.3^2}{11} + \frac{9.1^2}{17}\right)^{0.5}$$

- Next, we must evaluate the expression in brackets using the other rules.
- Exponents come first, $6.3^2 = 39.69$ and $9.1^2 = 82.81$ leaving

$$\frac{39.69}{11} + \frac{82.81}{17}$$

inside the brackets.
- Next comes the division, $39.69 \div 11 = 3.608$ and $82.81 \div 17 = 4.871$ leaving the addition $3.608 + 4.871 = 8.479$ as the term inside the brackets.
- This leaves us with $\sqrt{8.479} = 2.912$ (3dp).

e.g.

$$\frac{7 \times 3.2^2 + 9 \times 4.3^2}{8 + 10 - 2} = 14.88 \text{ (2dp)}$$

- First, we reformat to discover the hidden brackets, namely

$$(7 \times 3.2^2 + 9 \times 4.3^2) \div (8 + 10 - 2)$$

- Next, we must evaluate the expressions in brackets using the other rules.
- For the first, exponents come first, $3.2^2 = 10.24$ and $4.3^2 = 18.49$ leaving

$$7 \times 10.24 + 9 \times 18.49$$

inside the brackets.
- The multiplications take precedence over the addition, so $7 \times 10.24 = 71.68$ and $9 \times 18.49 = 166.41$ leaving the addition $71.68 + 166.41 = 238.09$ as the term inside the first brackets.
- The second brackets give $8+10-2 = 16$.
- This leaves $238.09 \div 16 = 14.88$ (2dp).

Inverse (opposite) operations

Addition and subtraction are inverse operations.

e.g.
If $x = y + z$ then $y = x - z$.

Multiplication and division are inverse operations.

e.g.
If $x = y \times z$ then $y = x/z$ (as long as $z \neq 0$).

For every non-zero number x we can find a number $\frac{1}{x}$ (or $1 \div x$ or $1/x$ or x^{-1}) which is called the inverse of x, with $x \times \frac{1}{x} = 1$.

e.g.
$2 \times \frac{1}{2} = 1$

Absolute value

$|x|$ reads the absolute value of x

$|x|$ is the *size* of x (ignoring the sign).

e.g.
$|3| = 3$ and $|-3| = 3$
$|0| = 0$

$|x - y|$ is the size of the difference between x and y.

e.g
$|3 - 1| = |2| = 2$ (notice the brackets)

e.g.
$|3 - 5| = |-2| = 2$

Summation notation

$\sum$ is an instruction to sum, i.e. add. (The symbol is the Greek letter Sigma, which is their capital "s").

Suppose we have a collection of data and we label each observation so that

$$x_1 = \text{1st observation}$$
$$x_2 = \text{2nd observation}$$
$$\vdots$$
$$x_i = i\text{th observation}$$

and so on.

The subscript i is called an index. It can be i or j or r or any other label. It takes each whole number value in turn, between the first and last shown below and above the $\sum$ symbol.

e.g.

$$\sum_{i=1}^{5} x_i = x_1 + x_2 + x_3 + x_4 + x_5$$

$$\sum_{j=2}^{4} x_j = x_2 + x_3 + x_4$$

with

i	1	2	3	4	5
x_i	1.3	2.4	2.1	3.2	4.0

then:

$$\sum_{i=1}^{5} x_i = x_1 + x_2 + x_3 + x_4 + x_5$$
$$= 1.3 + 2.4 + 2.1 + 3.2 + 4.0 = 13$$

$$\sum_{j=2}^{4} x_j = x_2 + x_3 + x_4$$
$$= 2.4 + 2.1 + 3.2 = 7.7$$

Symbol	Name	Translation
α	alpha	a
β	beta	b
ϵ	epsilon	e
μ	mu	m
ν	nu	n
σ	sigma	s
Σ	Sigma	S
χ	chi	ch

Similarly $\sum_{i=1}^{3} x_i^2$ means add up the squares of x_i for $i = 1$ to 3.

$$\sum_{i=1}^{3} x_i^2 = x_1^2 + x_2^2 + x_3^2$$
$$= 1.3^2 + 2.4^2 + 2.1^2$$
$$= 1.69 + 5.76 + 4.41 = 11.86$$

Commonly the sums are written without limits.

e.g.
$\sum x_i$ or $\sum x$ means add up all the x values.

Similarly, $\sum x_i^2$ is an instruction to add up all the squared x_i values. Note that BEDMAS ensures we do the squaring before the sum.

A final example combining various aspects of summation notation and BEDMAS:

e.g.
$$\sum_{i=2}^{4}(x_i^2 - 1)$$
$$= (x_2^2 - 1) + (x_3^2 - 1) + (x_4^2 - 1)$$
$$= (4^2 - 1) + (3^2 - 1) + (2^2 - 1)$$
$$= 15 + 8 + 3 = 26.$$

Greek letters

These are often a quick way to distinguish whether we are talking about populations or samples. The Greek letters found in this book, their names, and English equivalents are:

Appendix B

Calculator use

In this appendix, we demonstrate the statistical functions of the Casio *fx-82MS*. Output from the calculator is shown using `this` font and non-numerical keys are placed in a box.

The important keys for a statistical analysis are:

- MODE is used to put the calculator into a statistical mode.

- SHIFT is used to access additional functions in the calculator.

- CLR , accessed by typing SHIFT MODE , clears the statistical memories.

- DT , accessed by typing M+ , is used to enter data.

- ; , accessed by typing SHIFT , , is used when entering repeated observations.

- S-SUM , accessed by typing SHIFT 1 , stores various sums.

- S-VAR , accessed by typing SHIFT 2 , stores various sample statistics.

- REPLAY (the large button just below the screen) is used to move the cursor to access additional memories, and to "replay" the last input. In particular REPLAY ▷ is used to move right, and ◁ REPLAY is used to move left.

The calculator has two important modes: the standard deviation mode, and the regression mode. We first investigate the standard deviation mode. With either, as we enter data, the calculator keeps track of what has been entered by updating sums.

> To begin with, all sums must be set to zero. This is achieved by typing [SHIFT] [MODE] and choosing [3] corresponding to the choice ALL, and then [=]. Alternatively, in the standard deviation or regression modes, choose [CLR] [1] and [=], corresponding to the choice Scl.

When you clear the memory, the calculator should display a confirmation message: Reset All followed by ----------, or in the standard deviation or regression modes, Stat clear.

The standard deviation mode

Enter the standard deviation mode by typing [MODE] followed by [2] corresponding to the choice SD. A small SD will appear on the screen. Before entering any data, clear the memory using [CLR] [1] [=].

To begin, we assume we have the following data:

i	1	2	3	4	5
x_i	1.3	2.4	2.1	3.2	4.0

which we enter one by one into the calculator, using the [DT] key. The first observation is entered as follows: 1.3 [DT], after which the calculator reports n=1. Continuing, we enter:

$$2.4 \, [DT] \, 2.1 \, [DT] \, 3.2 \, [DT] \, 4.0 \, [DT]$$

after which the calculator should report n=5.

We can see from processing the data manually, that $\sum x = 1.3 + 2.4 + 2.1 + 3.2 + 4 = 13$, and this is confirmed by keying in:

$$[S\text{-}SUM] \, [2] \, [=] \text{ returning } \sum x = 13$$

where [S-SUM] is equivalent to [SHIFT] [1]. The sum $\sum x^2 = 1.3^2 + 2.4^2 + 2.1^2 + 3.2^2 + 4^2 = 38.1$ is more easily obtained using:

$$[S\text{-}SUM] \, [1] \, [=] \text{ returning } \sum x^2 = 38.1.$$

We also notice the option to retrieve the sample size n, by keying:

$$[S\text{-}SUM] \, [3] \, [=] \text{ returning } n = 5.$$

In addition to these sums, we can also obtain the mean and standard deviation of the data. These are obtained using:

[S-VAR][1][=] returning the mean $\bar{x} = \frac{1}{n}\sum x = \frac{1}{5} \times 13 = 2.6$

where [S-VAR] is equivalent to [SHIFT][2]. The standard deviation is calculated manually using

$$s = \sqrt{\frac{\sum x^2 - \frac{1}{n}(\sum x)^2}{n-1}} = \sqrt{\frac{38.1 - \frac{1}{5} \times 13^2}{4}} = \sqrt{\frac{4.3}{4}} = \sqrt{1.075} = 1.037 \text{ (3dp)}.$$

However, it is much easier to key in

[S-VAR][3][=] returning the standard deviation $s = 1.036822068$

which we round appropriately. Note that [S-VAR][2][=] returns the *population standard deviation*, using n in the denominator rather than the correct $n-1$. Throughout the examples and exercises in this book, we will never need to use this, and it is unlikely that we would need it in any other setting.

In many instances we have grouped or repeated data for which we need to approximate the mean and standard deviation. The calculator enables us to easily enter such data using the [;] key (this is equivalent to [SHIFT][,]). Consider the data set:

2.1 2.1 2.1 2.1 5.3 5.3 5.3 5.3 5.3 7.8 7.8 7.8

which includes 2.1 repeated four times, 5.3 five times and 7.8 three times. After clearing any previous data from the memory ([CLR][1][=] in SD mode), we enter the data as follows:

2.1[;]4[DT] 5.3[;]5[DT] 7.8[;]3[DT].

When we enter 2.1[;]4[DT], the calculator reports n=4 indicating four copies of 2.1 have been entered. Likewise, when we add 5.3[;]5[DT] it reports n=9. You should confirm $\bar{x} = 4.86$ and $s = 2.28$ (both to 2dp) for this data.

The regression mode

When we have paired data as in Chapter 4, we may find the calculator's regression mode useful. We enter the regression mode by keying [MODE] followed by [3] for the choice Reg and then [1] for the choice Lin, i.e. we select the *linear regression* mode. [The calculator is also capable of log-linear, and exponential regression, neither of which are discussed in this book.] As with the SD mode, a small REG should appear on the screen. Once again, the first step is to clear any data from the statistical memories. This achieved by keying: [CLR][1][=].

For this section we will add some y data to the x data we have already examined:

i	1	2	3	4	5
x_i	1.3	2.4	2.1	3.2	4.0
y_i	2.1	1.9	2.7	2.9	3.2

This time, we enter pairs of data using the [DT] key. The pairs are input using the [,] key to separate them, as follows:

$$1.3\,[,]\,2.1\,[DT]\quad 2.4\,[,]\,1.9\,[DT]\quad 2.1\,[,]\,2.7\,[DT]\quad 3.2\,[,]\,2.9\,[DT]\quad 4\,[,]\,3.2\,[DT]$$

after which the screen should report n=5, indicating we have entered the correct *number of pairs*. We can confirm the sums we found earlier using:

[S-SUM][1][=] giving $\sum x^2 = $ 38.1, [S-SUM][2][=] giving $\sum x = $ 13

and [S-SUM][3][=] giving $n = $ 5. This time, we notice the small arrow ➔ at the extreme right of the screen which indicates additional sums are available, in particular $\sum y$, $\sum y^2$, and $\sum xy$. These are accessed using [REPLAY ▷] and then [1], [2], or [3] as appropriate. The sum $\sum y = 2.1 + 1.9 + 2.7 + 2.9 + 3.2 = 12.8$ is obtained using:

[S-SUM][REPLAY ▷][2][=] returning $\sum y = $ 12.8.

The sum $\sum y^2 = 2.1^2 + 1.9^2 + 2.7^2 + 2.9^2 + 3.2^2 = 33.96$ is more easily obtained using:

[S-SUM][REPLAY ▷][1][=] returning $\sum y^2 = $ 33.96.

The final sum used in the regression-related formulae is $\sum xy = 1.3 \times 2.1 + \cdots + 4 \times 3.2 = 35.04$, which is obtained from the calculator using:

[S-SUM][REPLAY ▷][3][=] returning $\sum xy = $ 35.04.

The five sums: $\sum x, \sum x^2, \sum y, \sum y^2$ and $\sum xy$, as well as n are used to calculate Pearson's correlation (see Section 4.2.1), which has the formula

$$r = \frac{\sum xy - \frac{1}{n}\sum x \sum y}{\sqrt{\left(\sum x^2 - \frac{1}{n}(\sum x)^2\right)\left(\sum y^2 - \frac{1}{n}(\sum y)^2\right)}} = \frac{35.04 - \frac{1}{5} \times 13 \times 12.8}{\sqrt{(38.1 - \frac{1}{5} \times 13^2) \times (33.96 - \frac{1}{5} \times 12.8^2)}}$$

for which careful observation of the BEDMAS rules in Appendix A gives $r = 0.777$ (3dp). Once again it is much easier to use the calculator's functions. To access r, type:

[S-VAR][REPLAY ▷][REPLAY ▷][3][=] returning $r = $ 0.777392439

which can be rounded appropriately.

In obtaining r, we skipped some other useful statistics. The following are also available:

- [S-VAR][1][=] gives $\bar{x} = 2.6$
- [S-VAR][3][=] gives the standard deviation of the x data $s_x = 1.0368\ldots$
- [S-VAR][REPLAY ▷][1][=] gives $\bar{y} = 2.56$
- [S-VAR][REPLAY ▷][3][=] gives the standard deviation of the y data $s_y = 0.5458\ldots$

as well as the coefficients of the regression line $\hat{y} = a + bx$. Their formulae are

$$b = \frac{\sum xy - \frac{1}{n}\sum x \sum y}{\sum x^2 - \frac{1}{n}(\sum x)^2} = \frac{35.04 - \frac{1}{5} \times 13 \times 12.8}{38.1 - \frac{1}{5} \times 13^2} = 0.409 \text{ (3dp)}$$

and $a = \bar{y} - b\bar{x} = 2.56 - 0.409 \times 2.6 = 1.496$ (3dp). Once again, it is much easier to obtain these directly from the calculator, which labels them A and B instead of a and b. The constant a is obtained using

[S-VAR][REPLAY ▷][REPLAY ▷][1][=] returning $a = 1.4958\ldots$

and the slope b is obtained using

[S-VAR][REPLAY ▷][REPLAY ▷][2][=] returning $b = 0.4093\ldots$

both of which can be rounded appropriately.

The calculator performs one additional function for us, which is to compute *predictions*. Suppose we wanted to find out the value of the line at $x = 2.1$. To calculate this, we would insert $x = 2.1$ into the equation of the line, i.e.

$$\hat{y} = a + b \times 2.1 = 1.496 + 0.409 \times 2.1 = 2.3549.$$

We obtain a prediction directly from the calculator using

2.1 [S-VAR][REPLAY ▷][REPLAY ▷][REPLAY ▷][2][=] returning $\hat{y} = 2.35534\ldots$

which highlights our rounding errors when we calculated it manually. Note that we key in the x value first, and then instruct the calculator to give us the corresponding $\hat{y}$. Less useful is to obtain an x value for a given $\hat{y}$ value, which is done in a similar way, but choosing [1] instead of [2] at the final screen.

Additional features

The calculator has additional features which make data entry and management a breeze. Make sure you consult the calculator's manual to find out how to view the entered data using [REPLAY △] and [REPLAY ▽], and how to change incorrectly entered data.

Appendix C

Spreadsheet use

An increasingly common desktop tool is a spreadsheet – examples include Microsoft's Excel, or Openoffice.org's Calc. The latter is part of a free and open productivity suite and is available for download at www.openoffice.org.

Spreadsheets are an excellent way of managing small datasets, and can be used to perform many of the data operations in this book. In this appendix we list spreadsheet functions which may be directly relevant. We also include examples of how to manually perform some tasks.

We assume a basic knowledge of spreadsheets, in particular, data entry and basic function syntax.

C.1 Summary statistics

Consider a sample of 15 observations entered in cells A1 to A15 of a spreadsheet. You might like to enter 15 numbers in now, or, take advantage of the random number generator in the spreadsheet by typing the formula =RAND() in cell A1. Copy and paste this into cells A2 to A15 to get a sample of size 15.

This and other commands (also known as functions, formulas, or formulae) are shown in THIS FONT and should always be preceded by an "=" symbol in the spreadsheet. The things between the brackets "( )" are the *arguments* of the function, and should be separated by a semi-colon " ; " in openoffice.org's Calc, or a comma " , " in Microsoft's Excel.

The following functions only have one argument: the cell range of the data.

COUNT(A1:A15) counts the number of observations

SUM(A1:A15) adds the observations, i.e. $\sum x_i = x_1 + x_2 + \cdots + x_n$

SUMSQ(A1:A15) adds the squares of the observations, i.e. $\sum x_i^2 = x_1^2 + x_2^2 + \cdots + x_n^2$

AVERAGE(A1:A15) calculates the sample mean, i.e. $\bar{x} = \frac{1}{n}\sum x_i$

VAR(A1:A15) calculates the sample variance, i.e. $s^2 = \frac{1}{n-1}\sum(x_i - \bar{x})^2$

STDEV(A1:A15) calculates the sample standard deviation, i.e. $s = \sqrt{\frac{1}{n-1}\sum(x_i - \bar{x})^2}$. This is equivalent to the formula =SQRT(VAR(A1:A15))

MIN(A1:A15) returns the smallest value in the sample, the minimum

MAX(A1:A15) returns the largest value in the sample, the maximum

MEDIAN(A1:A15) returns the sample median

MAX(A1:A15)-MIN(A1:A15) returns the range of the sample

The next set of functions requires an argument in addition to the data range. Excel separates arguments by a comma, whereas openoffice.org separates them with a semicolon, as we do here.

In these instances, the extra argument is a number which can be entered into the function directly, or put in an empty cell and referenced, e.g. if the data are in cells A1 to A15, and 1 is entered in cell B1, then =QUARTILE(A1:A15,1) or =QUARTILE(A1:A15,B1) will both calculate the lower quartile of the data. The latter is particularly useful if you anticipate changing the number, e.g. to see a different quartile.

QUARTILE(A1:A15;B1) returns the specified quartile, according to the value in cell B1, as follows: 0 (minimum), 1 (lower quartile), 2 (median), 3 (upper quartile), 4 (maximum).

NB: The quartile function uses a method of interpolation, and will possibly give different values to the simpler method used in this book.

PERCENTILE(A1:A15;B1) returns the specified percentile, according to the value in cell B1. This value should be a decimal, between 0 and 1 (inclusive). The values 0, 0.25, 0.5, 0.75, 1 correspond to the quartiles.

NB: The percentile function uses a method of interpolation, and will possibly give different values to the simpler method used in this book.

QUARTILE(A1:A15;3)-QUARTILE(A1:A15;1) returns the interquartile range of the sample

The next functions provide bivariate statistics. We assume that the x data are entered in cells A1 to A15 and the y data are entered in cells B1 to B15.

CORREL(A1:A15;B1:B15) returns (Pearson's) correlation coefficient for the bivariate sample. This function is identical to PEARSON(A1:A15;B1:B15).

INTERCEPT(B1:B15;A1:A15) returns the intercept estimate a in the OLS regression line $\hat{Y} = a + bX$

NB: the y data come first in the function.

SLOPE(B1:B15;A1:A15) returns the slope estimate b in the OLS regression line $\hat{Y} = a + bX$

NB: the y data come first in the function.

C.1.1 Sorting data

The spreadsheet can be used to sort data, but this is done via a menu rather than using a function. Suppose you have data entered in cells A1 to B15, e.g. x data in A1 to A15 and y data in B1 to B15. To sort the x data, click the menu headed Data, followed by Sort. A dialogue box will open. Sort by Column A, select "ascending" (in Excel you are offered "smallest to largest"), and click "OK". Note that the x data are now in ascending order, while the y data have remained paired with the x data.

To sort the y data, repeat, choose instead to sort by Column B using the drop-down box.

C.1.2 Case study: Spearman's correlation

We assume, as before, that the x data are entered in cells A1 to A15 and the y data are entered in cells B1 to B15. We wish to evaluate Spearman's correlation coefficient

$$r_s = 1 - \frac{6\sum d^2}{n^3 - n}$$

1. In cell C1, enter =RANK(A1;A$1:A$15;1).

   NB: If you are using Excel, you will need to replace the two semi-colons with commas.

   The RANK formula returns the rank of the value in A1, from among the data in A1 to A15, in ascending order (the final 1). The $ are for "absolute" cell references and prevent the row number changing when the formula is copied and pasted elsewhere.

   Copy this formula, select cells C2 to C15 and paste the formula. While A1 changes in each cell, A$1:A$15 does not.

   NB: the function RANK does not correctly account for tied values. Manually check for these, and change the values by hand (overwrite the formulae with the appropriate rank values). You might like to use the sort feature described above.

2. Repeat step 1 for the y data in B1 to B15, using cells D1 to D15, i.e. enter =RANK(B1;B$1:B$15;1) in cell D1, and copy and paste. Correct for any ties as before.

3. In cell E1, enter =C1-D1. This calculates $d_1 = r(X_1) - r(Y_1)$, i.e. the difference in the ranks for the first observation. Copy and paste to cells E2 to E15.

   NB: as a check, in cell F1 enter =SUM(E1:E15). This is $\sum d_i$ and should equal zero if the ranks have been correctly calculated and ties correctly accounted for.

4. In cell F2, enter =SUMSQ(E1:E15). This returns $\sum d_i^2$.

5. In cell F3, enter =COUNT(E1:E15). This returns the sample size n, which is 15 in this example.

6. In cell F4, enter =1 - 6*F2/(F3^3 - F3). This evaluates Spearman's correlation as given by the formula above, and obeys the BEDMAS rules. F3^3 is equivalent to POWER(F3,3), and both calculate n^3. Alternatively, the denominator could be replaced by (15^3 - 15).

C.1.3 Case study: time series moving averages

Time series data are time consuming to process manually, but very easy to handle in a spreadsheet. We assume the time series observations are in cells A1 to A50, and will now calculate a 7-point centred moving average of these data.

1. In cell B4, enter the formula =AVERAGE(A1:A7). This calculates the average of the first seven observations, is centred on the 4th observation, and is equivalent to =SUM(A1:A7)/7.

   NB: we cannot calculate this moving average in cells B1 to B3 since there are not enough previous observations.

2. Copy the formula in cell B4 and paste it into cell B5. Click on this cell, and note that the formula has updated to =AVERAGE(A2:A8), i.e. the average has "moved", and is now centred on the 5th observation.

3. Copy the formula in cell B4 or B5, and paste into cells B6 to B97.

   NB: if your formula references empty cells, it will treat the contents as a zero. For example, if you copy the formula into cell B98, it will reference cell A101, and will now incorrectly include a 0 in the average.

4. The contents of cells B1 to B100 (including the three missing values at each end) are the 7-point moving average.

C.2 Probability distributions

Binomial probabilities are given by the function BINOMDIST. Enter x in cell A1, n in cell A2 and p in cell A3.

BINOMDIST(A1;A2;A3;0) returns $P(X = x)$ where $X \sim Bin(n, p)$, i.e. an individual term as in Table D.2 on page 325. The final argument 0 specifies the individual term.

BINOMDIST(A1;A2;A3;1) returns $P(X \leq x)$ where $X \sim Bin(n, p)$, i.e. a cumulative probability as in Table D.3 on page 327. The final argument 1 specifies the cumulative probability.

Normal probabilities are given by the functions NORMSDIST and NORMDIST. Enter z in cell A1, x in cell B1, μ in cell B2, and σ in cell B3.

NORMSDIST(A1) returns $P(Z < z)$ where $Z \sim N(0, 1)$ is the standard normal random variable, as in Table D.4 on page 329.

$P(Z > z)$ can be given by =1-NORMSDIST(A1).

NORMDIST(B1;B2;B3;1) returns $P(X < x)$ where $X \sim N(\mu, \sigma^2)$. The fourth argument 1 specifies a probability rather than the height of the "bell" curve.

NB: The third argument of the function is the standard deviation of X, not its variance.

Inverse normal probabilities are given by the functions NORMSINV and NORMINV. Enter p in cells A1 and B1, μ in cell B2, and σ in cell B3.

NORMSINV(A1) returns z such that $P(Z < z) = p$ where $Z \sim N(0, 1)$ is the standard normal random variable, as in Table D.5 on page 330.

NORMINV(B1;B2;B3) returns x such that $P(X < x) = p$ where $X \sim N(\mu, \sigma^2)$.

NB: The third argument of the function is the standard deviation of X, not its variance.

C.3 Inference

Spreadsheets are capable of performing some of the hypothesis tests in this book directly. We include only the basic (large sample) Z test and the two sample t test in this section.

We assume the large sample data are entered in cells A1 to A50 and the small sample data are entered in cells D1 to D15 and E1 to E15. Place μ from H_0 in cell B1. In cell B2 calculate the sample standard deviation of the large sample, i.e. use =STDEV(A1:A50) or enter the value for σ if you have it.

Use these functions with caution!

ZTEST(A1:A50;B1;B2) returns the probability of a test statistic greater than

$$\frac{\bar{X} - \mu}{\frac{s}{\sqrt{n}}}$$

where $\bar{X}$ is given by AVERAGE(A1:A50), μ is in B1, s is in B2 and n is given by COUNT(A1:A50) (and is 50 in this example). This is the p-value *in the case of a one-sided test with a greater-than alternative*. It can be doubled if the alternative is two-sided, or subtracted from one if the alternative is less-than.

NB: The help file in openoffice.org 3.2 suggests the p-value is two-tailed, but it is not! The normal distribution is always used for the p-value so this will be incorrect when n is small.

TTEST(D1:D15;E1:E15;F1;F2) returns the p-value of a test of $H_0 : \mu_1 = \mu_2$ where:

- sample 1 data are in D1 to D15
- sample 2 data are in E1 to E15
- the third argument (in F1) is 1 for a one-tailed test (and the data with the larger mean is sample 1) or 2 for a two-tailed test
- the fourth argument (in F2) is 1 for a paired analysis, 2 if the variances are equal, or 3 if the variances are not equal.

NB: when the variances are not equal, the degrees of freedom are calculated differently from the method used in the text.

TDIST(A1;A2;1) returns the probability that the t random variable, with degrees of freedom given in A2, exceeds A1. If the (small sample) test statistic is in A1, its degrees of freedom parameter is in A2, and the test has a greater-than alternative, this is the p-value of the test.

TDIST(A1;A2;2) returns the probability that the absolute value of a t random variable, with degrees of freedom given in A2, exceeds A1, where A1 contains a positive number. If the absolute value of the (small sample) test statistic is in A1, its degrees of freedom parameter is in A2, and the test has a two-sided alternative, this is the p-value of the test.

FDIST(A1;A2;A3) gives the probability that the F random variable, with degrees of freedom given in cells A2 and A3, exceeds the value in A1. If A1 contains the test statistic in a test of equality of variances, $n_\ell - 1$ is in A2 and $n_s - 1$ is in A3, then the value given is half the p-value of the test. If A1 contains an ANOVA test statistic, with $k - 1$ in A2 and $n - k$ in A3, then the value given is the p-value of the test.

CHIDIST(A1;A2) gives the probability that the χ^2 random variable, with degrees of freedom given in cell A2, exceeds the value in A1. If A1 contains the test statistic in a chi-square test, and its degrees of freedom parameter is in A2, then the value given is the p-value of the test. See also Section C.3.1.

C.3.1 Case study: contingency table test

The following instructions will perform the contingency table test of Chapter 12. We assume the observed frequencies are entered in cells A1 to C2, i.e. the table is 2×3. At the end of these instructions, we show you the completed spreadsheet for Example 12.4.

1. In cell D1 enter the formula =SUM(A1:C1). This is the first row total. Copy the formula to cell D2 to give the second row total.

2. In cell A3 enter the formula =SUM(A1:A2). This is the first column total. Copy the formula to cells B3 and B4 to give the remaining column totals.

3. In cell D3 enter the formula =SUM(A1:C2). This is the total sample size n. Check that SUM(A3:C3) and SUM(D1:D2) give the same number.

4. We now calculate the expected frequencies in cells A5 to C6. In cell A5 enter the formula =A$3*$D1/$D$3. This multiplies the first column total (in A3) by the first row total (in D1) and divides by n (in D3).

   We've put absolute cell references so the various references (row for column totals, column for row totals, and both for n) don't change when we copy and paste.

5. Copy the formula from A5 to cell A6 and B5 to C6. Select cell B6 and note the formula is =B$3*$D2/$D$3, i.e. the second column total (in B3) times the second row total (in D2) divided by n (in D3).

   Check that the expected frequencies have the same row and column totals as the observed frequencies by following steps 1 and 2.

6. We now calculate the contributions to the test statistic, i.e.

$$\frac{(o_i - e_i)^2}{e_i}$$

   in cells A9 to C10. In cell A9 enter the formula =(A1-A5)^2/A5. This subtracts the first expected frequency (in A5) from the first observed frequency (in A1), squares, then divides by the first expected frequency. The formula obeys the BEDMAS rules.

   Copy the formula from cell A9 and paste into cells A10 and B9 to C10.

7. In cell D11, enter the formula =SUM(A9:C10). This is the test statistic.

8. In cell D12, enter the formula =(COUNT(D1:D2)-1)*(COUNT(A3:C3)-1). This is the number of degrees of freedom, in this case, 2.

9. In cell D13, enter the formula =CHIDIST(D11;D12). This is the p-value of the test. Alternatively, skip this step, and compare the test statistic (in cell D11) with the appropriate critical value from Table D.7 on Page 332.

The completed spreadsheet for the data in Example 12.4 is shown in Figure C.1.

	A	B	C	D	E
1	10	8	6	24	Observed
2	6	16	18	40	frequencies
3	16	24	24	64	
4					
5	6	9	9	24	Expected
6	10	15	15	40	frequencies
7	16	24	24	64	
8					
9	2.66667	0.11111	1		
10	1.6	0.06667	0.6		
11				6.04444	Test statistic
12				2	Degrees of freedom
13				0.04869	p-value
14					

Cell D13: =CHIDIST(D11,D12)

Figure C.1 The completed spreadsheet for the chi-square test

Appendix D

Tables and table use

The following tables are provided in this appendix:

D.1 random number table
D.2 individual binomial probabilities
D.3 cumulative binomial probabilities
D.4 the standard normal distribution
D.5 the inverse standard normal distribution
D.6 Student's t-distribution
D.7 the χ^2 distribution
D.8 the F-distribution with 5% upper-tail probability
D.9 the F-distribution with $2\frac{1}{2}$% upper-tail probability
D.10 the F-distribution with 1% upper-tail probability
D.11 the Mann-Whitney U table, for 5% and 1%

Each table has a short introduction which indicates how the table should be used. Until you have had practice successfully using each table, you should always refer to the introductory notes before finding your critical values, or probabilities.

D.1 Random number table (2160 random digits, generated using statistical software R, downloaded from r-project.org.)

```
94880  47224  42145  26928  86390  17618  10021  27621  72460
53202  36605  32792  34116  39100  81670  41781  18689  74267
81863  12169  36687  96552  03374  13019  64104  51975  78822
93422  78146  22609  00411  30118  09252  61658  48593  64089
42508  61213  76201  60889  56451  24987  11992  33573  35755
55594  19503  12643  13315  92694  47245  18576  54596  12940
65320  66678  61659  14769  28469  44260  59913  17403  30609
47680  78658  19440  82270  27435  06636  08766  21013  27491
52224  76047  98316  65927  39075  46576  50595  34529  08816
59621  94478  18444  55082  35446  20242  12008  13452  82374
88817  93562  09253  97167  95526  94725  99285  49859  24367
61218  68182  21492  13418  28211  43431  58532  76110  81296
74533  92353  25016  40234  94022  42284  56337  09070  78095
60532  13991  28526  78582  80628  67106  80180  99344  28698
01959  38289  90588  04142  69326  29086  15989  90718  01321
50139  68374  19237  78002  97076  71528  81703  26410  47450
25179  95615  01663  43886  29673  01216  18049  76548  68060
56109  77820  13468  81512  09368  91660  93818  01600  64966
67593  80609  93287  00521  25659  75955  05702  34452  63655
32465  16923  44949  13736  20882  03868  72881  02999  20226
41133  59859  14163  30583  42817  15073  90437  58074  99111
90988  24026  29222  66827  53245  15121  01348  96707  35254
61986  08141  92994  90905  36906  82218  04501  21967  73425
88020  06493  90089  81958  16839  58095  78365  88936  62093
75514  97072  69020  03669  80944  59785  11604  17419  20658
90499  57200  01095  43428  66653  30950  61388  87809  60075
40468  88140  86377  51403  02249  18413  71589  38116  99495
49898  26851  80843  13528  72903  87946  85784  20411  40673
47930  52216  95419  05897  07400  81457  05816  99427  95937
16305  61479  84690  34731  70273  15250  26063  30299  96536
70274  51092  49175  03963  00215  78311  32355  10153  17044
05636  71779  49412  74922  38333  18423  73840  54292  87152
65518  64497  57270  00502  96252  50735  58021  99459  08078
38133  64762  23208  44029  62441  15050  24558  66962  46889
40887  51833  75910  94907  19634  92436  08720  43063  94742
35338  58986  84082  84246  19118  39728  57884  22040  50599
79450  16866  71535  79107  01990  15287  82985  03866  81076
48428  56359  91193  35927  34807  18735  27187  66924  50151
34632  53073  54109  05718  27987  67772  38115  39674  86107
52437  99672  44834  27754  31711  75654  78191  94765  20937
73751  35063  53400  67315  10186  77140  77134  32417  23800
13074  86641  00675  49579  81387  40789  51156  53360  63436
30577  15870  73134  28285  80994  12469  02963  41432  31294
16381  18037  82957  11573  04746  08740  02431  65528  84012
65170  30785  27685  77720  22065  19858  06538  83280  66647
15687  72681  39202  69837  36931  84195  54709  70430  47579
07460  14037  34831  59588  05704  54710  17833  12119  86452
34871  60309  01618  14964  13547  84942  46474  70851  72392
```

D.2 Binomial probabilities

(Tabulated values are $P(X = x)$ where X is $Bin(n,p)$. If $p > 0.5$ use $P(X = x) = P(Y = n - x)$, where $Y \sim Bin(n, 1 - p)$.)

n	x	0.01	0.05	0.10	0.15	0.20	0.25	0.30	0.35	0.40	0.45	0.50
2	0	0.9801	0.9025	0.8100	0.7225	0.6400	0.5625	0.4900	0.4225	0.3600	0.3025	0.2500
	1	0.0198	0.0950	0.1800	0.2550	0.3200	0.3750	0.4200	0.4550	0.4800	0.4950	0.5000
	2	0.0001	0.0025	0.0100	0.0225	0.0400	0.0625	0.0900	0.1225	0.1600	0.2025	0.2500
3	0	0.9703	0.8574	0.7290	0.6141	0.5120	0.4219	0.3430	0.2746	0.2160	0.1664	0.1250
	1	0.0294	0.1354	0.2430	0.3251	0.3840	0.4219	0.4410	0.4436	0.4320	0.4084	0.3750
	2	0.0003	0.0071	0.0270	0.0574	0.0960	0.1406	0.1890	0.2389	0.2880	0.3341	0.3750
	3		0.0001	0.0010	0.0034	0.0080	0.0156	0.0270	0.0429	0.0640	0.0911	0.1250
4	0	0.9606	0.8145	0.6561	0.5220	0.4096	0.3164	0.2401	0.1785	0.1296	0.0915	0.0625
	1	0.0388	0.1715	0.2916	0.3685	0.4096	0.4219	0.4116	0.3845	0.3456	0.2995	0.2500
	2	0.0006	0.0135	0.0486	0.0975	0.1536	0.2109	0.2646	0.3105	0.3456	0.3675	0.3750
	3		0.0005	0.0036	0.0115	0.0256	0.0469	0.0756	0.1115	0.1536	0.2005	0.2500
	4			0.0001	0.0005	0.0016	0.0039	0.0081	0.0150	0.0256	0.0410	0.0625
5	0	0.9510	0.7738	0.5905	0.4437	0.3277	0.2373	0.1681	0.1160	0.0778	0.0503	0.0313
	1	0.0480	0.2036	0.3281	0.3915	0.4096	0.3955	0.3602	0.3124	0.2592	0.2059	0.1563
	2	0.0010	0.0214	0.0729	0.1382	0.2048	0.2637	0.3087	0.3364	0.3456	0.3369	0.3125
	3		0.0011	0.0081	0.0244	0.0512	0.0879	0.1323	0.1811	0.2304	0.2757	0.3125
	4			0.0004	0.0022	0.0064	0.0146	0.0283	0.0488	0.0768	0.1128	0.1563
	5				0.0001	0.0003	0.0010	0.0024	0.0053	0.0102	0.0185	0.0313
6	0	0.9415	0.7351	0.5314	0.3771	0.2621	0.1780	0.1176	0.0754	0.0467	0.0277	0.0156
	1	0.0571	0.2321	0.3543	0.3993	0.3932	0.3560	0.3025	0.2437	0.1866	0.1359	0.0938
	2	0.0014	0.0305	0.0984	0.1762	0.2458	0.2966	0.3241	0.3280	0.3110	0.2780	0.2344
	3		0.0021	0.0146	0.0415	0.0819	0.1318	0.1852	0.2355	0.2765	0.3032	0.3125
	4		0.0001	0.0012	0.0055	0.0154	0.0330	0.0595	0.0951	0.1382	0.1861	0.2344
	5			0.0001	0.0004	0.0015	0.0044	0.0102	0.0205	0.0369	0.0609	0.0938
	6					0.0001	0.0002	0.0007	0.0018	0.0041	0.0083	0.0156
7	0	0.9321	0.6983	0.4783	0.3206	0.2097	0.1335	0.0824	0.0490	0.0280	0.0152	0.0078
	1	0.6590	0.2573	0.3720	0.3960	0.3670	0.3115	0.2471	0.1848	0.1306	0.0872	0.0547
	2	0.0020	0.0406	0.1240	0.2097	0.2753	0.3115	0.3177	0.2985	0.2613	0.2140	0.1641
	3		0.0036	0.0230	0.0617	0.1147	0.1730	0.2269	0.2679	0.2903	0.2918	0.2734
	4		0.0002	0.0026	0.0109	0.0287	0.0577	0.0972	0.1442	0.1935	0.2388	0.2734
	5			0.0002	0.0012	0.0043	0.0115	0.0250	0.0466	0.0774	0.1172	0.1641
	6				0.0001	0.0004	0.0013	0.0036	0.0084	0.0172	0.0320	0.0547
	7						0.0001	0.0002	0.0006	0.0016	0.0037	0.0078
8	0	0.9227	0.6634	0.4305	0.2725	0.1678	0.1001	0.0576	0.0319	0.0160	0.0084	0.0039
	1	0.0746	0.2793	0.3826	0.3847	0.3355	0.2670	0.1977	0.1373	0.0896	0.0548	0.0313
	2	0.0026	0.0515	0.1488	0.2376	0.2936	0.3115	0.2965	0.2587	0.2090	0.1569	0.1094
	3	0.0001	0.0054	0.0331	0.0839	0.1468	0.2076	0.2541	0.2786	0.2787	0.2568	0.2188
	4		0.0004	0.0046	0.0185	0.0459	0.0865	0.1361	0.1875	0.2322	0.2627	0.2734
	5			0.0004	0.0026	0.0092	0.0231	0.0467	0.0808	0.1239	0.1719	0.2188
	6				0.0002	0.0011	0.0038	0.0100	0.0217	0.0413	0.0703	0.1094
	7					0.0001	0.0004	0.0012	0.0033	0.0079	0.0164	0.0313
	8							0.0001	0.0002	0.0007	0.0017	0.0039
9	0	0.9135	0.6302	0.3874	0.2316	0.1342	0.0751	0.0404	0.0207	0.0101	0.0046	0.0020
	1	0.0830	0.2985	0.3874	0.3679	0.3020	0.2253	0.1556	0.1004	0.0605	0.0339	0.0176
	2	0.0034	0.0629	0.1722	0.2597	0.3020	0.3003	0.2668	0.2162	0.1612	0.1110	0.0703
	3	0.0001	0.0077	0.0446	0.1069	0.1762	0.2336	0.2668	0.2716	0.2508	0.2119	0.1641
	4		0.0006	0.0074	0.0283	0.0661	0.1168	0.1715	0.2194	0.2508	0.2600	0.2461
	5			0.0008	0.0050	0.0165	0.0389	0.0735	0.1181	0.1672	0.2128	0.2461
	6			0.0001	0.0006	0.0028	0.0087	0.0210	0.0424	0.0743	0.1160	0.1641
	7					0.0003	0.0012	0.0039	0.0098	0.0212	0.0407	0.0703
	8						0.0001	0.0004	0.0013	0.0035	0.0083	0.0176
	9								0.0001	0.0003	0.0008	0.0020
10	0	0.9044	0.5987	0.3487	0.1969	0.1074	0.0563	0.0282	0.0135	0.0060	0.0025	0.0010
	1	0.0914	0.3151	0.3874	0.3474	0.2684	0.1877	0.1211	0.0725	0.0403	0.0207	0.0098
	2	0.0042	0.0746	0.1937	0.2759	0.3020	0.2816	0.2335	0.1757	0.1209	0.0763	0.0439
	3	0.0001	0.0105	0.0574	0.1298	0.2013	0.2503	0.2668	0.2522	0.2150	0.1665	0.1172
	4		0.0010	0.0112	0.0401	0.0881	0.1460	0.2001	0.2377	0.2508	0.2384	0.2051
	5		0.0001	0.0015	0.0085	0.0264	0.0584	0.1029	0.1536	0.2007	0.2340	0.2461
	6			0.0001	0.0012	0.0055	0.0162	0.0368	0.0689	0.1115	0.1596	0.2051
	7				0.0001	0.0008	0.0031	0.0090	0.0212	0.0425	0.0746	0.1172
	8					0.0001	0.0004	0.0014	0.0043	0.0106	0.0229	0.0439
	9							0.0001	0.0005	0.0016	0.0042	0.0098
	10									0.0001	0.0003	0.0010

							p					
n	x	0.01	0.05	0.10	0.15	0.20	0.25	0.30	0.35	0.40	0.45	0.50
11	0	0.8953	0.5688	0.3138	0.1673	0.0859	0.0422	0.0198	0.0088	0.0036	0.0014	0.0005
	1	0.0995	0.3293	0.3835	0.3248	0.2362	0.1549	0.0932	0.0518	0.0266	0.0125	0.0054
	2	0.0050	0.0867	0.2131	0.2866	0.2953	0.2581	0.1998	0.1395	0.0887	0.0513	0.0269
	3	0.0002	0.0137	0.0710	0.1517	0.2215	0.2581	0.2568	0.2254	0.1774	0.1259	0.0806
	4		0.0014	0.0158	0.0536	0.1107	0.1721	0.2201	0.2428	0.2365	0.2060	0.1611
	5		0.0001	0.0025	0.0132	0.0388	0.0803	0.1321	0.1830	0.2207	0.2360	0.2256
	6			0.0003	0.0023	0.0097	0.0268	0.0566	0.0985	0.1471	0.1931	0.2256
	7				0.0003	0.0017	0.0064	0.0173	0.0379	0.0701	0.1128	0.1611
	8					0.0002	0.0011	0.0037	0.0102	0.0234	0.0462	0.0806
	9						0.0001	0.0005	0.0018	0.0052	0.0126	0.0269
	10								0.0002	0.0007	0.0021	0.0054
	11										0.0002	0.0005
12	0	0.8864	0.5404	0.2824	0.1422	0.0687	0.0317	0.0138	0.0057	0.0022	0.0008	0.0002
	1	0.1074	0.3413	0.3766	0.3012	0.2062	0.1267	0.0712	0.0368	0.0174	0.0075	0.0029
	2	0.0060	0.0988	0.2301	0.2924	0.2835	0.2323	0.1678	0.1088	0.0639	0.0339	0.0161
	3	0.0002	0.0173	0.0852	0.1720	0.2362	0.2581	0.2397	0.1954	0.1419	0.0923	0.0537
	4		0.0021	0.0213	0.0683	0.1329	0.1936	0.2311	0.2367	0.2128	0.1700	0.1208
	5		0.0002	0.0038	0.0193	0.0532	0.1032	0.1585	0.2039	0.2270	0.2225	0.1934
	6			0.0005	0.0040	0.0155	0.0401	0.0792	0.1281	0.1766	0.2124	0.2256
	7				0.0006	0.0033	0.0115	0.0291	0.0591	0.1009	0.1489	0.1934
	8				0.0001	0.0005	0.0024	0.0078	0.0199	0.0420	0.0762	0.1208
	9					0.0001	0.0004	0.0015	0.0048	0.0125	0.0277	0.0537
	10							0.0002	0.0008	0.0025	0.0068	0.0161
	11								0.0001	0.0003	0.0010	0.0029
	12										0.0001	0.0002
15	0	0.8601	0.4633	0.2059	0.0874	0.0352	0.0134	0.0047	0.0016	0.0005	0.0001	
	1	0.1303	0.3658	0.3432	0.2312	0.1319	0.0668	0.0305	0.0126	0.0047	0.0016	0.0005
	2	0.0092	0.1348	0.2669	0.2856	0.2309	0.1559	0.0916	0.0476	0.0219	0.0090	0.0032
	3	0.0004	0.0307	0.1285	0.2184	0.2501	0.2252	0.1700	0.1110	0.0634	0.0318	0.0139
	4		0.0049	0.0428	0.1156	0.1876	0.2252	0.2186	0.1792	0.1268	0.0780	0.0417
	5		0.0006	0.0105	0.0449	0.1032	0.1651	0.2061	0.2123	0.1859	0.1404	0.0916
	6			0.0019	0.0132	0.0430	0.0917	0.1472	0.1906	0.2066	0.1914	0.1527
	7			0.0003	0.0030	0.0138	0.0393	0.0811	0.1319	0.1771	0.2013	0.1964
	8				0.0005	0.0035	0.0131	0.0348	0.0710	0.1181	0.1647	0.1964
	9				0.0001	0.0007	0.0034	0.0116	0.0298	0.0612	0.1048	0.1527
	10					0.0001	0.0007	0.0030	0.0096	0.0245	0.0515	0.0916
	11						0.0001	0.0006	0.0024	0.0074	0.0191	0.0417
	12							0.0001	0.0004	0.0016	0.0052	0.0139
	13								0.0001	0.0003	0.0010	0.0032
	14										0.0001	0.0005
	15											
20	0	0.8179	0.3585	0.1216	0.0388	0.0115	0.0032	0.0008	0.0002			
	1	0.1652	0.3774	0.2702	0.1368	0.0576	0.0211	0.0068	0.0020	0.0005	0.0001	
	2	0.0159	0.1887	0.2852	0.2293	0.1369	0.0669	0.0278	0.0100	0.0031	0.0008	0.0002
	3	0.0010	0.0596	0.1901	0.2428	0.2054	0.1339	0.0716	0.0323	0.0123	0.0040	0.0011
	4		0.0133	0.0898	0.1821	0.2182	0.1897	0.1304	0.0738	0.0350	0.0139	0.0046
	5		0.0022	0.0319	0.1028	0.1746	0.2023	0.1789	0.1272	0.0746	0.0365	0.0148
	6		0.0003	0.0089	0.0454	0.1091	0.1686	0.1916	0.1712	0.1244	0.0746	0.0370
	7			0.0020	0.0160	0.0545	0.1124	0.1643	0.1844	0.1659	0.1221	0.0739
	8			0.0004	0.0046	0.0222	0.0609	0.1144	0.1614	0.1797	0.1623	0.1201
	9			0.0001	0.0011	0.0074	0.0271	0.0654	0.1158	0.1597	0.1771	0.1602
	10				0.0002	0.0020	0.0099	0.0308	0.0686	0.1171	0.1593	0.1762
	11					0.0005	0.0030	0.0120	0.0336	0.0710	0.1185	0.1602
	12					0.0001	0.0008	0.0039	0.0136	0.0355	0.0727	0.1201
	13						0.0002	0.0010	0.0045	0.0146	0.0366	0.0739
	14							0.0002	0.0012	0.0049	0.0150	0.0370
	15								0.0003	0.0013	0.0049	0.0148
	16									0.0003	0.0013	0.0046
	17										0.0002	0.0011
	18											0.0002
	19											
	20											

D.3 Cumulative binomial probabilities

(Tabulated values are $P(X \geq x)$ for certain values of n, p. If $p > 0.5$ use $P(Y \leq y) = 1 - P(Y \geq y+1)$, where $y = n - x$, and $Y \sim Bin(n, 1-p)$.)

n	x	0.01	0.05	0.10	0.15	0.20	p=0.25	0.30	0.35	0.40	0.45	0.50
2	0	1	1	1	1	1	1	1	1	1	1	1
	1	0.0199	0.0975	0.1900	0.2775	0.3600	0.4375	0.5100	0.5775	0.6400	0.6975	0.7500
	2	0.0001	0.0025	0.0100	0.0225	0.0400	0.0625	0.0900	0.1225	0.1600	0.2025	0.2500
3	0	1	1	1	1	1	1	1	1	1	1	1
	1	0.0297	0.1426	0.2710	0.3859	0.4880	0.5781	0.6570	0.7254	0.7840	0.8336	0.8750
	2	0.0003	0.0073	0.0280	0.0608	0.1040	0.1563	0.2160	0.2818	0.3520	0.4253	0.5000
	3		0.0001	0.0010	0.0034	0.0080	0.0156	0.0270	0.0429	0.0640	0.0911	0.1250
4	0	1	1	1	1	1	1	1	1	1	1	1
	1	0.0394	0.1855	0.3439	0.4780	0.5904	0.6836	0.7599	0.8215	0.8704	0.9085	0.9375
	2	0.0006	0.0140	0.0523	0.1095	0.1808	0.2617	0.3483	0.4370	0.5248	0.6090	0.6875
	3		0.0005	0.0037	0.0120	0.0272	0.0508	0.0837	0.1265	0.1792	0.2415	0.3125
	4			0.0001	0.0005	0.0016	0.0039	0.0081	0.0150	0.0256	0.0410	0.0625
5	0	1	1	1	1	1	1	1	1	1	1	1
	1	0.0490	0.2262	0.4095	0.5563	0.6723	0.7627	0.8319	0.8840	0.9222	0.9497	0.9688
	2	0.0010	0.0226	0.0815	0.1648	0.2627	0.3672	0.4718	0.5716	0.6630	0.7438	0.8125
	3		0.0012	0.0086	0.0266	0.0579	0.1035	0.1631	0.2352	0.3174	0.4069	0.5000
	4			0.0005	0.0022	0.0067	0.0156	0.0308	0.0540	0.0870	0.1312	0.1875
	5				0.0001	0.0003	0.0010	0.0024	0.0053	0.0102	0.0185	0.0313
6	0	1	1	1	1	1	1	1	1	1	1	1
	1	0.0585	0.2649	0.4686	0.6229	0.7379	0.8220	0.8824	0.9246	0.9533	0.9723	0.9844
	2	0.0015	0.0328	0.1143	0.2235	0.3446	0.4661	0.5798	0.6809	0.7667	0.8364	0.8906
	3		0.0022	0.0159	0.0473	0.0989	0.1694	0.2557	0.3529	0.4557	0.5585	0.6563
	4		0.0001	0.0013	0.0059	0.0170	0.0376	0.0705	0.1174	0.1792	0.2553	0.3438
	5			0.0001	0.0004	0.0016	0.0046	0.0109	0.0223	0.0410	0.0692	0.1094
	6					0.0001	0.0002	0.0007	0.0018	0.0041	0.0083	0.0156
7	0	1	1	1	1	1	1	1	1	1	1	1
	1	0.0679	0.3017	0.5217	0.6794	0.7903	0.8665	0.9176	0.9510	0.9720	0.9848	0.9922
	2	0.0020	0.0444	0.1497	0.2834	0.4233	0.5551	0.6706	0.7662	0.8414	0.8976	0.9375
	3		0.0038	0.0257	0.0738	0.1480	0.2436	0.3529	0.4677	0.5801	0.6836	0.7734
	4		0.0002	0.0027	0.0121	0.0333	0.0706	0.1260	0.1998	0.2898	0.3917	0.5000
	5			0.0002	0.0012	0.0047	0.0129	0.0288	0.0556	0.0963	0.1529	0.2266
	6				0.0001	0.0004	0.0013	0.0038	0.0090	0.0188	0.0357	0.0625
	7						0.0001	0.0002	0.0006	0.0016	0.0037	0.0078
8	0	1	1	1	1	1	1	1	1	1	1	1
	1	0.0773	0.3366	0.5695	0.7275	0.8322	0.8999	0.9424	0.9681	0.9832	0.9916	0.9961
	2	0.0027	0.0572	0.1869	0.3428	0.4967	0.6329	0.7447	0.8309	0.8936	0.9368	0.9648
	3	0.0001	0.0058	0.0381	0.1052	0.2031	0.3215	0.4482	0.5722	0.6846	0.7799	0.8555
	4		0.0004	0.0050	0.0214	0.0563	0.1138	0.1941	0.2936	0.4059	0.5230	0.6367
	5			0.0004	0.0029	0.0104	0.0273	0.0580	0.1061	0.1737	0.2604	0.3633
	6				0.0002	0.0012	0.0042	0.0113	0.0253	0.0498	0.0885	0.1445
	7					0.0001	0.0004	0.0013	0.0036	0.0085	0.0181	0.0352
	8							0.0001	0.0002	0.0007	0.0017	0.0039
9	0	1	1	1	1	1	1	1	1	1	1	1
	1	0.0865	0.3698	0.6126	0.7684	0.8658	0.9249	0.9596	0.9793	0.9899	0.9954	0.9980
	2	0.0034	0.0712	0.2252	0.4005	0.5638	0.6997	0.8040	0.8789	0.9295	0.9615	0.9805
	3	0.0001	0.0084	0.0530	0.1409	0.2618	0.3993	0.5372	0.6627	0.7682	0.8505	0.9102
	4		0.0006	0.0083	0.0339	0.0856	0.1657	0.2703	0.3911	0.5174	0.6386	0.7461
	5			0.0009	0.0056	0.0196	0.0489	0.0988	0.1717	0.2666	0.3786	0.5000
	6				0.0006	0.0031	0.0100	0.0253	0.0536	0.0994	0.1658	0.2539
	7					0.0003	0.0013	0.0043	0.0112	0.0250	0.0498	0.0898
	8						0.0001	0.0004	0.0014	0.0038	0.0091	0.0195
	9								0.0001	0.0003	0.0008	0.0020
10	0	1	1	1	1	1	1	1	1	1	1	1
	1	0.0956	0.4013	0.6513	0.8031	0.8926	0.9437	0.9718	0.9865	0.9940	0.9975	0.9990
	2	0.0043	0.0861	0.2639	0.4557	0.6242	0.7560	0.8507	0.9140	0.9536	0.9767	0.9893
	3	0.0001	0.0115	0.0702	0.1798	0.3222	0.4744	0.6172	0.7384	0.8327	0.9004	0.9453
	4		0.0010	0.0128	0.0500	0.1209	0.2241	0.3504	0.4862	0.6177	0.7340	0.8281
	5		0.0001	0.0016	0.0099	0.0328	0.0781	0.1503	0.2485	0.3669	0.4956	0.6230
	6			0.0001	0.0014	0.0064	0.0197	0.0473	0.0949	0.1662	0.2616	0.3770
	7				0.0001	0.0009	0.0035	0.0106	0.0260	0.0548	0.1020	0.1719
	8					0.0001	0.0004	0.0016	0.0048	0.0123	0.0274	0.0547
	9							0.0001	0.0005	0.0017	0.0045	0.0107
	10									0.0001	0.0003	0.0010

n	x	0.01	0.05	0.10	0.15	0.20	p 0.25	0.30	0.35	0.40	0.45	0.50
11	0	1	1	1	1	1	1	1	1	1	1	1
	1	0.1047	0.4312	0.6862	0.8327	0.9141	0.9578	0.9802	0.9912	0.9964	0.9986	0.9995
	2	0.0052	0.1019	0.3026	0.5078	0.6779	0.8029	0.8870	0.9394	0.9698	0.9861	0.9941
	3	0.0002	0.0152	0.0896	0.2212	0.3826	0.5448	0.6873	0.7999	0.8811	0.9348	0.9673
	4		0.0016	0.0185	0.0694	0.1611	0.2867	0.4304	0.5744	0.7037	0.8089	0.8867
	5		0.0001	0.0028	0.0159	0.0504	0.1146	0.2103	0.3317	0.4672	0.6029	0.7256
	6			0.0003	0.0027	0.0117	0.0343	0.0782	0.1487	0.2465	0.3669	0.5000
	7				0.0003	0.0020	0.0076	0.0216	0.0501	0.0994	0.1738	0.2744
	8					0.0002	0.0012	0.0043	0.0122	0.0293	0.0610	0.1133
	9						0.0001	0.0006	0.0020	0.0059	0.0148	0.0327
	10								0.0002	0.0007	0.0022	0.0059
	11										0.0002	0.0005
12	0	1	1	1	1	1	1	1	1	1	1	1
	1	0.1136	0.4596	0.7176	0.8578	0.9313	0.9683	0.9862	0.9943	0.9978	0.9992	0.9998
	2	0.0062	0.1184	0.3410	0.5565	0.7251	0.8416	0.9150	0.9576	0.9804	0.9917	0.9968
	3	0.0002	0.0196	0.1109	0.2642	0.4417	0.6093	0.7472	0.8487	0.9166	0.9579	0.9807
	4		0.0022	0.0256	0.0922	0.2054	0.3512	0.5075	0.6533	0.7747	0.8655	0.9270
	5		0.0002	0.0043	0.0239	0.0726	0.1576	0.2763	0.4167	0.5618	0.6956	0.8062
	6			0.0005	0.0046	0.0194	0.0544	0.1178	0.2127	0.3348	0.4731	0.6128
	7			0.0001	0.0007	0.0039	0.0143	0.0386	0.0846	0.1582	0.2607	0.3872
	8				0.0001	0.0006	0.0028	0.0095	0.0255	0.0573	0.1117	0.1938
	9					0.0001	0.0004	0.0017	0.0056	0.0153	0.0356	0.0730
	10							0.0002	0.0008	0.0028	0.0079	0.0193
	11								0.0001	0.0003	0.0011	0.0032
	12										0.0001	0.0002
15	0	1	1	1	1	1	1	1	1	1	1	1
	1	0.1399	0.5367	0.7941	0.9126	0.9648	0.9866	0.9953	0.9984	0.9995	0.9999	1.0000
	2	0.0096	0.1710	0.4510	0.6814	0.8329	0.9198	0.9647	0.9858	0.9948	0.9983	0.9995
	3	0.0004	0.0362	0.1841	0.3958	0.6020	0.7639	0.8732	0.9383	0.9729	0.9893	0.9963
	4		0.0055	0.0556	0.1773	0.3518	0.5387	0.7031	0.8273	0.9095	0.9576	0.9824
	5		0.0006	0.0127	0.0617	0.1642	0.3135	0.4845	0.6481	0.7827	0.8796	0.9408
	6		0.0001	0.0022	0.0168	0.0611	0.1484	0.2784	0.4357	0.5968	0.7392	0.8491
	7			0.0003	0.0036	0.0181	0.0566	0.1311	0.2452	0.3902	0.5478	0.6964
	8			0.0000	0.0006	0.0042	0.0173	0.0500	0.1132	0.2131	0.3465	0.5000
	9				0.0001	0.0008	0.0042	0.0152	0.0422	0.0950	0.1818	0.3036
	10					0.0001	0.0008	0.0037	0.0124	0.0338	0.0769	0.1509
	11						0.0001	0.0007	0.0028	0.0093	0.0255	0.0592
	12							0.0001	0.0005	0.0019	0.0063	0.0176
	13								0.0001	0.0003	0.0011	0.0037
	14										0.0001	0.0005
	15											
20	0	1	1	1	1	1	1	1	1	1	1	1
	1	0.1821	0.6415	0.8784	0.9612	0.9885	0.9968	0.9992	0.9998	1.0000	1.0000	1.0000
	2	0.0169	0.2642	0.6083	0.8244	0.9308	0.9757	0.9924	0.9979	0.9995	0.9999	1.0000
	3	0.0010	0.0755	0.3231	0.5951	0.7939	0.9087	0.9645	0.9879	0.9964	0.9991	0.9998
	4		0.0159	0.1330	0.3523	0.5886	0.7748	0.8929	0.9556	0.9840	0.9951	0.9987
	5		0.0026	0.0432	0.1702	0.3704	0.5852	0.7625	0.8818	0.9490	0.9811	0.9941
	6		0.0003	0.0113	0.0673	0.1958	0.3828	0.5836	0.7546	0.8744	0.9447	0.9793
	7			0.0024	0.0219	0.0867	0.2142	0.3920	0.5834	0.7500	0.8701	0.9423
	8			0.0004	0.0059	0.0321	0.1018	0.2277	0.3990	0.5841	0.7480	0.8684
	9			0.0001	0.0013	0.0100	0.0409	0.1133	0.2376	0.4044	0.5857	0.7483
	10				0.0002	0.0026	0.0139	0.0480	0.1218	0.2447	0.4086	0.5881
	11					0.0006	0.0039	0.0171	0.0532	0.1275	0.2493	0.4119
	12					0.0001	0.0009	0.0051	0.0196	0.0565	0.1308	0.2517
	13						0.0002	0.0013	0.0060	0.0210	0.0580	0.1316
	14							0.0003	0.0015	0.0065	0.0214	0.0577
	15								0.0003	0.0016	0.0064	0.0207
	16									0.0003	0.0015	0.0059
	17										0.0003	0.0013
	18											0.0002
	19											
	20											

D.4 Standard Normal Distribution

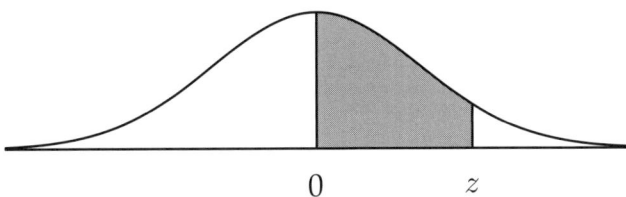

The tabulated values are the probability the $N(0, 1)$ random variable lies between the mean 0 and the listed values of z.

z	0	0.01	0.02	0.03	0.04	0.05	0.06	0.07	0.08	0.09
0.0	0.0000	0.0040	0.0080	0.0120	0.0160	0.0199	0.0239	0.0279	0.0319	0.0359
0.1	0.0398	0.0438	0.0478	0.0517	0.0557	0.0596	0.0636	0.0675	0.0714	0.0753
0.2	0.0793	0.0832	0.0871	0.0910	0.0948	0.0987	0.1026	0.1064	0.1103	0.1141
0.3	0.1179	0.1217	0.1255	0.1293	0.1331	0.1368	0.1406	0.1443	0.1480	0.1517
0.4	0.1554	0.1591	0.1628	0.1664	0.1700	0.1736	0.1772	0.1808	0.1844	0.1879
0.5	0.1915	0.1950	0.1985	0.2019	0.2054	0.2088	0.2123	0.2157	0.2190	0.2224
0.6	0.2257	0.2291	0.2324	0.2357	0.2389	0.2422	0.2454	0.2486	0.2517	0.2549
0.7	0.2580	0.2611	0.2642	0.2673	0.2704	0.2734	0.2764	0.2794	0.2823	0.2852
0.8	0.2881	0.2910	0.2939	0.2967	0.2995	0.3023	0.3051	0.3078	0.3106	0.3133
0.9	0.3159	0.3186	0.3212	0.3238	0.3264	0.3289	0.3315	0.3340	0.3365	0.3389
1.0	0.3413	0.3438	0.3461	0.3485	0.3508	0.3531	0.3554	0.3577	0.3599	0.3621
1.1	0.3643	0.3665	0.3686	0.3708	0.3729	0.3749	0.3770	0.3790	0.3810	0.3830
1.2	0.3849	0.3869	0.3888	0.3907	0.3925	0.3944	0.3962	0.3980	0.3997	0.4015
1.3	0.4032	0.4049	0.4066	0.4082	0.4099	0.4115	0.4131	0.4147	0.4162	0.4177
1.4	0.4192	0.4207	0.4222	0.4236	0.4251	0.4265	0.4279	0.4292	0.4306	0.4319
1.5	0.4332	0.4345	0.4357	0.4370	0.4382	0.4394	0.4406	0.4418	0.4429	0.4441
1.6	0.4452	0.4463	0.4474	0.4484	0.4495	0.4505	0.4515	0.4525	0.4535	0.4545
1.7	0.4554	0.4564	0.4573	0.4582	0.4591	0.4599	0.4608	0.4616	0.4625	0.4633
1.8	0.4641	0.4649	0.4656	0.4664	0.4671	0.4678	0.4686	0.4693	0.4699	0.4706
1.9	0.4713	0.4719	0.4726	0.4732	0.4738	0.4744	0.4750	0.4756	0.4761	0.4767
2.0	0.4772	0.4778	0.4783	0.4788	0.4793	0.4798	0.4803	0.4808	0.4812	0.4817
2.1	0.4821	0.4826	0.4830	0.4834	0.4838	0.4842	0.4846	0.4850	0.4854	0.4857
2.2	0.4861	0.4864	0.4868	0.4871	0.4875	0.4878	0.4881	0.4884	0.4887	0.4890
2.3	0.4893	0.4896	0.4898	0.4901	0.4904	0.4906	0.4909	0.4911	0.4913	0.4916
2.4	0.4918	0.4920	0.4922	0.4925	0.4927	0.4929	0.4931	0.4932	0.4934	0.4936
2.5	0.4938	0.4940	0.4941	0.4943	0.4945	0.4946	0.4948	0.4949	0.4951	0.4952
2.6	0.4953	0.4955	0.4956	0.4957	0.4959	0.4960	0.4961	0.4962	0.4963	0.4964
2.7	0.4965	0.4966	0.4967	0.4968	0.4969	0.4970	0.4971	0.4972	0.4973	0.4974
2.8	0.4974	0.4975	0.4976	0.4977	0.4977	0.4978	0.4979	0.4979	0.4980	0.4981
2.9	0.4981	0.4982	0.4982	0.4983	0.4984	0.4984	0.4985	0.4985	0.4986	0.4986
3.0	0.4987	0.4987	0.4987	0.4988	0.4988	0.4989	0.4989	0.4989	0.4990	0.4990
3.1	0.4990	0.4991	0.4991	0.4991	0.4992	0.4992	0.4992	0.4992	0.4993	0.4993
3.2	0.4993	0.4993	0.4994	0.4994	0.4994	0.4994	0.4994	0.4995	0.4995	0.4995
3.3	0.4995	0.4995	0.4995	0.4996	0.4996	0.4996	0.4996	0.4996	0.4996	0.4997
3.4	0.4997	0.4997	0.4997	0.4997	0.4997	0.4997	0.4997	0.4997	0.4997	0.4998
3.5	0.4998	0.4998	0.4998	0.4998	0.4998	0.4998	0.4998	0.4998	0.4998	0.4998
3.6	0.4998	0.4998	0.4999	0.4999	0.4999	0.4999	0.4999	0.4999	0.4999	0.4999
3.7	0.4999	0.4999	0.4999	0.4999	0.4999	0.4999	0.4999	0.4999	0.4999	0.4999
3.8	0.4999	0.4999	0.4999	0.4999	0.4999	0.4999	0.4999	0.4999	0.4999	0.4999
3.9	0.5000	0.5000	0.5000	0.5000	0.5000	0.5000	0.5000	0.5000	0.5000	0.5000

D.5 Inverse normal probability table

For a listed value of p, the table gives the value of z such that the standardised normal variate Z has a probability p of lying between 0 and z.

p	z	p	z	p	z	p	z	p	z	p	z
0.00	0.0000	0.10	0.2533	0.20	0.5244	0.30	0.8416	0.40	1.2816	0.470	1.8808
0.01	0.0251	0.11	0.2793	0.21	0.5534	0.31	0.8779	0.41	1.3408	0.471	1.8957
0.02	0.0502	0.12	0.3055	0.22	0.5828	0.32	0.9154	0.42	1.4051	0.472	1.9110
0.03	0.0753	0.13	0.3319	0.23	0.6128	0.33	0.9542	0.43	1.4758	0.473	1.9268
0.04	0.1004	0.14	0.3585	0.24	0.6433	0.34	0.9945	0.44	1.5548	0.474	1.9431
0.05	0.1257	0.15	0.3853	0.25	0.6745	0.35	1.0364	0.45	1.6449	0.475	1.9600
0.06	0.1510	0.16	0.4125	0.26	0.7063	0.36	1.0803	0.46	1.7507	0.476	1.9774
0.07	0.1764	0.17	0.4399	0.27	0.7388	0.37	1.1264			0.477	1.9954
0.08	0.2019	0.18	0.4677	0.28	0.7722	0.38	1.1750			0.478	2.0141
0.09	0.2275	0.19	0.4959	0.29	0.8064	0.39	1.2265			0.479	2.0335

p	z	p	z	p	z	p	z	p	z
0.480	2.0537	0.485	2.1701	0.490	2.3263	0.495	2.5758	0.4995	3.2905
0.481	2.0749	0.486	2.1973	0.491	2.3656	0.496	2.6521	0.4999	3.7190
0.482	2.0969	0.487	2.2262	0.492	2.4089	0.497	2.7478	0.49995	3.8906
0.483	2.1201	0.488	2.2571	0.493	2.4573	0.498	2.8782	0.49999	4.2649
0.484	2.1444	0.489	2.2904	0.494	2.5121	0.499	3.0902		

D.6 Student's t-distribution

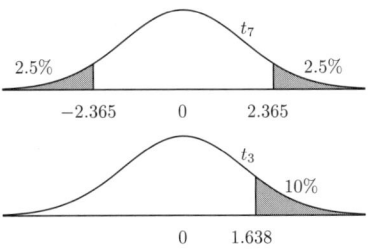

The row labels are the degrees of freedom.

The two-tail column headings give the probability that T will exceed the tabulated value c in absolute value, i.e. $P(T < -c \text{ or } T > c)$, e.g. for $\nu = 7$, $P(T < -2.365 \text{ or } T > 2.365) = 0.05$.

The one-tail column headings give the probability that T will exceed the tabulated value, e.g. for $\nu = 3$, $P(T > 1.638) = 0.10$.

	Two-tail probability values						
	0.20	0.10	0.05	0.02	0.01	0.002	0.001
	One-tail probability values						
ν (df)	0.10	0.05	0.025	0.01	0.005	0.001	0.0005
1	3.078	6.314	12.706	31.821	63.656	318.289	636.578
2	1.886	2.920	4.303	6.965	9.925	22.328	31.600
3	1.638	2.353	3.182	4.541	5.841	10.214	12.924
4	1.533	2.132	2.776	3.747	4.604	7.173	8.610
5	1.476	2.015	2.571	3.365	4.032	5.893	6.869
6	1.440	1.943	2.447	3.143	3.707	5.208	5.959
7	1.415	1.895	2.365	2.998	3.499	4.785	5.408
8	1.397	1.860	2.306	2.896	3.355	4.501	5.041
9	1.383	1.833	2.262	2.821	3.250	4.297	4.781
10	1.372	1.812	2.228	2.764	3.169	4.144	4.587
11	1.363	1.796	2.201	2.718	3.106	4.025	4.437
12	1.356	1.782	2.179	2.681	3.055	3.930	4.318
13	1.350	1.771	2.160	2.650	3.012	3.852	4.221
14	1.345	1.761	2.145	2.624	2.977	3.787	4.140
15	1.341	1.753	2.131	2.602	2.947	3.733	4.073
16	1.337	1.746	2.120	2.583	2.921	3.686	4.015
17	1.333	1.740	2.110	2.567	2.898	3.646	3.965
18	1.330	1.734	2.101	2.552	2.878	3.610	3.922
19	1.328	1.729	2.093	2.539	2.861	3.579	3.883
20	1.325	1.725	2.086	2.528	2.845	3.552	3.850
21	1.323	1.721	2.080	2.518	2.831	3.527	3.819
22	1.321	1.717	2.074	2.508	2.819	3.505	3.792
23	1.319	1.714	2.069	2.500	2.807	3.485	3.767
24	1.318	1.711	2.064	2.492	2.797	3.467	3.745
25	1.316	1.708	2.060	2.485	2.787	3.450	3.725
26	1.315	1.706	2.056	2.479	2.779	3.435	3.707
27	1.314	1.703	2.052	2.473	2.771	3.421	3.690
28	1.313	1.701	2.048	2.467	2.763	3.408	3.674
29	1.311	1.699	2.045	2.462	2.756	3.396	3.659
30	1.310	1.697	2.042	2.457	2.750	3.385	3.646
40	1.303	1.684	2.021	2.423	2.704	3.307	3.551
60	1.296	1.671	2.000	2.390	2.660	3.232	3.460
120	1.289	1.658	1.980	2.358	2.617	3.160	3.373
∞	1.282	1.645	1.960	2.326	2.576	3.090	3.290

D.7 The χ^2 distribution

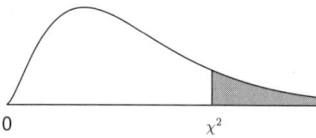

The figure at the top of each column gives the probability that the chi-square random variable with ν degrees of freedom lies above the tabulated value,

e.g. $P(\chi^2_{13} > 22.362) = 5\%$.

p \ ν	0.995	0.99	0.975	0.95	0.10	0.05	0.025	0.01	0.005	0.001
1	0.000	0.000	0.001	0.004	2.706	3.841	5.024	6.635	7.879	10.827
2	0.010	0.020	0.051	0.103	4.605	5.991	7.378	9.210	10.597	13.815
3	0.072	0.115	0.216	0.352	6.251	7.815	9.348	11.345	12.838	16.266
4	0.207	0.297	0.484	0.711	7.779	9.488	11.143	13.277	14.860	18.466
5	0.412	0.554	0.831	1.145	9.236	11.070	12.832	15.086	16.750	20.515
6	0.676	0.872	1.237	1.635	10.645	12.592	14.449	16.812	18.548	22.457
7	0.989	1.239	1.690	2.167	12.017	14.067	16.013	18.475	20.278	24.321
8	1.344	1.647	2.180	2.733	13.362	15.507	17.535	20.090	21.955	26.124
9	1.735	2.088	2.700	3.325	14.684	16.919	19.023	21.666	23.589	27.877
10	2.156	2.558	3.247	3.940	15.987	18.307	20.483	23.209	25.188	29.588
11	2.603	3.053	3.816	4.575	17.275	19.675	21.920	24.725	26.757	31.264
12	3.074	3.571	4.404	5.226	18.549	21.026	23.337	26.217	28.300	32.909
13	3.565	4.107	5.009	5.892	19.812	22.362	24.736	27.688	29.819	34.527
14	4.075	4.660	5.629	6.571	21.064	23.685	26.119	29.141	31.319	36.124
15	4.601	5.229	6.262	7.261	22.307	24.996	27.488	30.578	32.801	37.698
16	5.142	5.812	6.908	7.962	23.542	26.296	28.845	32.000	34.267	39.252
17	5.697	6.408	7.564	8.672	24.769	27.587	30.191	33.409	35.718	40.791
18	6.265	7.015	8.231	9.390	25.989	28.869	31.526	34.805	37.156	42.312
19	6.844	7.633	8.907	10.117	27.204	30.144	32.852	36.191	38.582	43.819
20	7.434	8.260	9.591	10.851	28.412	31.410	34.170	37.566	39.997	45.314
21	8.034	8.897	10.283	11.591	29.615	32.671	35.479	38.932	41.401	46.796
22	8.643	9.542	10.982	12.338	30.813	33.924	36.781	40.289	42.796	48.268
23	9.260	10.196	11.689	13.091	32.007	35.172	38.076	41.638	44.181	49.728
24	9.886	10.856	12.401	13.848	33.196	36.415	39.364	42.980	45.558	51.179
25	10.520	11.524	13.120	14.611	34.382	37.652	40.646	44.314	46.928	52.619
26	11.160	12.198	13.844	15.379	35.563	38.885	41.923	45.642	48.290	54.051
27	11.808	12.878	14.573	16.151	36.741	40.113	43.195	46.963	49.645	55.475
28	12.461	13.565	15.308	16.928	37.916	41.337	44.461	48.278	50.994	56.892
29	13.121	14.256	16.047	17.708	39.087	42.557	45.722	49.588	52.335	58.301
30	13.787	14.953	16.791	18.493	40.256	43.773	46.979	50.892	53.672	59.702
40	20.707	22.164	24.433	26.509	51.805	55.758	59.342	63.691	66.766	73.403
50	27.991	29.707	32.357	34.764	63.167	67.505	71.420	76.154	79.490	86.66
60	35.534	37.485	40.482	43.188	74.397	79.082	83.298	88.379	91.952	99.608
70	43.275	45.442	48.758	51.739	85.527	90.531	95.023	100.425	104.215	112.317
80	51.172	53.540	57.153	60.391	96.578	101.879	106.629	112.329	116.321	124.839
90	59.196	61.754	65.647	69.126	107.565	113.145	118.136	124.116	128.299	137.208
100	67.328	70.065	74.222	77.929	118.498	124.342	129.561	135.807	140.170	149.449

D.8 One-tail 5% critical values for the F-distribution

The tabulated value is the value of F with ν_1, ν_2 degrees of freedom, which is exceeded with a probability of 5%.

ν_1 = degrees of freedom for the numerator mean square

ν_2 = degrees of freedom for the denominator mean square

ν_2 \ ν_1	1	2	3	4	5	6	7	8	9	10	12	15	20	30	60	∞
1	161.45	199.50	215.71	224.58	230.16	233.99	236.77	238.88	240.54	241.88	243.90	245.95	248.02	250.10	252.20	254.32
2	18.513	19.000	19.164	19.247	19.296	19.329	19.353	19.371	19.385	19.396	19.412	19.429	19.446	19.463	19.479	19.496
3	10.128	9.5521	9.2766	9.1172	9.0134	8.9407	8.8867	8.8452	8.8123	8.7855	8.7447	8.7028	8.6602	8.6166	8.5720	8.5265
4	7.7086	6.9443	6.5914	6.3882	6.2561	6.1631	6.0942	6.0410	5.9988	5.9644	5.9117	5.8578	5.8025	5.7459	5.6878	5.6281
5	6.6079	5.7861	5.4094	5.1922	5.0503	4.9503	4.8759	4.8183	4.7725	4.7351	4.6777	4.6188	4.5581	4.4957	4.4314	4.3650
6	5.9874	5.1432	4.7571	4.5337	4.3874	4.2839	4.2067	4.1468	4.0990	4.0600	3.9999	3.9381	3.8742	3.8082	3.7398	3.6689
7	5.5915	4.7374	4.3468	4.1203	3.9715	3.8660	3.7871	3.7257	3.6767	3.6365	3.5747	3.5107	3.4445	3.3758	3.3043	3.2298
8	5.3176	4.4590	4.0662	3.8379	3.6875	3.5806	3.5005	3.4381	3.3881	3.3472	3.2839	3.2184	3.1503	3.0794	3.0053	2.9276
9	5.1174	4.2565	3.8625	3.6331	3.4817	3.3738	3.2927	3.2296	3.1789	3.1373	3.0729	3.0061	2.9365	2.8637	2.7872	2.7067
10	4.9646	4.1028	3.7083	3.4780	3.3258	3.2172	3.1355	3.0717	3.0204	2.9782	2.9130	2.8450	2.7740	2.6996	2.6211	2.5379
11	4.8443	3.9823	3.5874	3.3567	3.2039	3.0946	3.0123	2.9480	2.8962	2.8536	2.7876	2.7186	2.6464	2.5705	2.4901	2.4045
12	4.7472	3.8853	3.4903	3.2592	3.1059	2.9961	2.9134	2.8486	2.7964	2.7534	2.6866	2.6169	2.5436	2.4663	2.3842	2.2962
13	4.6672	3.8056	3.4105	3.1791	3.0254	2.9153	2.8321	2.7669	2.7144	2.6710	2.6037	2.5331	2.4589	2.3803	2.2966	2.2064
14	4.6001	3.7389	3.3439	3.1122	2.9582	2.8477	2.7642	2.6987	2.6458	2.6022	2.5342	2.4630	2.3879	2.3082	2.2229	2.1307
15	4.5431	3.6823	3.2874	3.0556	2.9013	2.7905	2.7066	2.6408	2.5876	2.5437	2.4753	2.4034	2.3275	2.2468	2.1601	2.0658
16	4.4940	3.6337	3.2389	3.0069	2.8524	2.7413	2.6572	2.5911	2.5377	2.4935	2.4247	2.3522	2.2756	2.1938	2.1058	2.0096
17	4.4513	3.5915	3.1968	2.9647	2.8100	2.6987	2.6143	2.5480	2.4943	2.4499	2.3807	2.3077	2.2304	2.1477	2.0584	1.9604
18	4.4139	3.5546	3.1599	2.9277	2.7729	2.6613	2.5767	2.5102	2.4563	2.4117	2.3421	2.2686	2.1906	2.1071	2.0166	1.9168
19	4.3808	3.5219	3.1274	2.8951	2.7401	2.6283	2.5435	2.4768	2.4227	2.3779	2.3080	2.2341	2.1555	2.0712	1.9795	1.8780
20	4.3513	3.4928	3.0984	2.8661	2.7109	2.5990	2.5140	2.4471	2.3928	2.3479	2.2776	2.2033	2.1242	2.0391	1.9464	1.8432
21	4.3248	3.4668	3.0725	2.8401	2.6848	2.5727	2.4876	2.4205	2.3661	2.3210	2.2504	2.1757	2.0960	2.0102	1.9165	1.8117
22	4.3009	3.4434	3.0491	2.8167	2.6613	2.5491	2.4638	2.3965	2.3419	2.2967	2.2258	2.1508	2.0707	1.9842	1.8894	1.7831
23	4.2793	3.4221	3.0280	2.7955	2.6400	2.5277	2.4422	2.3748	2.3201	2.2747	2.2036	2.1282	2.0476	1.9605	1.8648	1.7570
24	4.2597	3.4028	3.0088	2.7763	2.6207	2.5082	2.4226	2.3551	2.3002	2.2547	2.1834	2.1077	2.0267	1.9390	1.8424	1.7330
25	4.2417	3.3852	2.9912	2.7587	2.6030	2.4904	2.4047	2.3371	2.2821	2.2365	2.1649	2.0889	2.0075	1.9192	1.8217	1.7110
26	4.2252	3.3690	2.9752	2.7426	2.5868	2.4741	2.3883	2.3205	2.2655	2.2197	2.1479	2.0716	1.9898	1.9010	1.8027	1.6906
27	4.2100	3.3541	2.9603	2.7278	2.5719	2.4591	2.3732	2.3053	2.2501	2.2043	2.1323	2.0558	1.9736	1.8842	1.7851	1.6717
28	4.1960	3.3404	2.9467	2.7141	2.5581	2.4453	2.3593	2.2913	2.2360	2.1900	2.1179	2.0411	1.9586	1.8687	1.7689	1.6541
29	4.1830	3.3277	2.9340	2.7014	2.5454	2.4324	2.3463	2.2782	2.2229	2.1768	2.1045	2.0275	1.9446	1.8543	1.7537	1.6376
30	4.1709	3.3158	2.9223	2.6896	2.5336	2.4205	2.3343	2.2662	2.2107	2.1646	2.0921	2.0148	1.9317	1.8409	1.7396	1.6223
40	4.0847	3.2317	2.8387	2.6060	2.4495	2.3359	2.2490	2.1802	2.1240	2.0773	2.0035	1.9245	1.8389	1.7444	1.6373	1.5089
60	4.0012	3.1504	2.7581	2.5252	2.3683	2.2541	2.1665	2.0970	2.0401	1.9926	1.9174	1.8364	1.7480	1.6491	1.5343	1.3893
120	3.9201	3.0718	2.6802	2.4472	2.2899	2.1750	2.0868	2.0164	1.9588	1.9105	1.8337	1.7505	1.6587	1.5543	1.4290	1.2539
∞	3.8414	2.9957	2.6049	2.3719	2.2141	2.0986	2.0096	1.9384	1.8799	1.8307	1.7522	1.6664	1.5705	1.4591	1.3180	1.0000

D.9 One-tail $2\frac{1}{2}$% critical values for the F-distribution

The tabulated value is the value of F with ν_1, ν_2 degrees of freedom, which is exceeded with a probability of $2\frac{1}{2}$%.

ν_1 = degrees of freedom for the numerator mean square

ν_2	1	2	3	4	5	6	7	8	9	10	12	15	20	30	60	∞
1	647.79	799.50	864.160	899.58	921.85	937.11	948.22	956.66	963.28	968.63	976.71	984.87	993.10	1001.4	1009.8	1018.3
2	38.506	39.000	39.165	39.248	39.298	39.331	39.355	39.373	39.387	39.398	39.415	39.431	39.448	39.465	39.481	39.498
3	17.443	16.044	15.439	15.101	14.885	14.735	14.624	14.540	14.473	14.419	14.337	14.253	14.167	14.081	13.992	13.902
4	12.218	10.649	9.9792	9.6045	9.3645	9.1973	9.0741	8.9796	8.9047	8.8439	8.7512	8.6565	8.5599	8.4613	8.3604	8.2573
5	10.007	8.4336	7.7636	7.3879	7.1464	6.9777	6.8531	6.7572	6.6811	6.6192	6.5245	6.4277	6.3286	6.2269	6.1225	6.0153
6	8.8131	7.2599	6.5988	6.2272	5.9876	5.8198	5.6955	5.5996	5.5234	5.4613	5.3662	5.2687	5.1684	5.0652	4.9589	4.8491
7	8.0727	6.5415	5.8898	5.5226	5.2852	5.1186	4.9949	4.8993	4.8232	4.7611	4.6658	4.5678	4.4667	4.3624	4.2544	4.1423
8	7.5709	6.0595	5.4160	5.0526	4.8173	4.6517	4.5286	4.4333	4.3572	4.2951	4.1997	4.1012	3.9995	3.8940	3.7844	3.6702
9	7.2093	5.7147	5.0781	4.7181	4.4844	4.3197	4.1970	4.1020	4.0260	3.9639	3.8682	3.7694	3.6669	3.5604	3.4493	3.3329
10	6.9367	5.4564	4.8256	4.4683	4.2361	4.0721	3.9498	3.8549	3.7790	3.7168	3.6209	3.5217	3.4185	3.3110	3.1984	3.0798
11	6.7241	5.2559	4.6300	4.2751	4.0440	3.8807	3.7586	3.6638	3.5879	3.5257	3.4296	3.3299	3.2261	3.1176	3.0035	2.8828
12	6.5538	5.0959	4.4742	4.1212	3.8911	3.7283	3.6065	3.5118	3.4358	3.3736	3.2773	3.1772	3.0728	2.9633	2.8478	2.7249
13	6.4143	4.9653	4.3472	3.9959	3.7667	3.6043	3.4827	3.3880	3.3120	3.2497	3.1532	3.0527	2.9477	2.8372	2.7204	2.5955
14	6.2979	4.8567	4.2417	3.8919	3.6634	3.5014	3.3799	3.2853	3.2093	3.1469	3.0502	2.9493	2.8437	2.7324	2.6142	2.4872
15	6.1995	4.7650	4.1528	3.8043	3.5764	3.4147	3.2934	3.1987	3.1227	3.0602	2.9633	2.8621	2.7559	2.6437	2.5242	2.3953
16	6.1151	4.6867	4.0768	3.7294	3.5021	3.3406	3.2194	3.1248	3.0488	2.9862	2.8890	2.7875	2.6808	2.5678	2.4471	2.3163
17	6.0420	4.6189	4.0112	3.6648	3.4379	3.2767	3.1556	3.0610	2.9849	2.9222	2.8249	2.7230	2.6158	2.5020	2.3801	2.2474
18	5.9781	4.5597	3.9539	3.6083	3.3820	3.2209	3.0999	3.0053	2.9291	2.8664	2.7689	2.6667	2.5590	2.4445	2.3214	2.1869
19	5.9216	4.5075	3.9034	3.5587	3.3327	3.1718	3.0509	2.9563	2.8801	2.8172	2.7196	2.6171	2.5089	2.3937	2.2696	2.1333
20	5.8715	4.4613	3.8587	3.5147	3.2891	3.1283	3.0074	2.9128	2.8365	2.7737	2.6758	2.5731	2.4645	2.3486	2.2234	2.0853
21	5.8266	4.4199	3.8188	3.4754	3.2501	3.0895	2.9686	2.8740	2.7977	2.7348	2.6368	2.5338	2.4247	2.3082	2.1819	2.0422
22	5.7863	4.3828	3.7829	3.4401	3.2151	3.0546	2.9338	2.8392	2.7628	2.6998	2.6017	2.4984	2.3890	2.2718	2.1446	2.0032
23	5.7498	4.3492	3.7505	3.4083	3.1835	3.0232	2.9023	2.8077	2.7313	2.6682	2.5699	2.4665	2.3567	2.2389	2.1107	1.9677
24	5.7166	4.3187	3.7211	3.3794	3.1548	2.9946	2.8738	2.7791	2.7027	2.6396	2.5411	2.4374	2.3273	2.2090	2.0799	1.9353
25	5.6864	4.2909	3.6943	3.3530	3.1287	2.9685	2.8478	2.7531	2.6766	2.6135	2.5149	2.4110	2.3005	2.1816	2.0516	1.9055
26	5.6586	4.2655	3.6697	3.3289	3.1048	2.9447	2.8240	2.7293	2.6528	2.5896	2.4908	2.3867	2.2759	2.1565	2.0257	1.8781
27	5.6331	4.2421	3.6472	3.3067	3.0828	2.9228	2.8021	2.7074	2.6309	2.5676	2.4688	2.3644	2.2533	2.1334	2.0018	1.8527
28	5.6096	4.2205	3.6264	3.2863	3.0626	2.9027	2.7820	2.6872	2.6106	2.5473	2.4484	2.3438	2.2324	2.1121	1.9797	1.8291
29	5.5878	4.2006	3.6072	3.2674	3.0438	2.8840	2.7633	2.6686	2.5919	2.5286	2.4295	2.3248	2.2131	2.0923	1.9591	1.8072
30	5.5675	4.1821	3.5894	3.2499	3.0265	2.8667	2.7460	2.6513	2.5746	2.5112	2.4120	2.3072	2.1952	2.0739	1.9400	1.7867
40	5.4239	4.0510	3.4633	3.1261	2.9037	2.7444	2.6238	2.5289	2.4519	2.3882	2.2882	2.1819	2.0677	1.9429	1.8028	1.6371
60	5.2856	3.9253	3.3425	3.0077	2.7863	2.6274	2.5068	2.4117	2.3344	2.2702	2.1692	2.0613	1.9445	1.8152	1.6668	1.4821
120	5.1523	3.8046	3.2269	2.8943	2.6740	2.5154	2.3948	2.2994	2.2217	2.1570	2.0548	1.9450	1.8249	1.6899	1.5299	1.3104
∞	5.0239	3.6889	3.1161	2.7858	2.5665	2.4082	2.2875	2.1918	2.1136	2.0483	1.9447	1.8326	1.7085	1.5660	1.3883	1.0000

ν_2 = degrees of freedom for the denominator mean square

D.10 One-tail 1% critical values for the F-distribution

The tabulated value is the value of F with ν_1, ν_2 degrees of freedom, which is exceeded with a probability of 1%.

ν_2 \ ν_1	1	2	3	4	5	6	7	8	9	10	12	15	20	30	60	∞
1	4052.2	4999.5	5403.4	5624.6	5763.6	5859.0	5928.4	5981.1	6022.5	6055.8	6106.3	6157.3	6208.7	6260.6	6313.0	6365.9
2	98.503	99.000	99.166	99.249	99.299	99.333	99.356	99.374	99.388	99.399	99.416	99.433	99.449	99.466	99.482	99.499
3	34.116	30.817	29.457	28.710	28.237	27.911	27.672	27.489	27.345	27.229	27.052	26.872	26.690	26.505	26.316	26.125
4	21.198	18.000	16.694	15.977	15.522	15.207	14.976	14.799	14.659	14.546	14.374	14.198	14.020	13.838	13.652	13.463
5	16.258	13.274	12.060	11.392	10.967	10.672	10.456	10.289	10.158	10.051	9.8883	9.7222	9.5526	9.3793	9.2020	9.0204
6	13.745	10.925	9.7795	9.1483	8.7459	8.4661	8.2600	8.1017	7.9761	7.8741	7.7183	7.5590	7.3958	7.2285	7.0567	6.8800
7	12.246	9.5466	8.4513	7.8466	7.4604	7.1914	6.9928	6.8400	6.7188	6.6201	6.4691	6.3143	6.1554	5.9920	5.8236	5.6495
8	11.259	8.6491	7.5910	7.0061	6.6318	6.3707	6.1776	6.0289	5.9106	5.8143	5.6667	5.5151	5.3591	5.1981	5.0316	4.8588
9	10.561	8.0215	6.9919	6.4221	6.0569	5.8018	5.6129	5.4671	5.3511	5.2565	5.1114	4.9621	4.8080	4.6486	4.4831	4.3105
10	10.044	7.5594	6.5523	5.9943	5.6363	5.3858	5.2001	5.0567	4.9424	4.8491	4.7059	4.5581	4.4054	4.2469	4.0819	3.9090
11	9.6460	7.2057	6.2167	5.6683	5.3160	5.0692	4.8861	4.7445	4.6315	4.5393	4.3974	4.2509	4.0990	3.9411	3.7761	3.6024
12	9.3302	6.9266	5.9525	5.4120	5.0643	4.8206	4.6395	4.4994	4.3875	4.2961	4.1553	4.0096	3.8584	3.7008	3.5355	3.3608
13	9.0738	6.7010	5.7394	5.2053	4.8616	4.6204	4.4410	4.3021	4.1911	4.1003	3.9603	3.8154	3.6646	3.5070	3.3413	3.1654
14	8.8616	6.5149	5.5639	5.0354	4.6950	4.4558	4.2779	4.1399	4.0297	3.9394	3.8001	3.6557	3.5052	3.3476	3.1813	3.0040
15	8.6831	6.3589	5.4170	4.8932	4.5556	4.3183	4.1415	4.0045	3.8948	3.8049	3.6662	3.5222	3.3719	3.2141	3.0471	2.8684
16	8.5310	6.2262	5.2922	4.7726	4.4374	4.2016	4.0259	3.8896	3.7804	3.6909	3.5527	3.4089	3.2587	3.1007	2.9330	2.7528
17	8.3997	6.1121	5.1850	4.6690	4.3359	4.1015	3.9267	3.7910	3.6822	3.5931	3.4552	3.3117	3.1615	3.0032	2.8348	2.6530
18	8.2854	6.0129	5.0919	4.5790	4.2479	4.0146	3.8406	3.7054	3.5971	3.5082	3.3706	3.2273	3.0771	2.9185	2.7493	2.5660
19	8.1849	5.9259	5.0103	4.5003	4.1708	3.9386	3.7653	3.6305	3.5225	3.4338	3.2965	3.1533	3.0031	2.8442	2.6742	2.4893
20	8.0960	5.8489	4.9382	4.4307	4.1027	3.8714	3.6987	3.5644	3.4567	3.3682	3.2311	3.0880	2.9377	2.7785	2.6077	2.4212
21	8.0166	5.7804	4.8740	4.3688	4.0421	3.8117	3.6396	3.5056	3.3981	3.3098	3.1730	3.0300	2.8796	2.7200	2.5484	2.3603
22	7.9454	5.7190	4.8166	4.3134	3.9880	3.7583	3.5867	3.4530	3.3458	3.2576	3.1209	2.9779	2.8274	2.6675	2.4951	2.3055
23	7.8811	5.6637	4.7649	4.2636	3.9392	3.7102	3.5390	3.4057	3.2986	3.2106	3.0740	2.9311	2.7805	2.6202	2.4471	2.2558
24	7.8229	5.6136	4.7181	4.2184	3.8951	3.6667	3.4959	3.3629	3.2560	3.1681	3.0316	2.8887	2.7380	2.5773	2.4035	2.2107
25	7.7698	5.5680	4.6755	4.1774	3.8550	3.6272	3.4568	3.3239	3.2172	3.1294	2.9931	2.8502	2.6993	2.5383	2.3637	2.1694
26	7.7213	5.5263	4.6366	4.1400	3.8183	3.5911	3.4210	3.2884	3.1818	3.0941	2.9578	2.8150	2.6640	2.5026	2.3273	2.1315
27	7.6767	5.4881	4.6009	4.1056	3.7848	3.5580	3.3882	3.2558	3.1494	3.0618	2.9256	2.7827	2.6316	2.4699	2.2938	2.0965
28	7.6356	5.4529	4.5681	4.0740	3.7539	3.5276	3.3581	3.2259	3.1195	3.0320	2.8959	2.7530	2.6017	2.4397	2.2629	2.0642
29	7.5977	5.4204	4.5378	4.0449	3.7254	3.4995	3.3303	3.1982	3.0920	3.0045	2.8685	2.7256	2.5742	2.4118	2.2344	2.0342
30	7.5625	5.3903	4.5097	4.0179	3.6990	3.4735	3.3045	3.1726	3.0665	2.9791	2.8431	2.7002	2.5487	2.3860	2.2079	2.0062
40	7.3141	5.1785	4.3126	3.8283	3.5138	3.2910	3.1238	2.9930	2.8876	2.8005	2.6648	2.5216	2.3689	2.2034	2.0194	1.8047
60	7.0771	4.9774	4.1259	3.6490	3.3389	3.1187	2.9530	2.8233	2.7185	2.6318	2.4961	2.3523	2.1978	2.0285	1.8363	1.6006
120	6.8509	4.7865	3.9491	3.4795	3.1735	2.9559	2.7918	2.6629	2.5586	2.4721	2.3363	2.1915	2.0346	1.8600	1.6557	1.3805
∞	6.6349	4.6052	3.7816	3.3192	3.0173	2.8020	2.6393	2.5113	2.4073	2.3209	2.1847	2.0385	1.8783	1.6964	1.4730	1.0000

ν_1 = degrees of freedom for the numerator mean square

ν_2 = degrees of freedom for the denominator mean square

D.11 Mann-Whitney U two-sample test

Critical region: $\mathcal{U} \leqslant$ tabulated value. n_ℓ is the larger sample size, and n_s is the smaller sample size.

5% Table

	n_ℓ 2	3	4	5	6	7	8	9	10	11	12	13	14	15	16	17	18	19	20	21	22	23	24	n_ℓ 25	
n_ℓ 2							0	0	0	0	1	1	1	1	1	2	2	2	2	3	3	3	3	3	2 n_s
3				0	1	1	2	2	3	3	4	4	5	5	6	6	7	7	8	8	9	9	10	10	3
4				1	2	3	4	4	5	6	7	8	9	10	11	11	12	13	14	15	16	17	17	18	4
5					3	5	6	7	8	9	11	12	13	14	15	17	18	19	20	22	23	24	25	27	5
6			0	1		6	8	10	11	13	14	16	17	19	21	22	24	25	27	29	30	32	33	35	6
7			0	1	3		10	12	14	16	18	20	22	24	26	28	30	32	34	36	38	40	42	44	7
8			1	2	4	6		15	17	19	22	24	26	29	31	34	36	38	41	43	45	48	50	53	8
9		0	1	3	5	7	9		20	23	26	28	31	34	37	39	42	45	48	50	53	56	59	62	9
10		0	2	4	6	9	11	13		26	29	33	36	39	42	45	48	52	55	58	61	64	67	71	10
11		0	2	5	7	10	13	16	18		33	37	40	44	47	51	55	58	62	65	69	73	76	80	11
12		1	3	6	9	12	15	18	21	24		41	45	49	53	57	61	65	69	73	77	81	85	89	12
13		1	3	7	10	13	17	20	24	27	31		50	54	59	63	67	72	76	80	85	89	94	98	13
14		1	4	7	11	15	18	22	26	30	34	38		59	64	69	74	78	83	88	93	98	102	107	14
15		2	5	8	12	16	20	24	29	33	37	42	46		70	75	80	85	90	96	101	106	111	117	15
16		2	5	9	13	18	22	27	31	36	41	45	50	55		81	86	92	98	103	109	115	120	126	16
17		2	6	10	15	19	24	29	34	39	44	49	54	60	65		93	99	105	111	117	123	129	135	17
18		2	6	11	16	21	26	31	37	42	47	53	58	64	70	75		106	112	119	125	132	138	145	18
19	0	3	7	12	17	22	28	33	39	45	51	57	63	69	74	81	87		119	126	133	140	147	154	19
20	0	3	8	13	18	24	30	36	42	48	54	60	67	73	79	86	92	99		134	141	149	156	163	20
21	0	3	8	14	19	25	32	38	44	51	58	64	71	78	84	91	98	105	112		150	157	165	173	21
22	0	4	9	14	21	27	34	40	47	54	61	68	75	82	89	96	104	111	118	125		166	174	182	22
23	0	4	9	15	22	29	35	43	50	57	64	72	79	87	94	102	109	117	125	132	140		183	192	23
24	0	4	10	16	23	30	37	45	52	60	68	75	83	91	99	107	115	123	131	139	147	155		201	24
n_ℓ 25	0	5	10	17	24	32	39	47	55	63	71	79	87	96	104	112	121	129	138	146	155	163	172		25 n_s
	n_s 2	3	4	5	6	7	8	9	10	11	12	13	14	15	16	17	18	19	20	21	22	23	24	n_s 25	

1% Table

Equal sample sizes

n	1	2	3	4	5	6	7	8	9	10	11	12	13	14	15	16	17	18	19	20	21	22	23	24	25
5%				0	2	5	8	13	17	23	30	37	45	55	64	75	87	99	113	127	142	158	175	192	211
1%					0	2	4	7	11	16	21	27	34	42	51	60	70	81	93	105	118	133	148	164	180

n	26	27	28	29	30	31	32	33	34	35	36	37	38	39	40	41	42	43	44	45	46	47	48	49	50
5%	230	250	272	294	317	341	365	391	418	445	473	503	533	564	596	628	662	697	732	769	806	845	884	924	965
1%	198	216	235	255	276	298	321	344	369	394	420	447	475	504	533	564	595	627	660	694	729	765	802	839	877

Appendix E

Solutions to selected exercises

A set of supplementary exercises (all exercises from the first edition of this book), are available, with full solutions, at http://www.pearsoned.co.nz.

Chapter 2

2.1 (a) continuous, (b) nominal, (e) discrete, (f) ordinal.

2.6 (e) self disclosed mode and distance – are these accurate? (g) an unlikely event has gained attention *after* six rounds, but these six continue to feature.

2.17 (c) relative frequency to compare samples of different size.

2.20 (a) the data are asymmetric, with a mode in the 31-35 age range. (b) We do not have the actual ages.

2.27 (b) the percentages add to more than 100%.

2.28 (a) we need to know the sample size before being able to interpret the frequencies. (b) "Political preference" could be classified as ordinal. It is typically done on a two-dimensional scale (Left-Right and Libertarian-Authoritarian).

Chapter 3

3.2 (a) $\bar{x} = 10.54$, $s = 3.57$. (b) $\bar{x} = 9.18$, $s = 2.87$.

3.11 (b) $\bar{x} = 29.5$, $s = 8.6$. (c) LQ = 23, median = 28, UQ = 37. No outliers, no observations outside (2, 58).

3.18 (b) LQ = 85, median = 89, UQ = 95. Four outliers less than 70: 63, 63, 63, 64. (c) Stemplot gives more information. (d) The outliers.

3.19 $\bar{x} = 39.5$, $s = 18.8$

3.20 (a) Data missing, and not sorted. (b) Median = 1992 (c) LQ = 1979 and UQ = 2004, so outliers below 1941.5. Austria is an outlier, but Denmark is not.

3.24 (a) The mode is 2 plays. (b) $\bar{x} = 2.28$ and $s = 1.28$.

3.30 (a) midpoints: $5, 15, \ldots, 95$. (b) $\bar{x} = 42.4$ and $s = 18.7$. (c) Mean and standard deviation very similar, though different because midpoints are used rather than observations.

3.33 (a) midpoints: $0.5, 1.5, 3.5, 8, 15.5, 75.5$. For those using a bike, $\bar{x} = 6.41$ and $s = 9.50$. (b) Would need share of each mode of transport, i.e. how many people in each category.

Chapter 4

4.5 (a) Clear outlier at (46,9), with positive relationship between rest of sample. (b) $r = 0.4218$. (d) $r = 0.5478$.

4.9 (a) Clear outlier at (82,30.28), with positive relationship between rest of sample. Should use Spearman's, or remove outlier. (b) Count has ranks 1.5, 1.5, 10.5, 3, 15, 12, 10.5, 14, 13, 16, 5.5, 9, 5.5, 5.5, 8, 5.5, $\sum d_i^2 = 79$ and $r_s = 0.8828$.

4.18 (b) $r = 0.9850$. Men's ranks 1, 2, 3, 4, 5, 6 and women's ranks 1, 2, 3, 4, 5, 6 with $\sum d_i^2 = 0$ and $r_s = 1$. We should not use the "ranks" given in the table. (c) Pearson's probably reasonable in this instance. Spearman's is misleading due to the missing categories.

4.20 (b) Pearson's not suitable since "Rating" is ordinal. (c) Ratings have ranks 1, 2.5, 2.5, 4, 5.5, 5.5, 7.5, 7.5, 10, 10, 10. Excess yields have ranks 11, 3, 8, 10, 5, 6.5, 2, 6.5, 9, 1, 4. $\sum d_i^2 = 317$, $r_s = -0.4409$. (c) Correlation is negative as expected.

4.26 (b) Both variables are measurements, linear relationship appears reasonable. (c) $r = 0.5354$ (d) $\hat{y} = -0.49 + 0.73x$

4.28 (a) Strong linear relationship. $r = 0.9974$. $\hat{y} = 0.559 + 0.00116x$. The intercept implies approximately 34 minutes for a trip of 0 km, i.e. safety messages, taxiing. The slope implies a speed of approximately 860 km/h.

4.39 (b) $y = 2x$ represents the cost of buying two "singles" instead of one "double". (c) The regression line is $\hat{y} = 1.275 + 1.493x$. (d) $e_{13} = 6.74 - 5.53 = 1.21$ so the biscuits are more expensive than expected. $e_2 = 14.65 - 17.07 = -2.42$ so the cheese is less expensive than expected.

4.40 (b) $\hat{y} = -0.3222 + 1.0593x$ (c) Fitted values are: 13.4, 14.5, 20.9, 19.8, 23, 23, 21.9, 19.8, 17.7. Residuals are: 0.6, 0.5, -1.9, 1.2, 4, 0, -1.9, -2.8, 0.3.

Chapter 5

5.6 (b) $\bar{y} = 1.544$ and $s = 0.172$. (c) Series does not seem stationary.

5.7 Consider plotting number of bronze medals, number of bronze + number of silver, total number.

5.11 (c) Wind direction is a periodic variable, i.e. $370° = 10°$. (d) Wind direction changed from near 0 to near 180, wind speed rose, it started to rain, and the temperature fell. The phenomenon was a southerly storm passing through.

5.14 (b) $\bar{y} = 137.5$ (c) The moving average is NA, 131.5, 128.8, 131.9, 149, 145.3, 154.4, 153.6, 143, 140.1, 116, NA, where "NA" denotes the end effects. (d) These data appear stationary.

5.18 (b) $\hat{y}_t = -8154.0 + 4.088t$ where $t = 1995, \ldots, 2004$. (c) Rapid increase in 1998 (RTDs?). Positive growth throughout period.

5.24 (b) $\hat{y}_t = -2437.9 + 1.296t$, seems reasonable during sample period. (c) Not sure if relationship will hold outside sample period. (d) Could be better recording, e.g. data from countries previously inaccessible to USGS.

5.30 $r = 0.2685$ which is approximately equal to $r_1 = 0.245$.

5.31 The first 30 digits of the first row have $r = -0.1277$ and $r_1 = -0.118$. These are both close to zero, indicating an approximately random sequence.

Chapter 6

6.2 (a) probability of reaching a person, given he/she is a main household shopper, is 0.622
(b) probability of reaching a person, given he/she has an income over $120,000 p.a. is 71%

6.4 (b) $P(D) = 0.58 \times 0.75 + 0.42 \times 0.6 = 0.687$ (c) $P(M|D) = 0.42 \times 0.6/0.687 = 0.367$.

6.9 (b) $0.65 \times 0.97 = 0.6305$ (c) $0.35 \times 0.65 + 0.65 \times 0.03 \times 0.67 = 0.2406$
(d) $0.35 \times 0.65/0.2406 = 0.9457$

6.13 (b) $P(D,Y) = 0.182 \times 0.151 = 0.0275$; $P(D',Y) = 0.182 \times 0.849 = 0.1545$;
$P(D,P) = 0.349 \times 0.116 = 0.0405$; $P(D',P) = 0.349 \times 0.884 = 0.3085$; $P(D,M) = 0.316 \times 0.093 = 0.0294$; $P(D',M) = 0.316 \times 0.907 = 0.2866$; $P(D,O) = 0.153 \times 0.027 = 0.0041$; $P(D',O) = 0.153 \times 0.973 = 0.1489$.
(c) $P(D) = 0.0275 + 0.0405 + 0.0294 + 0.0041 = 0.1015$.
(d) $P(Y|D) = 0.0275/0.1015 = 0.2709$; $P(P|D) = 0.0405/0.1015 = 0.3990$;
$P(M|D) = 0.0294/0.1015 = 0.2897$; $P(O|D) = 0.0041/0.1015 = 0.0404$.

6.18 (b) $P(W_{ab}) = 0.999 \times 0.307 + 0.001 \times 0.614 = 0.3073$
(c) $P(W_{ab}, W_{aw}) = 0.001 \times 0.614 = 0.00061$
(d) $P(W'_{aw}|W_{ab}) = 0.999 * 0.307/0.3073 = 0.9995$

6.22 (a) $P(F,W) = 691/2484 = 0.2782$
(b) 1032 males; 1452 females. $P(M) = 1032/2484 = 0.4155$, $P(F) = 1452/2484 = 0.5845$.
(c) $P(B|M) = 746/1032 = 0.7229$, $P(W|M) = 286/1032 = 0.2771$
(d) $P(B|F) = 761/1452 = 0.5241$, $P(W|F) = 691/1452 = 0.4759$.
(e) $n(W) = 691 + 286 = 977$, $P(M|W) = 286/977 = 0.2927$, $P(F|W) = 691/977 = 0.7073$
(f) $n(B) = 761+746 = 1507$, $P(M|B) = 746/1507 = 0.4950$, $P(F|B) = 761/1507 = 0.5050$.

Chapter 7

7.1 (a) 0.064 (e) 0.0004

7.2 (a) 1 (e) 0.0005

7.3 (c) 0.5000 (e) $P(X = 0)$ is excluded.

7.8 $E(X) = 100 \times 0.32 = 32$, $var(X) = 100 \times 0.32 \times (1-0.32) = 21.76$, $sd(X) = \sqrt{21.76} = 4.66$

7.12 (a) $n = 12$ is fixed, $p = 0.15$ is fixed, trials are independent and identical, each individual is either scammed or not, and X counts the number scammed. (b) 0.1422 (c) 0.8578 is the probability that at least one in the sample has been scammed.

7.18 (a) $n = 20$ is fixed, $p = 0.75$ is fixed, trials are independent and identical, each invididual either speaks English only or not, and X counts the number who use English only.
(b) $P(X = 20) = 0.0032$ (c) $P(X \geq 10) = 0.9961$ (d) $E(X) = 20 \times 0.75 = 15$.

Chapter 8

8.1 (a) 0.4265

8.2 (a) 0.4830

8.3 (b) $0.4793 + 0.3599 = 0.8392$

8.4 (a) $0.4192 - 0.2939 = 0.1253$

8.5 (a) $0.5 - 0.4115 = 0.0885$

8.7 (b) $P(X < 8) = P(Z < (8-12)/\sqrt{6}) = P(Z < -1.63) = 0.5 - 0.4484 = 0.0516$

8.10 (a) $P(X < 0) = P(Z < -5.34/3.7) = P(Z < -1.44) = 0.5 - 0.4251 = 0.0749$

8.12 (b) $P(\bar{X} < 8) = P(Z < (8-12)/(\sqrt{6/5})) = P(Z < -3.65) = 0.5 - 0.4999 = 0.0001$. (e) Much smaller probability since $\bar{X}$ much closer to μ than X. (f) CLT not needed since X is normal. It wouldn't apply anyway since $n < 30$.

8.14 We cannot determine the probability since we do not know the distribution of X, only its mean and standard deviation. It will be 0.5 if X is symmetric.

8.16 (a) 1.6449 (b) -0.5244 (e) 1.2816

8.18 $P(0 < Z < 0.6745) = 0.25$, so $P(Z > 0.6745) = 0.25$. 0.6745 is the upper quartile of Z.

8.21 (a) $P(Z > 0.5244) = 0.3$, so $x = 3.2 + 0.5244 \times \sqrt{4.6} = 4.32$. (b) $P(Z > -0.5244) = 0.7$, so $x = 3.2 - 0.5244 \times \sqrt{4.6} = 2.08$.

8.27 (b) $(10-320)/240 = -1.29$ and $P(Z < -1.29) = 0.5 - 0.4015$. Not a problem since X is not normal. (c) By the CLT, $\bar{X} \sim N(320, 24^2)$ since $240/\sqrt{100} = 24$. $P(290 < \bar{X} < 350) = P(-1.25 < Z < 1.25) = 2 \times 0.3944 = 0.7888$. (d) Probably not. Depends if the suburb is representative of the city.

8.30 X has mean $np = 76.8$ and standard deviation $\sqrt{np(1-p)} = 7.23$. $P(60 \leqslant X \leqslant 80) \approx P(59.5 < X' < 80.5)$ where X' is normal. This is $P(-2.39 < Z < 0.51) = 0.4916 + 0.1950 = 0.6866$. The actual binomial probability is 0.6904.

Chapter 9

9.3 (a) $0.6 \pm 1.96 \times 2.3/\sqrt{40} = 0.6 \pm 0.71$ (b) 0 lies in CI, but need a one-sided test. Test $H_0 : \mu = 0$ against $H_0 : \mu > 0$. Test statistic is $(0.6 - 0)/(2.3/\sqrt{40}) = 1.65$ and this is $N(0,1)$. Rejection region is values > 1.645 so reject H_0. Yes, evidence of understatement on average.

9.7 (a) CLT applies with n large. (b) $1.97 \pm 1.96 \times 1.60/\sqrt{104} = 1.97 \pm 0.31$. (c) Test $H_0 : \mu = 1.7$ against $H_0 : \mu > 1.7$. Test statistic is $(1.97 - 1.7)/(1.6/\sqrt{104}) = 1.72$ and this is $N(0,1)$. Rejection region is values > 1.645 so reject H_0. (d) Test $H_0 : \mu = 2.3$ against $H_0 : \mu < 2.3$. Test statistic $Z = (1.97 - 2.3)/(1.6/\sqrt{104}) = -2.10$ and this is $N(0,1)$. Rejection region is values < -1.645 so reject H_0. (e) Would need to know how often donations are made, i.e. calculate average donation *per visitor*, not per donor.

9.9 Assume All Black back heights are normally distributed. $\bar{x} = 182.1$, $s = 2.15$, $n = 17$. Test $H_0 : \mu = 177.3$ against $H_0 : \mu > 1.773$. Test statistic is $(182.1 - 177.3)/(2.15/\sqrt{17}) = 9.21$ and this is t_{16}. Rejection region is values > 1.746 so reject H_0.

9.13 (a) 95% CI is $1.86 \pm 2.262 \times 0.58/\sqrt{10} = 1.86 \pm 0.41$ (b) Need the (average) log-concentrations to be normally distributed. (c) Test $H_0 : \mu = 1.43$ against $H_0 : \mu > 1.43$. Test statistic is $(1.86 - 1.43)/(0.58/\sqrt{10}) = 2.34$ and this is t_9. Rejection region is values > 1.833 so reject H_0. (d) Test $H_0 : \mu = 1.59$ against $H_0 : \mu > 1.59$. Test statistic is $(1.86 - 1.59)/(0.58/\sqrt{10}) = 1.47$ and this is t_9. Rejection region is values > 1.833 so cannot reject H_0. (e) Theobromine seems a reasonable cough-suppressant, on a par with codeine.

9.18 (a) Test $H_0 : \mathcal{M} = 29.9$ vs $H_a : \mathcal{M} < 29.9$. Test statistic is 3 observations greater than 29.9. p-value is $P(X \leqslant 3) = 0.1719 > 0.05$ so cannot reject H_0. (b) 41.2 years is an outlier.

9.21 (a) Test $H_0 : \mathcal{M} = 1000$ vs $H_a : \mathcal{M} > 1000$. Test statistic is 7 observations (out of 10) greater than 29.9. p-value is $P(X \geqslant 3) = 0.1719 > 0.05$ so cannot reject H_0. (b) Test $H_0 : \mu = 1000$ against $H_0 : \mu > 1000$. $\bar{x} = 1018.33$ and $s = 63.9$ with $n = 12$. Test statistic is $(1018.33 - 1000)/(63.9/\sqrt{12}) = 0.99$ and this is t_{11}. Rejection region is values > 1.796 so cannot reject H_0. (c) No outliers, but data are asymmetric. Same decision, so choice probably not critical.

9.23 (a) $\hat{p} = 108/250 = 0.432$. 95% CI is $0.432 \pm 1.96\sqrt{0.432 \times 0.568/250} = 0.432 \pm 0.061$. (b) Test $H_0 : p = 0.29$ against $H_a : p > 0.29$. Test statistic is $(0.432 - 0.29)/\sqrt{0.29 \times 0.71/250} = 4.95$ and this is $N(0,1)$. Rejection region is values > 1.645 so reject H_0.

9.37 (a) Test $H_0 : p = 0.5$ against $H_a : p > 0.5$. Test statistic is 8 correct (successes) out of 8. p-value is $P(X \geqslant 8) = 0.0039 < 0.05$ so reject H_0. (b) Test $H_0 : p = 0.5$ against $H_a : p > 0.5$. Test statistic is 2 correct (successes) out of 2. p-value is $P(X \geqslant 2) = 0.2500 > 0.05$ so cannot reject H_0. (c) The second test is more reasonable. Assume there were many critters picking the outcome of the games. Paul is the one who chose the first six correctly and made the news.

9.43 (a) Can't use $p = 0$ to obtain a required sample size of zero. (b) Sample evidence suggests p is small. (c) Using $p = 0.1$, $n = (1.96/0.005)^2 0.1 \times 0.9 = 13829.76$ so need 13830. Note that using 0.5 gives 38416, almost three times as many.

9.45 (a) $\hat{p} = 0.0267$, with standard error $\sqrt{\hat{p}(1-\hat{p})/600} = 0.00658$. $\hat{p} - 3se(\hat{p}) = 0.007 > 0$. Normal approximation is reasonable. (b) 95% CI is $0.0267 \pm 1.96 \times 0.00658 = 0.0267 \pm 0.0129$. (c) $n = (1.96/0.01)^2 0.0267 \times (1 - 0.0267) = 998.32$, so need at least 999 people.

Chapter 10

10.1 (a) 95% CI is $6.8 - 2.4 \pm 1.96\sqrt{6.6^2/291 + 4.5^2/519} = 4.4 \pm 0.85$. (b) Test $H_0 : \mu_1 = \mu_2$ against $H_a : \mu_1 > \mu_2$ (males = 1). Test statistic is $(6.8 - 2.4 - 0)/\sqrt{6.6^2/291 + 4.5^2/519} = 10.13$ and this is $N(0,1)$. Rejection region is values $Z > 1.645$ so reject H_0. 0 was not in CI in (a). (c) $P(Z > 10.13) = 0.5 - 0.5000 = 0.0000 < 0.05$ so reject H_0.

10.5 (a) Not important since CLT applies to both samples. (b) Test $H_0 : \mu_1 = \mu_2$ against $H_a : \mu_1 \neq \mu_2$ (Metallica = 1). Test statistic is $(6.18 - 5.53 - 0)/\sqrt{1.58^2/95 + 3.08^2/94} = 1.82$ and this is $N(0,1)$. Rejection region is values $|Z| > 1.96$ so cannot reject H_0.

10.14 The two samples have $\bar{x}_1 = 5.49$, $s_1 = 0.56$, $n_1 = 4$ and $\bar{x}_2 = 3.82$, $s_2 = 0.67$, and $n_2 = 9$. (a) $s_p^2 = (3 \times 0.56^2 + 8 \times 0.67^2)/11 = 0.412$. (b) 95% CI is $5.49 - 3.82 \pm 2.201\sqrt{0.412/4 + 0.412/9} = 1.67 \pm 0.85$. (c) Test $H_0 : \mu_1 = \mu_2$ against $H_a : \mu_1 \neq \mu_2$. Test statistic is $(5.49 - 3.82 - 0)/\sqrt{0.412/4 + 0.412/9} = 4.33$ and this is t_{11}. Rejection region is values $|T| > 2.201$ so reject H_0 (also, 0 not in the CI). (d) Normality of prices, equal variance. No outliers in either sample, but very small sample sizes. s_1 and s_2 similar.

10.22 Test $H_0 : \sigma_1^2 = \sigma_2^2$ against $H_a : \sigma_1^2 \neq \sigma_2^2$. $s_1^2 = 4.61$ for $n_1 = 17$ backs. $s_2^2 = 46.38$ for $n_2 = 23$ forwards. Test statistic is $46.38/4.61 = 10.06$ and this is $F(22, 16)$. Rejection region from $2\frac{1}{2}$% table is values > 2.6808 (use $F(20, 16)$ as $F(22, 16)$ not in tables). We reject H_0.

10.28 The Metallica ranks are: 3, 4, 10, 11, 12, 13, 15, 17, 18, 19, 20, 22 with $S_1 = 164$. The Pink Floyd ranks are: 1, 2, 5, 6, 7, 8, 9, 14, 16, 21, 23, 24 with $S_2 = 136$. Test $H_0 : \mathcal{M}_1 = \mathcal{M}_2$ against $H_0 : \mathcal{M}_1 \neq \mathcal{M}_2$. $U_1 = 164 - 12 \times 13/2 = 86$ and $U_2 = 136 - 12 \times 13/2 = 58$. The test stat is $\mathcal{U} = \min(86, 58) = 58$. Rejection region is values $\leqslant 37$ so cannot reject H_0.

10.36 (a) The data are for the same years. (b) Differences (Mauritius − Tunisia) are: 0, -0.1, -0.5, -0.8, -0.3, -0.5, -0.9, -0.7, 0.5, 0.5, 1.1, 1.2. These do not appear very symmetric, and possible bimodal with division according to time. (c) Test $H_0 : \mu_d = 0$ against $H_0 : \mu_d > 0$. Test statistic is $(-0.041 - 0)/(0.718/\sqrt{12}) = -0.201$ and this is t_{11}. Rejection region is values > 1.796 so cannot reject H_0. (d) Test $H_0 : \mathcal{M}_d = 0$ against $H_0 : \mathcal{M}_d < 0$. Test statistic is number of positive differences 4 (out of 11). p-value is $P(X \geq 4) = 0.8867 > 0.05$ so again cannot reject H_0. (e) Notice that early evidence suggested Mauritius was less corrupt (negative differences) while the more recent evidence suggests the opposite.

10.42 (a) The percentages are for the same products. (b) The differences (premium − home) are: 7.8, 6, 16, 0, 29, 54, 45, 32, -4, -3, 0, 21, -7, 10, 10, -3, 0, 5. 54 (corned beef) is an outlier, so the sign test should be used. (c) Test $H_0 : \mathcal{M}_d = 0$ against $H_0 : \mathcal{M}_d > 0$. Test statistic is number of positive differences 11 (out of 15, 3 discarded). p-value is $P(X \geq 11) = 0.0592 > 0.05$ so cannot reject H_0.

10.44 Test $H_0 : p_1 = p_2$ against $H_0 : p_1 \neq p_2$. $p^* = (4800 \times 0.19 + 25800 \times 0.15)/(4800 + 25800) = 0.156$ with $p^*(1-p^*) = 0.1312$ and standard error $\sqrt{0.132/4800 + 0.132/25800} = 0.00571$. Test statistic is $(0.19 - 0.15 - 0)/0.00571 = 7.00$ and this is $N(0,1)$. Rejection region is values $|Z| > 1.96$ so reject H_0.

10.56 (a) 95% CI is $0.308 - 0.182 \pm 1.96 \times \sqrt{0.308 \times 0.692/496 + 0.182 \times 0.818/98} = 0.126 \pm 0.087$. (b) Test $H_0 : p_1 = p_2$ against $H_0 : p_1 > p_2$. $p^* = (496 \times 0.308 + 98 \times 0.182)/(496 + 98) = 0.287$ with $p^*(1-p^*) = 0.205$ and standard error $\sqrt{0.205/496 + 0.205/98} = 0.050$. Test statistic is $(0.308 - 0.182 - 0)/0.050 = 2.52$ and this is $N(0,1)$. Rejection region is values > 1.645 so reject H_0.

10.59 Standard error becomes $\sqrt{1.58^2/95 \times 28/122 + 3.08^2/94 \times 68/162} = 0.220$ instead of 0.357, reflecting a large proportion of each population in the sample. Test $H_0 : \mu_1 = \mu_2$ against $H_a : \mu_1 \neq \mu_2$ (Metallica = 1). Test statistic is $(6.18 - 5.53 - 0)/0.220 = 2.95$ and this is $N(0,1)$. Rejection region is values $|Z| > 1.96$ so reject H_0.

Chapter 11

11.3 (a) MST $= 3925.18/2 = 1962.59$, MSE $= 14323.25/24 = 596.8021$. $F = 1962.59/596.8021 = 3.29$. (b) H_0 : all three population group means are equal, vs H_a : at least one is different. The test stat is $F = 3.29$ and this is $F(2,24)$. The rejection region is values > 3.4028, so we cannot reject H_0. (c) All three populations should be normally distributed, with a common variance.

11.5 (a) $k - 1 = 3$. (b) $n - k = 128$ and $k = 4$, so $n = 132$. (c) The test statistic is $F(3,128)$. The rejection region is values > 3.2269 (using $F(3,120)$ instead of $F(3,128)$) so we can reject H_0. (d) Yes, since p-value < 0.05.

11.12 (a) The totals are: 193.2, 166, 314.9 with grand total 674.1. The sample variances are: 178.5, 144.7, 48.2. SST $= 193.2^2/7 + 166^2/9 + 314.9^2/8 - 674.1^2/24 = 1855.565$, SSE $= 6 \times 178.5 + 8 \times 144.7 + 7 \times 48.2 = 2566.541$. MST $= 1855.565/2 = 927.78$, MSE $= 2566.541/21 = 122.2162$. Test statistic is $F = 927.78/122.2162 = 7.591$ and this is $F(2,21)$. Rejection region is values > 3.4668, so reject H_0. (c) Sample means are: 27.6, 18.4, 39.4. $s_p = \sqrt{122.2162} = 11.06$. Half confidence intervals are: $27.6 \pm 11.06/\sqrt{7} = 27.6 \pm 4.2$, $18.4 \pm 11.06/\sqrt{9} = 18.4 \pm 3.7$ and $39.4 \pm 11.06/\sqrt{8} = 39.4 \pm 3.9$. (d) The three means all appear to be different. We can conclude kids' cereals in NZ have more sugar than adults' cereals in NZ, and that US kids' cereals have more sugar than NZ kids' cereals.

11.15 (a) The totals are: 680.15, 765.60, 518.40 with grand total 1964.15. SST $= 680.15^2/305 + 765.6^2/319 + 518.40^2/216 - 1964.15^2/840 = 4598.334$, SSE $= 304 \times 0.52^2 + 318 \times 0.61^2 + 215 \times 0.52^2 = 258.67$. MST $= 5.614/2 = 2.807$, MSE $= 258.67/837 = 0.3090$. Test statistic is $F = 2.807/0.3090 = 9.084$ and this is $F(2,837)$. Rejection region is values > 2.9957 (using $F(2,\infty)$ since $F(2,837)$ is not in the tables), so reject H_0. (c) $s_p = \sqrt{0.3090} = 0.556$. Half confidence intervals are: $2.23 \pm 0.556/\sqrt{305} = 2.23 \pm 0.03$, $2.40 \pm 0.556/\sqrt{319} = 2.40 \pm 0.03$ and $2.40 \pm 0.556/\sqrt{216} = 2.40 \pm 0.04$. The car interval does not overlap with the train and bus intervals, indicating the car population mean is lower than the other two. This is consistent with rejecting H_0.

11.18 Test H_0 : all three boxes have same median donations vs H_a : at least one median differs. Notes has ranks: 23, 26, 23, 16.5, 9.5, 16.5, 27, 23 with $R_1 = 168.5$; coins has ranks: 4, 9.5, 1, 7, 12, 16.5, 4, 9.5, 23 with $R_2 = 86.5$; empty has ranks: 16.5, 16.5, 4, 9.5, 23, 16.5, 16.5, 4, 16.5 with $R_3 = 123.0$. $\sum R_i^2/n_i = 5667.056$ and $K = 12 \times 5667.056/(27 \times 28) - 3 \times 27 = 8.953$ and this is χ_2^2. The rejection region is values > 5.991, so reject H_0.

Chapter 12

12.1 (a) $n = \sum f_i = 300$. $e_i = np_i = 300/6 = 50$. (b) Test H_0 : $p_i = 1/6$ vs H_a : at least one differs. First term is $(43 - 50)^2/50 = 0.98$. Test statistic is $\chi^2 = 0.98 + 1.28 + 1.62 + 0 + 0.5 + 0.5 = 4.88$ and this is χ_5^2. Rejection region is values > 11.070 so cannot reject H_0.

12.4 (a) 7 out of 14 = 50% of the *expected* frequencies are less than 5. (b) 40+ has observed frequency = 8 and expected frequency = 7.9. (c) First term is $(58 - 50.7)^2/50.7 = 1.051$. Test statistic is $\chi^2 = 1.051 + 3.299 + 0.002 + 0.177 + 1.289 + 0.337 + 0.008 + 0.1 + 0.001 = 6.264$ and this is $\chi_{9-1-1}^2 = \chi_7^2$. Rejection region is values > 14.067 so cannot reject H_0.

12.7 Test H_0 : gender and seating position are independent, against H_a : they are not independent. Number of males is 259, number of "front seat" is 314, $n = 711$. Expected number of males in front seat is $259 \times 314/711 = 114.4$. Expected frequencies are (top to bottom, L to R): 114.4, 18.6, 126, 199.6, 32.4, 220. First term is $(163 - 114.4)^2/114.4 = 20.6$. Test statistic is $\chi^2 = 20.7 + 0.7 + 16.1 + 11.8 + 0.4 + 9.2 = 58.9$ and this is χ_2^2. Rejection region is values > 5.991 so reject H_0.

12.13 (a) Number of Māori is 63, number of "Nicotine replacement" is 222, $n = 380$. Expected number of Māori using nicotine replacement is $63 \times 222/380 = 36.8$. Expected frequencies are (top to bottom, L to R): 36.8, 6.5, 14.3, 5.5, 185.2, 32.5, 71.7, 27.5. These are all above 5, so no need for amalgamation. (b) Test H_0 : race and method are independent, against H_a : they are not independent. First term is $(37 - 36.8)^2/36.8 = 0.001$. Test statistic is $\chi^2 = 0 + 0.94 + 2.31 + 2.2 + 0 + 0.19 + 0.46 + 0.44 = 6.54$ and this is χ_3^2. Rejection region is values > 7.815 so cannot reject H_0. (c) Since cannot reject H_0, no valid patterns.

12.26 (a) Number of "Labour" is 1083, number of "no" is 2352, $n = 3094$. Expected number of Labour "no" responses is $1083 \times 2352/3094 = 823.3$. Expected frequencies are (top to bottom, L to R): 823.3, 1001.2, 339, 36.5, 119.3, 12.9, 19.8, 259.7, 315.8, 107, 11.5, 37.7, 4.1, 6.2. (b) Amalgamation is not necessary, since 13 out of 14 expected frequencies exceed 5. (c) Test H_0 : political party and answer are independent, against H_a : they are not independent. First term is $(804 - 823.3)^2/823.3 = 0.45$. Test statistic is $\chi^2 = 0.45 + 0.04 + 0.01 + 0.17 + 7.37 + 0.73 + 0.39 + 1.43 + 0.12 + 0.04 + 0.54 + 23.35 + 2.32 + 1.23 = 38.19$ and this is χ_6^2. Rejection region is values > 12.592 so reject H_0. The proportion of each party who say "Yes" is 25.8, 24.4, 24.4, 29.2, 5.1, 5.9, 34.6. These proportions are high for NZ First and the Māori Party, and low for Act and United Future.

12.34 (a) Donors enter into both "donors" and "visitors" columns, whereas we must assume each individual appears only once in the table. (b) The table is (top to bottom, L to R): 10, 10, 443, 249. (c) Test H_0 : day and donation are independent, against H_a : they are not independent. Number of donors is 20, number of visitors on Saturday is 453, $n = 712$. First expected frequency is $20 \times 453/712 = 12.7$. Expected frequencies are (top to bottom, L to R): 12.7, 7.3, 440.3, 251.7. First term is $(|10 - 12.7| - 0.5)^2/12.7 = 0.38$. Test statistic is

$\chi^2 = 0.38 + 0.67 + 0.01 + 0.02 = 1.08$ and this is χ^2_1. Rejection region is values > 3.841 so cannot reject H_0. (d) The claim above is not supported by the data.

Chapter 13

13.7 Test $H_0 : \beta = 0$ vs $H_a : \beta \neq 0$. Have $\hat{\sigma} = 3.618$. Test statistic is $7.342/(3.618/\sqrt{11.489}) = 10.34$ and this is t_{23}. The rejection region is values $|T| > 2.069$, so reject H_0.

13.11 (a) $\hat{y} = 1.2751 + 1.4925x$ (b) Test $H_0 : \beta = 2$ vs $H_a : \beta < 2$. Have $\hat{\sigma} = 1.33$. Test statistic is $(1.4925 - 2)/(1.33/\sqrt{126.70}) = -4.295$ and this is t_{18}. The rejection region is values < -1.734, so reject H_0.

13.14 (b) $\hat{y} = 14.78 + 0.9844x$ (c) The new scales give weights that are too high by about 15 kg. This diminishes slightly as x increases, since the slope is less than one. (d) Test $H_0 : \beta = 1$ vs $H_a : \beta \neq 1$. Have $\hat{\sigma} = 1.598$. Test statistic is $(0.9844 - 1)/(1.598/\sqrt{5250}) = -0.707$ and this is t_{16}. The rejection region is values $|T| > 2.120$, so cannot reject H_0. (e) The intercept estimate should be deducted from the readings. Further data should be collected to establish whether or not the slope is one.

13.20 When $x = 2.50$, $\hat{y} = 5.01$. $\hat{\sigma} = 1.33$, $n = 19$, $\bar{x} = 5.13$ and $S_{XX} = 196.9846$. The 95% CI is $5.01 \pm 2.110 \times 1.33 \times \sqrt{1 + 1/19 + (2.5 - 5.13)^2/196.9846} = 5.01 \pm 2.93$. $x = 13$ is well outside the range of the x data, and the prediction is possibly meaningless.

13.22 When $x = 235$, $\hat{y} = 246.1$. $\hat{\sigma} = 1.598$, $n = 18$, $\bar{x} = 225$ and $S_{XX} = 5250$. The 95% CI is $246.1 \pm 2.120 \times 1.598 \times \sqrt{1 + 1/18 + (235 - 225)^2/5250} = 246.1 \pm 3.5$. $x = 270$ is well outside the range of the x data, and the prediction is possibly meaningless.

Index

*Page numbers in bold indicate the primary reference, usually boxed in the text.

a, see regression, intercept
α, see hypothesis test, level
alternative hypothesis, see hypothesis test
amalgamation, 272
ANOVA, 248–253
    assumptions, 258–259
    interpretation, 256–258
    model-based approach, 254–256
    sums of squares
        calculation, **250**
        interpretation, **254**
    table, 251

b, see regression, slope
back-to-back stemplot, 15
bar chart, 20–22
    stacked, 21, 276
barplot, 269
Bayes' rule, 123–124
bell curve, see normal, curve
bimodal, 15, 19
$Bin(n,p)$, **136**
binomial, 134–142
    CLT application, 162
    conditions, **135**
    distribution, 137
    mean, **141**
    normal approximation, 161, **163**
    notation, **136**
    probabilities, 137–141, 319
    random variable, **135**
    variance, **141**
box-and-whisker plot, see boxplot
boxplot, 42–46
    checking for normality, 185
    modified, 45–46
    side-by-side, 43

categorical data, 269
categorical variable, **8**
census, 9
central limit theorem, see CLT
chance, 116
CI, see confidence interval
class
    boundaries, 17
    limits, 17
CLT, **156**, 156–159
    binomial, 161, 162
    Mann-Whitney U-test, 227
    proportions, 191
    sample mean, 171
coefficient of determination, 72
complete model, **255**
conclusion, see hypothesis test
confidence interval, 171–173
    ANOVA
        difference in means, **257**
        individual mean, **256**
    half, **257**
    interpretation, 173
    large sample
        difference in means, **214**
        difference in proportions, 230
        for μ, **172**
        proportion, 191–193

small sample
    difference in means, **219**
    for μ, 186
contingency table, 277
    spreadsheet analysis, 320–322
continuity correction
    binomial, 162, 163
continuous variable, 8
correlation, 57–66
    Pearson's, **58**, 57–62, 316
    sign, **59**
    Spearman's, **63**, 62–66, 317–318
    strength, **60**
count variable, **8**
critical value, 177

data dredging, 13
degrees of freedom
    ANOVA, 251–252
    chi-squared distribution, 261
    F-distribution, 221
    for two sample t-test, 218
    Kruskal-Wallis, 261
    one-way chi-square test, 271
    t-distribution, 185
    t-test, **185**
dependent variable, see variable, dependent
difference in means, 212–214
difference in proportions, 229–230
discrete variable, 8
distribution, 13, see probability distribution
    chi-squared, 261
    F, 221, 248
    sampling, 159
    skewed, 34
    t, 185
dotplot, 16

estimation, **170**
event, **117**
expected frequency
    contingency table, **278**
    one-way chi-square, **270**
explanatory variable, see variable, explanatory

F-distribution, see distribution, F
finite population correction factor, see FPCF
FPCF, 197, 233
frequency, **16**
frequency polygon, 20

Gaussian distribution, see normal distribution
goodness-of-fit test, 272–275
grand mean, 249
grand total, 249
grouped data, 40–42

H_0, see hypothesis test, null hypothesis
H_a, see hypothesis test, alternative hypothesis
half confidence intervals, **257**
histogram, 16–20
hypothesis test, **170**, 171, 173–178
    alternative hypothesis, 174, **174**
    calculating a p-value, **181**
    chi-square
        one-way, 269–271
    conclusion, **178**, 178
        using p-value, **180**, 180–183
    difference in variances, 221–223
    goodness-of-fit, 272–275
    hypotheses, 174–175
    large sample
        difference in means, 215–216
        difference in proportions, 231–232
        for μ, 174–178
        proportion, 193–194
    level, 177
    null hypothesis, 174
    rejection region, **176**, 176–177
    small sample
        difference in distributions, see Mann-Whitney U-test
        difference in means, 220
        for μ, 187
        median, 188–190
        proportion, 195–196
    steps, **174**
    test statistic, 175–176
hypothesis tests
    ANOVA, 248–253

incidence, 116
independence, 118, 156
independent variable, *see* variable, explanatory
interquartile range, **37**, 37, 316
interval estimate, *see* confidence interval
inverse normal, *see* normal, inverse
IQR, *see* interquartile range

jittering, 56

Kruskal-Wallis test, 260–261

large sample, **158**, 171
level, *see* hypothesis test
line chart, 21
line of best fit, *see* regression
lower quartile, **34**, 43, 316
LQ, *see* lower quartile

Mann-Whitney U-test, 223–225
    normal approximation, 227
    test statistic, **225**
margin of error, 199
maximum, 43
mean, **32**, 32, 43, 311, 316
    from grouped data, **41**
    normal, 148
mean square
    treatments, *see* MST
mean squared error, *see* MSE
measurement variable, **8**
measurements, 146
median, **33**, 33–34, 316
    test for, 188
minimum, 43
modified boxplot, 45–46
MSE, 250
    pooled variance, 250
MST, 250
mutually exclusive, **117**

$N(\mu, \sigma^2)$, **148**
nominal variable, 8
non-parametric, 188
non-parametric test, 223, 260
normal
    curve, 147

    distribution, 146–149
    inverse, 153–155, 319
    notation, **148**
    probability, 150–153, 319
    random variable, **146**
    standard distribution, **148**
normal distribution
    as a t-distribution, 185
null hypothesis, *see* hypothesis test

observed frequency, **270**
observed significance level, *see* p-value
one-way chi-square test, 269–271
order, 43
ordinal variable, 8
outcome, 115, 116
outlier, **44**, 44–45
outliers, 156

p-value, **180**, 180
parameters
    binomial, 136
    normal, 148
Pearson's linear correlation coefficient, *see* correlation, Pearson's
percentiles, 34–35, 316
point estimate, 170
pooled sample proportion, **231**
pooled variance
    ANOVA, 248, 250
    difference in means, **218**
population, **6**
    mean, **32**
    standard deviation, **39**
    variance, **38**
probability, 115, 116, **116**
    conditional, 123
    distribution, 124
    normal, *see* normal probability
    rules, 117–119, 123
    trees, 120–123
proportion, 116, **116**
    sample, *see* sample proportion, **142**
proportions, 191
p-value, 183
    sign test, 189

qualitative variable, **8**

quantitative variable, **8**
quartile
    lower, *see* lower quartile
    upper, *see* upper quartile

random sample, 10
random variable, 124
    binomial, *see* binomial random variable
    normal, **146**
range, **36**
ranks, 33, 63, 223
reduced model, **254**
regression, 66–75
    assumptions, 73–75
    calculating, 312–314
    estimated coefficients, **69**
    fitted values, **69**
    intercept, **69**, 317
    linear time trend, **98**
    model, **68**
    prediction, 71–72, 314
    residual, **69**
    RSS, 67
    slope, **69**, 317
rejection region, *see* hypothesis test, rejection region
relative frequency, **16**
residual
    ANOVA, **258**
    regression, **69**
residual sum of squares, *see* regression, RSS, 255
robust, 34, 223
R^2, *see* coefficient of determination

sample, **9**
    mean, **32**, 311, 316
    random, 10
    simple random, 10
    size, **9**
    standard deviation, **39**, 311, 316
    stratified, 11–12
    variance, **38**, 316
sample mean
    standardise, 161
sample proportion, **142**, 142–143, 194
    mean and variance, **143**

scatterplot, 55–57
    repeated points, 56
side-by-side boxplots, 43
sign test, 188–190
significance level, *see* hypothesis test, level
    observed, *see* p-value
skewed distribution, 34
small sample, 184
sorting, 15, 317
Spearman's rank-order correlation coefficient, *see* correlation, Spearman's
split-stem stemplot, 15
spreadsheet use, 315–322
stacked bar chart, 21, 276
standard deviation, 38–40, 311
    from grouped data, **41**, 312
    normal, 148
    population, **39**
    sample, **39**
standard error, **160**, 199
standard normal distribution, *see* normal, standard distribution, 185
standardisation, 148, **148**, 153
stem and leaf plot, *see* stemplot
stemplot, 13–16, 33
    back-to-back, 15
    split-stem, 15
Student's t-distribution, *see* t-distribution

t-test, *see* hypothesis test, small sample, for μ
t-distribution, 185
test statistic
    ANOVA, **251**
    chi-square
        contingency table, 279
        one-way, **271**
    difference in variances, **222**
    Kruskal-Wallis, **261**
    large sample
        difference in means, **216**
        difference in proportions, **232**
        for μ, **175**
        for p, **194**
    Mann-Whitney U-test, **225**
    small sample
        difference in means, **220**

for μ, **187**
for median, **189**
for p, **195**
ties, 64, 224, 261
time series, **91**, 90–106
    autocorrelation, 104–106
    components, 94–96
    frequency, 91
    irregular, **94**, 96
    linear time trend, **98**, 98–99
        parameter estimates, **99**
    model, 96, 98–99
    moving average, 100–103, 318
        centred, **101**
        end effect, 100
        seasonal, **102**
        weighted, 102
    non-stationary, 98
    plot, 21, **92**
    sample period, 91, 94
    seasonal, 93, **94**, 96, 101
    stationary, 98
    trend, **94**, 96, 101
        linear time, 98–99
tree diagrams, *see* probability trees
two-sample t-test, **220**
type I and II errors, 183–184

upper quartile, **34**, 43, 316
UQ, *see* upper quartile

variable
    categorical, **8**
    continuous, **8**
    dependent, 55
    discrete, **8**
    explanatory, 55
    nominal, **8**
    ordinal, **8**
    qualitative, **8**
    quantitative, **8**
variance, 38–40
    from grouped data, **41**
    population, **38**
    sample, **38**

Z, 148, **148**
z-scores, 155